THE ASCENT OF MAN

Sources and Interpretations

THE ASCENT OF MAN

Sources and Interpretations

John F. Henahan, Editor

Little, Brown and Company Boston/Toronto

Preface

Sixth Printing

Library of Congress Catalog Card Number 74-27176
Published simultaneously in Canada by Little, Brown and Company (Canada) Limited
Manufactured in the United States of America

Project Manager: Judith E. Plitt
Designer: Richard N. Carter
Assistant Designer: Kitty R. Anderson

Man is unique among the species. That much is clear, although the specific characteristics that set him apart have long been the subject of controversy. We can, however, draw some fairly safe conclusions.

If man isn't unique for his creativity, he certainly is singular in the abundance of creative products and artifacts that he leaves behind. Man alone *records* his ideas and his history. These records take many forms—works of art, books, films, nonbiodegradable waste, etc.—and serve many purposes—to inform, to gain fame and fortune, to please. All have one common characteristic: they pose as many questions as they answer. Each generation is faced with solving the puzzle presented by an increasing store of artifacts left by his predecessors. The progression of print, magnetic audiotape, videotape, and holograms reflects man's increasingly sophisticated attempts to create accurate and complete records, yet they are at best, imperfect reflections of the original ideas, and they reveal very little about the mental processes that conceived them.

This book brings together a vast assortment of man's recorded attempts to understand himself, the world around him, and his place in cultural and biological evolution. Whether his story is construed as a process of ascent, descent, or mere roaming about on a plane of chance is simply a measure of an individual's optimism.

Thus, the selections included here can only reflect their authors' subjective interpretations of an accumulation of "objective truth." To examine scientific theories, philosophic doctrines, industrial technologies, and poetic musings all in a place, as we have done here, is to perceive the richness of their interconnections and mutual influences. Their boundaries become muddled and their common qualities begin to emerge. Such juxtaposition offers a natural milieu for cultural anthropologists, philosophers, and historians of science to bring to the material their respective interpretations. It is indeed a fruitful and insightful process to tease an idea from its historical and cultural setting, only to see it spring back from the tension of its connection.

Editor's Introduction

With the sudden death of Jacob Bronowski in August 1974, our modern era lost one of its true renaissance men. Mathematician, poet, playwright, and humanist—Dr. Bronowski was convinced that man's unique facilities have allowed him to go beyond the limits of biological evolutionary processes to carve out a culture that embraces the arts, the sciences, and all that distinguishes man from animal. That conviction was a major theme of *The Ascent of Man* and he was obviously pleased that the television series and the book gave him a vehicle for communicating with a larger audience than had ever been available to him in the past. Dr. Bronowski was intrigued with the idea of using the films and the book as educational tools, and in the weeks before his death he participated enthusiastically in developing this anthology for that purpose.

The readings compiled in this book are intended to serve a variety of functions. In some cases, they provide documentation for specific points covered in *The Ascent of Man*. Other selections convey some of the historical background underlying the steps in man's ascent. Still others were selected not only for their relevance, but in some cases, for their entertainment value. In all cases, we hope the selections do justice to the spirit of Jacob Bronowski and his masterful treatment of *The Ascent of Man*.

Thanks are due to all who participated in the preparation of this anthology. A special "thank you" is extended to Sylvia Fitzgerald, Dr. Bronowski's invaluable assistant.

At Miami Dade Community College, Virginia Gentle coordinated the efforts of staff members in helping to compile this anthology. Her assistants are Roberta Odum, Pat Miller, Rose Quaranta, and Betty Little. Members of the Miami Dade faculty who gave their time to the project include Mildred Lunaas Bain, Dean of Academic Affairs; William J. Brackmann, Assistant Professor of Physics; Roy Jervis, Professor of Biology; John A. Jones, Associate Professor of Biology; Michael Nicklanovich, Assistant Professor of Biology; Margaret Pelton, Chairman of Fine Arts; Janet Rich, Chairman of Mathematics; Winston Richter, Director of Natural Sciences Division; Donald Roach, Chairman of Chemistry; S.T. Swirsky, Associate Professor of Archaeology and History; and Ena Williams, Associate Professor of English.

At the University of California Extension, San Diego, I had the valuable assistance and guidance of Martin Chamberlain, Dean; Mary Lindenstein Walshok, Project Director; Darrell Icenogle, UCSD Project Coordinator; and Bobbie Savitz, writer.

In addition, I acknowledge the help and cooperation of the library personnel at the University of California, San Diego; The Salk Institute for Biological Studies; Miami Dade Community College, and the University of Miami. In all phases of the anthology's preparation, Leila Wopschall was a very helpful assistant and Judy Besta cheerfully performed many typing chores. Michael Rowbotham's research assistance is also greatly appreciated.

The Editor

Contents

v Preface
vii Editor's Introduction

1
Lower than the Angels

1 The Descent of Man—an Excerpt by Charles Darwin
3 *Australopithecus africanus*: The Man-Ape of South Africa by Raymond A. Dart
6 Man the Toolmaker by L.S.B. Leakey
9 The Emergence of *Homo erectus* by Edmund White
13 *Homo neandertalis* Meets *Homo sapiens* by William Golding
15 The Brain Grows Up by Phillip V. Tobias
17 The 'Meaning' of Cave Paintings by J.Z. Young
19 The Impact of the Concept of Culture on the Concept of Man by Clifford Geertz
21 Change in the Modern World from *Cultural Anthropology Today* (CRM Books, Publisher)

2
The Harvest of the Seasons

23 Domestication of Animals by F.E. Zeuner
28 The Domestication of Maize by Paul C. Manglesdorf
31 The First Jericho by Kathleen M. Kenyon
38 Plate Tectonics by D.P. McKenzie
40 The Invention of the Plow by E. Cecil Curwin
45 The Equestrian Revolution of the Plains Indian by Peter Farb
47 Dialogues of Plato—an Excerpt
48 Gulliver in Brobdingnag by Jonathan Swift
49 Reflections on the Manhattan Project by Lt. Gen. Leslie R. Groves

3
The Grain in the Stone

53 Some Prehistoric Connections Between Siberia and America by James B. Griffin
57 What Are Blood Groups? by The American Red Cross
59 The Anasazi of Arizona by the Faculty of the University of Arizona
62 The Masonry of the Incas by John Hemming
63 The Quipu by J. Alden Mason
64 The Roman Architectural Contribution by William Fleming
67 The Rise of Freemasonry by Jean Gimpel
70 On Creativity
72 Simon Rodia by Robert Bryan

4

The Hidden Structure

77 A Chinese Alchemical Recipe by Nathan Sivin
79 Greek Alchemy: Pluses and Minuses by Charles A. Browne
81 The Beginning of Metallurgy by Cyril Stanley Smith
84 Origin of Metals by Lucretius
85 Paracelsus by Henry M. Pachter
89 The Discovery of Oxygen by F.W. Gibbs
93 Atomic Theory by John Dalton
96 The Police Constable's Atomic Theory by Flann O'Brien

5

The Music of the Spheres

101 The Greek Rationalization of Nature by Morris Kline
108 Mayan Mathematics by Charles Gallenkamp
112 Proposition 47 by Euclid
113 The Celestial Chorus by Johannes Kepler
114 Space Lattices and Packing of Atoms by Michell J. Sienko and Robert A. Plane
118 The Calculus by Benjamin David (and the Staff Editors of the *Life Science Library, Mathematics*)

6

The Starry Messenger

129 Babylonian Stargazers by Rudolf Thiel
132 The Problem of the Planets by Thomas S. Kuhn
134 Ptolemy's System of Planetary Motion by Gerald Holton and Duane Roller
136 The Copernican Revolution by Robert T. Lagemann
140 Galileo and the Starry Messenger by Galileo Galilei (translated by Stillman Drake)

7

The Majestic Clockwork

147 Tycho Meets Kepler by Arthur Koestler
153 The Astronomical Partnership of Tycho and Kepler by Robert T. Lagemann
157 Newton and the Pipes of Pan by J.E. McGuire and P.M. Rattansi
159 Goethe's Theory of Light by Dietrich Schroeer
163 Michelson and the Speed of Light by Bernard Jaffe
167 The Theory of Relativity by Francis W. McCarthy, Joan H. Brenner, and Maurice Temple

8

The Drive for Power

177 Coketown by Charles Dickens
178 Letter to the Earl of Liverpool by Robert Owen
180 William Blake and the Age of Revolution by J. Bronowski
184 On the Industrial Revolution by Marc Sequin
186 Composed upon Westminster Bridge by William Wordsworth
186 Composed a Few Miles Above Tintern Abbey by William Wordsworth
188 Frankenstein—an Excerpt by Mary Shelley
190 Heat Is Motion by George Gamow
192 James Watt and the Steam Engine by J.D. Bernal
195 Science, Technology, and Values by Joseph B. Platt
197 The Model T by Allan Nevins and Frank Ernest Hill

9

The Ladder of Creation

201 First Essay on Population—an Excerpt by Thomas R. Malthus
203 On *The Origin of the Species* by Charles Darwin
207 Science, Religion, and the Future by Canon C.E. Raven
212 Spontaneous Generation by A.I. Oparin
215 Louis Pasteur by René Dubos
217 The Origin of Life by Stanley M. Miller
220 Directed Panspermia by F.H.C. Crick and L.E. Orgel

10

World Within World

227 A Letter to My Mother by H.G.J. Moseley
228 Moseley, X-Rays, and the Structure of the Atom by E.N. da C. Andrade
231 Seurat and Science by John Rewald
234 The Quantum Theory of Max Planck by Barbara Lovett Cline
241 The Genesis of the Nuclear Energy Project by Enrico Fermi
248 Starfolk by Carl Sagan
252 What Is Entropy? by Isaac Asimov
253 Natural Changes Create Disorder by Isaac Asimov
254 Is the Universe Running Down? by Isaac Asimov

11

Knowledge or Certainty

257 The Edge of Precision by Stanley L. Jaki
262 The Uncertainty Principle by Paul G. Hewitt
266 Heisenberg and Bohr: A Conversation by Werner Heisenberg
269 Patent for a Chain Reaction by Leo Szilard

271 My Trial as a War Criminal by Leo Szilard
277 Particles of Light by James S. Perlman
279 Physical Optics by Charlotte R. Ward
280 Science Is Not Enough by Vannevar Bush

12

Generation upon Generation

283 The Discovery of Units of Heredity by Ruth Moore
288 The Language of the Genes by G.W. Beadle
295 The Double Helix by J.D. Watson and F.H.C. Crick
296 *Brave New World*—an Excerpt by Aldous Huxley
299 Ethical Aspects of the New Perspectives in Biomedical Research by Salvador E. Luria
301 Nature Does Not Always Know Best by Peter B. Medawar
303 Sexual Reproduction in Bacteria from *Geology Today* (CRM Books, Publisher)

13

The Long Childhood

309 Language and Brain Deficiency by H.J. Jerison
311 The Speech Center by Dean E. Wooldridge
315 Mind, Brain, and Humanistic Values by Roger W. Sperry
324 Physical Manipulation of the Brain from *The Hastings Center Report* (May 1973)
326 Is Man Still Ascending? by Julian Huxley
328 *Genesis* According to the Badger by T.H. White

330 Bibliography

THE ASCENT OF MAN

Sources and Interpretations

1
Lower than the Angels

What a piece of work is man!
how noble in reason!
how infinite in faculty!
in form and moving
how express and admirable!
in action how like an angel!
in apprehension how like a god!
the beauty of the world!
the paragon of animals!

William Shakespeare

In the *ascent* of man it is his faculties that place him somewhat "lower than the angels" in the scheme of things. His wit and imagination have allowed man to rise far above other animals, to develop magnificent cultures in a few thousand years. Other animal species have flourished on Earth unchanged for millions of years, stuck on an evolutionary dead-end street. It is man's uniqueness that has made the difference, a theme that Dr. Bronowski elaborates poetically and dramatically throughout *The Ascent of Man*. From a slightly different perspective, Charles Darwin saw tantalizing biological similarities in man and beast that allowed him to develop his world-shaking theory of the origin of the species. Darwin discussed those similarities a little more than one hundred years ago in his own great work *The Descent of Man*.

The Descent of Man—an Excerpt

Charles Darwin

He who wishes to decide whether man is the modified descendant of some pre-existing form, would probably first enquire whether man varies, however slightly, in bodily structure and in mental faculties; and if so, whether the variations are transmitted to his offspring in accordance with the laws which 1
prevail with the lower animals. Again, are the variations the result, as far as our ignorance permits us to judge, of the same general causes, and are they governed by the same general laws, as in the case of other organisms; for instance, by correlation, the inherited effects of use and disuse, etc.? Is man subject to similar malconformations, the result of arrested development, of reduplication of parts, etc., and does he display in any of his anomalies reversion to some former and ancient type of structure? It might also naturally be enquired whether man, like so many other animals, has given rise to varieties and sub-races, differing but slightly from each other, or to races differing so much that they must be classed as doubtful species? How are such races distributed over the world; and how, when crossed, do they react on each other in the first and succeeding generations? And so with many other points.

The enquirer would next come to the important point, whether man tends to increase at so rapid a rate, as to lead to occasional severe struggles for existence; and consequently to beneficial variations, whether in body or mind, being preserved, and injurious ones eliminated. Do the races or species of men, whichever term may be applied, encroach on and replace one another, so that some finally become extinct? We shall see that all these questions, as indeed is obvious in respect to most of them, must be answered in the affirmative, in the same manner as with the lower animals. But the several considerations just referred to may be conveniently deferred for a time: and we will first see how far the bodily structure of man shows traces, more or less plain, of his descent from some lower form. In succeeding chapters the mental powers of man, in comparison with those of the lower animals, will be considered.

The Bodily Structure of Man

It is notorious that man is constructed on the same general type or model as other mammals. All the bones in his skeleton can be compared with corresponding bones in a monkey, bat, or seal. So it is with his muscles, nerves, blood-vessels and internal viscera. The brain, the most important of all the organs, follows the same law, as shewn by Huxley and other anatomists. Bischoff, who is a hostile witness, admits that every chief fissure and fold in the brain of man has its analogy in that of the orang; but he adds that at no period of development do their brains perfectly agree; nor could perfect agreement be expected, for otherwise their mental powers would have been the same. . . . But it would be superfluous here to give further details on the correspondence between man and the higher mammals in the structure of the brain and all other parts of the body.

It may, however, be worth while to specify a few points, not directly or obviously connected with structure, by which this correspondence or relationship is well shewn.

Man is liable to receive from the lower animals, and to communicate to them, certain diseases, as hydrophobia, variola, the glanders, syphilis, cholera, herpes, etc.; and this fact proves the close similarity of their tissues and blood, both in minute structure and composition, far more plainly than does their comparison under the best microscope, or by the aid of the best chemical analysis. Monkeys are liable to many of the same non-contagious diseases as we are; thus Rengger, who carefully observed for a long time the *Cebus azarae* in its native land, found it liable to catarrh, with the usual symptoms, and which, when often recurrent, led to consumption. These monkeys suffered also from apoplexy, inflammation of the bowels, and cataract in the eye. The younger ones when shedding their milk-teeth often died from fever. Medicines produced the same effect on them as on us. Many kinds of monkeys have a strong taste for tea, coffee, and spiritous liquors: they will also, as I have myself seen, smoke tobacco with pleasure. Brehm asserts that the natives of north-eastern Africa catch the wild baboons by exposing vessels with strong beer, by which they are made drunk. He has seen some of these animals, which he kept in confinement, in this state; and he gives a laughable account of their behavior and strange grimaces. On the following morning they were very cross and dismal; they held their aching heads with both hands, and wore a most pitiable expression: when beer or wine was offered them, they turned away with disgust, but relished the juice of lemons. An American monkey, an Ateles, after getting drunk on brandy, would never touch it again, and thus was wiser than many men. These trifling facts prove how similar the nerves of taste must be in monkeys and man, and how similarly their whole nervous system is affected. . . .

Thus we can understand how it has come to pass that man and all other vertebrate animals have been constructed on the same general model, why they pass through the same early stages of development, and why they retain certain rudiments in common. Consequently we ought frankly to admit their community of descent; to take any other view, is to admit that our own structure, and that of all the animals around us, is a mere snare laid to entrap our judgment. This conclusion is greatly strengthened, if we look to the members of the whole animal series, and consider the evidence derived from their affinities or classification, their geographical distribution and geological succession. It is only our natural prejudice, and that arrogance which made our forefathers declare that they were descended from demigods, which leads us to demur to this conclusion. But the time will before long come, when it will be thought wonderful that naturalists, who were well acquainted with the comparative structure and development of man, and other mammals, should have believed that each was the work of a separate act of creation. . . .

It has often been objected to such views as the foregoing, that man is one of the most helpless and defenceless creatures in the world; and that during his early and less well-developed condition, he would have been still more helpless. The Duke of Argyll, for instance, insists that "the human frame has diverged from the structure of brutes, in the direction of greater physical helplessness and weakness. That is to say, it is a divergence which of all others it is most impossible to ascribe to mere natural selection." He adduces the naked and unprotected state of the body, the absence of great teeth or claws for defence, the small strength and speed of man, and his slight power of discovering food or of avoiding danger by smell. To these deficiencies there might be added one still more serious, namely, that he cannot climb quickly, and so escape from enemies. The loss of hair would not have been a great injury to the inhabitants of a warm country. For we know that the unclothed Fuegians can exist under a wretched climate. When we compare the defenceless state of man with that of apes, we must remember that the great canine teeth with which the latter are provided, are possessed in their full development by the males alone, and are chiefly used by them for fighting with their rivals; yet the females, which are not thus provided, manage to survive.

In regard to bodily size or strength, we do not know whether man is descended from some small species, like the chimpanzee, or from one as powerful as the gorilla; and therefore, we cannot say whether man has become larger and stronger, or smaller and weaker, than his ancestors. We should, however, bear in mind that an animal possessing great size, strength, and ferocity, and which, like the gorilla, could defend itself from all enemies, would not perhaps have become social: and this would most effectually have checked the acquirement of the higher mental qualities, such as sympathy and the love of his fellows. Hence it might have been an immense advantage to man to have sprung from some comparatively weak creature.

The small strength and speed of man, his want of natural weapons, etc., are more than counterbalanced, firstly, by his intellectual powers, through which he has formed for himself weapons, tools, etc., though still remaining in a barbarous state, and, secondly, by his social qualities which lead him to give and receive aid from his fellow-men. No country in the world abounds in a greater degree with dangerous beasts than southern Africa; no country presents more fearful physical hardships than the arctic regions; yet one of the puniest of races, that of the bushmen, maintains itself in southern Africa, as do the dwarfed Esquimaux [Eskimos] in the arctic regions. The ancestors of man were, no doubt, inferior in intellect, and probably in social disposition, to the lowest existing savages; but it is quite conceivable that they might have existed, or even flourished, if they had advanced in intellect whilst gradually losing their brute-like powers such as that of climbing trees, etc. But these ancestors would not have been exposed to any special danger, even if far more helpless and defenceless than any existing savages, had they inhabited some warm continent or large island, such as Australia, New Guinea, or Borneo, which is now the home of the orang. And natural selection arising from the competition of tribe with tribe, in some such large area as one of these, together with the inherited effects of habit, would, under favourable conditions, have sufficed to raise man to his present high position in the organic scale.

In "Lower Than the Angels," Dr. Bronowski notes that even though "it is tempting—very tempting to a scientist—to hope that the most original achievements of the mind are also the most recent," yesterday's discoveries are no less important to man's cultural and scientific ascent. He speaks of discoveries that may have occurred long ago. It is also true that although the recent discoveries of the Leakeys in Africa certainly are monumental in that they have helped place man the toolmaker on an anthropological timetable of sorts, no less monumental was R.A. Dart's painstaking peeling away of sand and lime that encased the Taung baby skull he found in an African quarry more than fifty years ago. Dart's account of this event conveys his sense of excitement:

No diamond cutter ever worked more lovingly or with such care on a priceless jewel—nor, I am sure, with such inadequate tools. . . . But on the seventy-third day, December 23, the rock parted. I could view the face from the front, although the right side was still imbedded. The creature which had contained this massive brain was no giant anthropoid such as a gorilla. What emerged was a baby's face, an infant with a full set of milk teeth and its permanent molars just in the process of erupting. I doubt if there was any parent prouder of his offspring than I was of my Taungs baby on that Christmas. [1]

In reporting his discovery in *Nature* in 1925, Dart explained why he had decided that the small skull represented "a creature well advanced beyond modern anthropoids [apes]." His report follows in edited form.

Australopithecus africanus: The Man-Ape of South Africa

Raymond A. Dart

In the first place, the whole cranium displays *humanoid* rather than anthropoid lineaments. . . .

In the second place, the dentition is *humanoid* rather than anthropoid. The specimen is juvenile, for the first permanent molar tooth only has erupted in both jaws on both sides of the face; *i.e.* it corresponds anatomically with a human child of six years of age. . . .

In the third place, the mandible [jawbone] itself is *humanoid* rather than anthropoid. . . .

Excerpted from Raymond Dart, "Australopithecus africanus: The Man-Ape of South Africa," *Nature,* Vol. 115, No. 2884, February 7, 1923, pp. 195–199. Reprinted by permission of Macmillan Journals Ltd

[1] John E. Pfeiffer, *The Emergence of Man*, New York, Harper & Row, 1969, p. 63.

That hominid characters were not restricted to the face in this extinct primate group is borne out by the relatively forward situation of the foramen magnum. The position of the basion can be assessed within a few millimetres of error, because a portion of the right exoccipital is present alongside the cast of the basal aspect of the cerebellum. Its position is such that the basi-prosthion measurement is 89 mm., while the basi-inion measurement is at least 54 mm. This relationship may be expressed in the form of a "head-balancing" index of 60.7. The same index in a baboon provides a value of 41.3, in an adult chimpanzee 50.7, in Rhodesian man 83.7, in a dolichocephalic European 90.0, and in a brachycephalic European 105.8. It is significant that this index, which indicates in a measure the poise of the skull upon the vertebral column, points to the assumption by this fossil group of an attitude appreciably more erect than that of modern anthropoids. The improved poise of the head, and the better posture of the whole body framework which accompanied this alteration in the angle at which its dominant member was supported, is of great significance. It means that a greater reliance was being placed by this group upon the feet as organs of progression, and that the hands were being freed from their more primitive function of accessory organs of locomotion. Bipedal [two-legged] animals, their hands were assuming a higher evolutionary role not only as delicate tactual, examining organs which were adding copiously to the animal's knowledge of its physical environment, but also as instruments of the growing intelligence in carrying out more elaborate, purposeful, and skilled movements, and as organs of offence and defence. The latter is rendered the more probable, in view, first, of their failure to develop massive canines and hideous features, and secondly, of the fact that even living baboons and anthropoid apes can and do use sticks and stones as implements and as weapons of offence. . . .

Whatever the total dimensions of the adult brain may have been, there are not lacking evidences that the brain in this group of fossil forms was distinctive in type and was an instrument of greater intelligence than that of living anthropoids. The face of the endocranial cast is scarred unfortunately in several places. . . . It is evident that the relative proportion of cerebral to cerebellar matter in this brain was greater than in the gorilla's. The brain does not show that general pre- and post-Rolandic flattening characteristic of the living anthropoids, but presents a rounded and well-filled-out contour, which points to a symmetrical and balanced development of the faculties of associative memory and intelligent activity. . . .

It is obvious, meanwhile, that it represents a fossil group distinctly advanced beyond living anthropoids in those two dominantly human characters of facial and dental recession on one hand, and improved quality of the brain on the other. Unlike Pithecanthropus, it does not represent an ape-like man, a caricature of precocious hominid failure, but a creature well advanced beyond modern anthropoids in just those characters, facial and cerebral, which are to be anticipated in an extinct link between man and his simian ancestor. At the same time, it is equally evident that a creature with anthropoid brain capacity, and lacking the distinctive, localized temporal expansions which appear to be concomitant with and necessary to articulate man, is no true man. It is therefore logically regarded as a man-like ape. I propose tentatively, then, that a new family of *Homo-simiadae* be created for the reception of the group of individuals which it represents, and that the first known species of the group be designated *Australopithecus africanus*, in commemoration, first, of the extreme southern and unexpected horizon of its discovery, and secondly, of the continent in which so many new and important discoveries connected with the early history of man have recently been made, thus vindicating the Darwinian claim that Africa would prove to be the cradle of mankind.

In anticipating the discovery of the true links between the apes and man in tropical countries, there has been a tendency to overlook the fact that, in the luxuriant forests of the tropical belts, Nature was supplying with profligate and lavish hand an easy and sluggish solution, by adaptive specialisation, of the problem of existence in creatures so well equipped mentally as living anthropoids are. For the production of man a different apprenticeship was needed to sharpen the wits and quicken the higher manifestations of intellect—a more open veldt country where competition was keener between swiftness and stealth, and where adroitness of thinking and movement played a preponderating role in the preservation of the species. Darwin has said, "no country in the world abounds in a greater degree with dangerous beasts than Southern Africa," and, in my opinion, Southern Africa, by providing a vast open country with occasional wooded belts and a relative scarcity of water, together with a fierce and bitter mammalian competition, furnished a laboratory such as was essential to this penultimate phase of human evolution.

It was indeed *Australopithecus*, the toolmaker, who prepared a pebble, stored it for later use, and "released the brake which the environment imposes on all other creatures." Once released, man's prowess with stone and tool developed simultaneously with his biological and cultural evolution. The importance of stone tools in man's ascent is illustrated in this selection by Louis S.B. Leakey. It was written years before the noted anthropologist found markings on pebbles that indicated that *Australopithecus* was making stone tools more than two million years ago.

Man the Toolmaker

L.S.B. Leakey

The making of weapons and implements of stone, bone, and wood doubtless goes back to the time when man first became man. One of the most widely accepted descriptive phrases to designate man as he is today is 'the toolmaker'. As more and more of the fossilized remains of the stages in human evolution come to light, it becomes increasingly difficult to draw the line between human and pre-human beings. Nevertheless, such creatures become recognizable as having human qualities when members of their stock have started to make tools as distinct from merely using natural objects as tools. . . .

. . . We often think and speak of prehistoric chipped stones as crude instruments. To remove any possible misunderstanding, it is well to say at once that there are multitudes of stone implements the manufacture of which has involved technical skill of the very highest order. The attainment of that skill implies traditions extending over ages, and an accompanying high degree of inventive skill combined with endless patience. The technical achievements of primitive man are thus worthy of the greatest respect. The earliest achievements of men of the Stone Age must be accepted as fully comparable, in their place, time, and circumstance, to those of the greatest modern inventors and engineers.

The term 'stone' is comprehensive, covering such widely different materials as flint and obsidian on the one hand and coarsegrained granites and quartzites on the other. Naturally, early man chose, when he could, the finer-grained and more homogeneous rocks from which to make his artifacts; though there is scarcely any form of stone which he did not use at one time or another. Contrary to common belief, the nature of the stone available made very little difference to the techniques employed in working it. From the earliest period until about 3000 B.C., techniques for working stone were in effect confined to various methods of chipping and flaking. We therefore begin by discussing these techniques.

Hammerstone Technique

At first, only the simplest hammerstone technique of flaking was employed. This consisted of taking a water-worn pebble of suitable size and shape, holding it in the hand, and using it as a hammer for knocking off flakes from another stone in order to shape it roughly into form. This sounds fairly easy, but in fact the worker had to master a knowledge of the correct angle at which to strike a blow before he could be certain of detaching a flake at the desired point and in the desired direction. The acquisition of such a technique was by no means easy, and there is ample evidence that it involved periods of time that are vast when measured by later standards.

When a blow is struck upon a piece of stone with the convex surface of a water-worn pebble, fracture does not take place in a continuation of the direction of the blow. As a result of stresses built up near the surface during the period of contact, the material breaks under tension round the periphery of the area of contact, and a crack spreads rapidly downwards and outwards. Theoretically—and indeed actually, if the blow is of sufficient force, and if the struck stone is of suitable texture—a strong vertical blow on a piece of fairly flat stone will punch out a complete cone, just as a bullet hitting a pane of plate-glass forms a conical, and not a cylindrical, hole in the glass. Ordinarily, however, a vertical blow with a hammerstone on another piece of stone causes only shattering, either because the blow is not strong enough to punch out a perfect cone, or because most stones have inherent lines of weakness and faults which interfere with the symmetrical distribution of the stress.

In the course of time, Stone Age man learned by experience that, in order to detach a flake in a desired direction from a piece of stone, he had to direct his blow at an angle of roughly 120° to the direction in which he desired to remove the flake, and also at a point near the edge of the stone from which the flake was to be detached. In this way only a part of the cone of force penetrated the stone, the rest being dissipated outside. Thus a flake was detached with a clearly defined semi-cone—or bulb of percussion as it is sometimes called—marking the point of impact of the hammerstone. Once Stone Age man had mastered the hammerstone technique of detaching flakes from other stones, he gradually and slowly used the knowledge to make new and better types of stone implements, and he also evolved certain modifications of this first technique.

Anvil Technique

To meet this difficulty the so-called anvil technique was at some point invented. In this technique, the 'hammer' was a fixed block of stone, and the stone from which it was desired to detach a really large flake (from which a big implement could be made) was swung against this fixed anvil. Here again, the essential for success was knowledge of the correct angle at which the stone should be struck against the anvil. The angle is fundamentally the same as in the simpler hammerstone technique.

A main disadvantage of the hammerstone and anvil techniques of flaking was that the scar from which the flake had been struck had a deep negative bulb of percussion at the point of impact of the original blow. Thus—since several flakes were detached in succession—the cutting edges of tools made by this technique were irregular and jagged. Massive as well as small instruments of this type with jagged cutting edges are associated with what is known as the Chellean stage of the great hand-axe culture. During the first half of this period only the hammerstone technique was employed, although naturally even with this technique there was a slow and gradual improvement in the types of tool that were made.

Cylinder-Hammer Technique

Towards the second half of the period a new technique of flaking—the so-called cylinder-hammer technique—was discovered or invented, and this new development is taken to mark the beginning of the Acheulian stage (from St. Acheul, near Amiens) of the hand-axe culture. The older hammerstone technique was still used for making rough tools as well as for the initial preparation of finer tools, but the new cylinder-hammer technique was used for the finishing-off process.

The new technique consists of knocking off very thin flakes with very shallow bulbs of percussion—thus producing a nearly even surface—by means of a round-edged hammer of some comparatively soft material such as bone, hard wood, or weathered stone. The essential factors in this technique are: (a) that the force of the blow is directed at the actual edge of the stone struck and not at a little distance in from the edge, and (b) that although one particular point of the hammer hits the stone first, the fact that the hammer is comparatively soft means that, almost instantaneously, force is applied also from other points along the rounded surface of the hammer. Thus, instead of the crack spreading from one point and giving rise to a marked bulb of percussion, it spreads from a larger area of contact through a flattened arc. This results in the removal of a flake that is very flat. A series, or rather the intersections of a series, of these flat flakes produces a nearly straight cutting edge.

It is probable, though the evidence is of an indirect character, that the commonest form of soft cylinder-hammer used by Stone Age man was a branch of wood. The evidence must be indirect, since we can hardly hope to find such perishable articles as wooden hammers on actual Stone Age sites.

At about the time that one group of Stone Age men discovered the cylinder-hammer technique for removing flat flakes, another group was evolving what is called the prepared-core technique to obtain flakes of a very special shape. This technique—sometimes called tortoise-core—differs greatly from other early Stone Age techniques in that the shape of the flake to be finally struck off, to serve as a tool with little or no subsequent retouching, was blocked out on the core or lump of stone beforehand and then removed by one blow. Previously it had been customary to take either a nodule, a pebble, a lump of stone, or a large flake already struck from a piece of stone, and strike off a series of flakes to trim the specimen into a tool.

The prepared-core technique appears in South Africa and in north-west Europe at about the same time.

Homo erectus was far from the dumb-brute caricature of the cave man who became extinct when modern man appeared on the scene. Today's theory makes *Homo erectus* a complex, social, direct ancestor of *Homo sapiens* who left a scattered record from England to China and from Java to South Africa. His wooden spear with fire-hardened tip, together with the stone tools from which it was made has been found in Clacton, England, in deposits that date back more than 400,000 years. Mankind's long childhood provided the cultural base whereby, after a million years of evolution, modern man appeared ready to move through the adolescence of another quarter million years. His maturity is symbolized by the arts and culture of Altamira Caves in Santander, Spain. He was ready for the explosive post-glacial development that we call civilization. The following excerpt describes this period in the cultural growth of *Homo erectus*.

The Emergence of *Homo erectus*

Edmund White

On a late spring day 400,000 years ago, a band of about 25 craggy-visaged men, women and children stopped at a sandy cove on the Mediterranean coast. They were looking for a place to stay, and they chose a spot, perched on a sand dune and protected by a limestone cliff, at the mouth of the Paillon Valley. Today, the city of Nice, France, rises around their ancient campsite, but archeologists have unearthed the place, called Terra Amata, where these 25 primitive visitors once made a brief stopover. From the assortment of fossil bones, stone tools and imprints they left and from the consistency of the sand in which they left them, the archeologists have been able to determine that they stayed only about three days and can describe, in extraordinary detail, just what the visitors did there before moving on. . . .

What sort of man visited this cove on the coast of Europe 400,000 years ago? Who was he? Although he came each spring for many years, the fossils he left behind him included no human bones—only a single human footprint in the hard sand. Fortunately, there are other sites of similar age where fossil bones of this prehistoric creature have been found, along with similar tools and animal bones and in two instances, hearths. Indeed, when the evidence from all these archeological finds is pieced together, the picture that emerges is remarkably complete, revealing not only what this man looked like, but where he came from. In some ways, with all the advantages of hindsight, scientists probably know more about him than he knew about himself.

He was the first man. He is known, in the scheme of evolution, as *Homo erectus*, or upright man. He was the direct descendant of Australopithecus, a creature considered the missing link between the apes and man. Physically, Homo erectus was quite distinct from his more apelike ancestors. While Australopithecus had evolved a larger brain, relative to his size, than modern apes have and was able to make primitive tools, he was still comparatively small—Pygmy-sized in many cases—and not yet as well adapted as true men are to totally erect posture and walking long distances.

Homo erectus stood straight-backed and walked with a springy stride. He was around five and a half feet tall and owed his manly posture and gait to several developments in the anatomy of his pelvis and foot—changes that had already been apparent in Australopithecus, but that in Homo erectus came close to their modern form. Homo erectus' pelvis was more bowl-shaped, so that the socket at which the femur connects to the pelvis was farther inward than in Australopithecus. This shift gave Erectus a straight-legged stance as opposed to the slightly bowed stance of Australopithecus. In addition, Erectus' foot, arched to support his whole weight, had lost all trace of the ability to grasp that ape feet possess. As a result of these changes, Homo erectus was the first primate who could walk upright for long periods of time without strain.

No complete set of Erectus' hand bones has been found, but his skill at making tools suggests that he had hands like ours, capable of a unique type of grip that sets man apart from other living primates. All primates can use a simple hand grip, with the fingers held together as a hook over some object like a branch. This "power grip" serves well enough for such activities as swinging from a tree limb or grabbing a bunch of greens. Homo erectus' hand was obviously adapted to use the more complicated "precision grip," with the thumb and other fingers pressing from opposite sides so they can hold a needle or throw a spear. Australopithecus had undoubtedly begun to use a version of this grip but because his fingers were thick and his thumb relatively short he would not have been able to use it as effectively as Homo erectus, who must have had thinner and straighter fingers and a longer and more flexible thumb, similar to modern man's; with the use of what is termed an opposable thumb, his manual dexterity must have been considerable.

While Homo erectus was not built quite like his forebears, neither did he resemble us exactly. His bones were heavier and thicker than a modern man's, and bigger bones required thicker muscles to move them. These skeletal differences, however, were not particularly noticeable. "Below the neck," one expert has noted, "the differences between Homo erectus and today's man could only be detected by an experienced anatomist." From the neck up, however, Erectus looked much more primitive. He had a low, sloping forehead, thick, jutting brows and a massive jaw with only a hint of a chin. Yet, with the exception of his eyebrows, these primitive aspects were less pronounced than they had been in his forebears, and the basic proportions of his head had begun to change. As his brain expanded, his head became higher domed than that of Australopithecus and his forehead receded less. In addition, his jaw became smaller as his demands on it lessened: His diet began to include more easily chewed, cooked food. Taken together these changes tended to humanize his looks; nevertheless his face was probably the least modern thing about him.

In the long progression of primates that led from apelike creatures to man, the changes marking Homo erectus as the first men are matters of degree. Many early anthropologists regarded Australopithecus as very bestial, but the modern view is that man's forebears were far more human than the first apish reconstructions suggested. It follows that the distinctions between Australopithecus and Homo erectus were more subtle than they were once thought to be. Somewhere between early Australopithecus and the first men there were numerous intermediate forms, becoming more and more like humans with succeeding generations. In fact, some classifiers have proposed a separate species, *Homo habilis*, for a creature whom others regard as the last of the Australopithecines. Habilis' brain was slightly larger than that of other Australopithecines, he was a shade taller (about four and a half to five feet tall), and his stance may have been a little more erect. Whether he should be lumped with the Australopithecines or distinguished from them as a separate species of primitive man is an argument best left to the taxonomists. But if we could have seen him with our own eyes, Homo habilis would have looked to us like a creature with a rather apelike head on top of a remarkably human-looking body. His successors, Homo erectus, were different. As one anthropologist has remarked, "they were the first people of whom, if you met them walking across a field, you would say, 'these are definitely not apes, they're men.' "

Far more than increased height, erect stature, manly gait or improved grip, the remarkable change in Homo erectus' brain and behavior clearly marked him as the first of men. The sharpest difference between Homo erectus and all the primates who preceded him was the large size of his brain, and the complexity of behavior this made possible. While the size of the brain case cannot always be relied on as an accurate measure of brain power, it is generally true that among animals of similar size, species with large brains are more intelligent than those with small brains. . . .

Homo erectus' brain was not only larger than earlier primate brains, but it was probably far more complex in internal organization. There are no fossil brains to show what the inside of his brain was like, of course, but there is considerable indirect evidence. A comparison of the brains of modern apes and man, for example, shows that the larger human brain has quite different proportions, with larger and more complex individual cells, enlarged centers for such functions as speech, hearing and sight, and many more internal interconnections between brain cells. Scientists consider the internal pattern of organization as important as size in making the human brain what it is. . . .[2]

. . . This restructuring of the brain was already evident in Australopithecus, for he made tools and showed other signs of manlike behavior, but the great change came with Homo erectus and gave him the mental equipment to distinguish him as a man. The proof lies in what he did with his better brain. In some ways, he simply went further along paths his forebears had begun to tread. In other ways, his accomplishments were unique. He was much more successful than Australopithecus at making tools and at planning and coordinating his hunts.

From the bones of the animals that the first men brought back to their primeval campsites, it is clear they were patient and premeditating hunters, able to lie in wait where they knew their prey would pass or to spread out stealthily and encircle some swift and leaping beast. By such means, the first men brought down gazelles and antelopes, tangled with saber-toothed cats, and at last could hunt down even the goliaths of the grasslands, the elephants.

Homo erectus' increasing concentration on hunting also led to the development of a social organization that is clearly human, being based on a distinct division of labor between men who hunted and women who picked and gathered food. . . .

But the creation of human society is only one of the great accomplishments that give Homo erectus his place as the first among men. He spread out over much of the world, making his way into the wintry climates of the north where there is no evidence that a monkey or ape had ever survived before. He mastered fire, using it to cook, to warm himself and even to trap big game. And he learned to talk. Australopithecus must have been capable of making expressive sounds that others around him could comprehend. But although his system of communication was considerably more advanced than that of the apes, he still did not use words. From the development of Homo erectus' brain, and from changes that probably occurred in his vocal tract, it seems that he was capable of some rudimentary human speech. . . .

The story of Homo erectus—who he was, where and when he lived, and what he did in his time on earth—can now be told because a small band of determined scientists spent their lives discovering these facts. Parts of the tale came to light in widely scattered corners of the world—fossils, artifacts and living sites were found at different times by men who sometimes did not

[2]See "The Brain Grows Up" by Phillip V. Tobias in this chapter and selections in Chapter 13.—Ed.

recognize the real meaning of their discoveries. Some of the discoveries, indeed, provoked bitter and prolonged controversies among the experts. But in recent years the quantity of evidence accumulated by the fossil hunters has come to be recognized as the portrait of a distinct species—a record left behind by the first men and assembled by his modern ancestors in an extraordinary tale of scientific detection.

There is still no clear evidence that *Homo sapiens* and *Homo neandertalis*, or Neanderthal, existed at the same time in Europe. No more clear is whether Neanderthal was effectively extinct when the long-faced, narrow-skulled *Homo sapiens* crossed the evolutionary bridge or *Homo sapiens* slew *Homo neandertalis*. Dr. Bronowski strongly doubted the likelihood of the latter possibility. However, it is a romantic concept put forth with considerable passion in *The Inheritors* by William Golding. Here in Golding's mind's eye is the confrontation of Lok and Fa, a Neanderthal couple, and the "new people," who ride magically over the water in dug-out logs.

Homo neandertalis* Meets *Homo sapiens 13

William Golding

At last they saw the new people face to face and in sunlight. They were incomprehensibly strange. Their hair was black and grew in the most unexpected ways. The bone-face in the front of the log had a pine-tree of hair that stood straight up so that his head, already too long, was drawn out as though something were pulling it upward without mercy. The other bone-face had hair in a huge bush that stood out on all sides like the ivy on the dead tree.

There was hair growing thickly over their bodies about the waist, the belly and the upper part of the leg so that this part of them was thicker than the rest. Yet Lok did not look immediately at their bodies; he was far too absorbed in the stuff round their eyes. A piece of white bone was placed under them, fitting close, and where the broad nostrils should have shown were narrow slits and between them the bone was drawn out to a point. Under that was another slit over the mouth, and their voices came fluttering through it. There was a little dark hair jutting out under the slit. The eyes of the face that peered through all this bone were dark and busy. There were eyebrows above them, thinner than the mouth or the nostrils, black, curving out and up so that the men looked menacing and wasp-like. Lines of teeth and seashells hung round their necks, over grey, furry skin. Over the eyebrows, the bone bulged up and swept back to be hidden under the hair. As the log came closer, Lok could see that the colour was not really bone white and shining but duller. It was more the colour of the big fungi, the ears that the people ate, and something like them in texture. Their legs and arms were stick-thin so that the joints were like the nodes in a twig.

Now that Lok was looking almost into the log he saw that it was much broader than before; or rather that it was the two logs moving side by side.

There were more bundles and curious shapes in this log and a man lay among them on his side. His body and bone was like the others but his hair grew on his head in a mass of sharp points and glistened and looked hard as the points on a chestnut case. He was doing something to one of the sharp twigs and his curved stick lay beside him.

The logs sidled right into the bank. The man at the back—Lok thought of him as Pine-tree—spoke softly. Bush laid down his wooden leaf and caught hold of the grass of the bank. Chestnut-head took his curved stick and twig and stole across the logs until he was crouched on the earth itself. Lok and Fa were almost directly above him. They could smell his individual scent, a sea-smell, meat-smell, fearsome and exciting. He was so close that any moment he might wind them for all he was below them and Lok inhibited his own scent in sudden fear, though he did not know what he did. He reduced his breathing till it was the merest surface and the very leaves were more lively.

Chestnut-head stood under them in the sun pattern. The twig was across the curved stick. He looked this way and that round the dead tree, he inspected the ground, he looked forward again into the forest. He spoke sideways to the others in the boat out of his slit; soft twittering speech; the whiteness quivered.

Lok felt the shock of a man who has trusted to a bough that is not there. He understood in a kind of up-side-down sensation that there was no Mal face, Fa face, Lok face concealed under the bone. It was skin.

Bush and Pine-tree had done something with strips of hide that joined the logs to the bushes. They got quickly out of the log and ran forward out of sight. Presently there was the sound of someone striking stone against wood. Chestnut-head crept forward too and was hidden.

There was nothing of interest now but the logs. They were very smooth and shiny inside where the wood could be seen and outside there were long smears like the whiteness on a rock when the sea has gone back and the sun has dried it. The edges were rounded, depressed in places where the hands of the bone-faces had rested. The shapes inside them were too various and numerous to be sorted. There were round stone, sticks, hides, there were bundles bigger than Lok, there were patterns of vivid red, bones that had grown into live shapes, the very ends of the brown leaves where the men held them were shaped like brown fish, there were smells, there were questions and no answers. Lok looked without seeing and the picture slid apart and came together again. Across the water there was no movement on the island.

Fa touched him on the hand. She was turning herself in the tree. Lok followed her carefully and they made themselves spy-holes that looked down into the clearing.

Already the familiar had altered. The tangle of bush and stagnant water to the left of the clearing was the same and so was the impenetrable marsh to the right. But where the trail through the forest touched the clearing thorn bushes were now growing thickly. There was a gap in these bushes and as they watched they saw Pine-tree come through the gap with another thorn bush over his shoulder. The stem was clean white and pointed. In the forest behind him the noise of chopping went on.

Fear was coming from Fa. It was not a shared picture but a general sense, a bitter smell, a dead silence and agonized attention, a motionlessness and tensed awareness that began to call forth the same in him. Now, more clearly than ever before there were two Loks, outside and inside. The inner Lok could look for ever. But the outer that breathed and heard and smelt and was awake always, was insistent and tightening on him like another skin. It forced the knowledge of its fear, its sense of peril on him long before his brain could understand the picture. He was more frightened than ever before in his life, more than when he had crouched on a rock with Fa and a cat had paced to and fro by a drained kill, looking up and wondering whether they were worth the trouble.

Fa's mouth crept to his ear.

"We are shut in."

As Dr. Bronowski suggests, man's cultural evolution has been built on "a richer and more intricate conjunction of human faculties" than other animals possess. Certainly one of the most refined and useful of man's faculties is his brain, the evolution of which is traced in more detail in Chapter 13. Here, Phillip V. Tobias points out that bigger skulls do not necessarily mean better brains in an evolutionary sense. Neanderthal man, presumed to be less of an intellectual innovator than *Homo sapiens*, had a mean cranial capacity of 1470 cubic centimeters; man's mean cranial capacity is only 1370 cubic centimeters. 15

The Brain Grows Up

Phillip V. Tobias

It seems that the trend toward increased brain size, which marked the first 2 or 3 million years of human evolution, had spent itself—and had done so as recently as the late Pleistocene. The wave of brain expansion had passed its peak. Some of our Stone Age men carried the twin processes of reduction of teeth and jaws, on the one hand, and expansion of brain on the other, so far that they seem to represent an ancient foreshadowing of the popular idea of the man of the future. The matter has been most vividly portrayed by Loren Eiseley, Benjamin Franklin Professor at the University of Pennsylvania. In one of the essays in his delightful florilegium of anthropological sketches, *The Immense Journey*, he writes:

One . . . skull lies in the lockers of a great metropolitan museum. It is labelled simply: Strandloper, South Africa. I have never looked longer into any human face than I have upon the features of that skull. I come there often, drawn in spite of myself. It is a face that would lend reality to the fantastic tales of our childhood. There is a hint of Wells's Time Machine folk in it—

From pp. 100–103, Phillip V. Tobias, *The Brain in Hominid Evolution*, Columbia University Press, 1971. Reprinted by permission of the publisher.

those pathetic, childlike people whom Wells pictures as haunting earth's autumnal cities in the far future of the dying planet.

Yet this skull has not been spirited back to us through future eras by a Time Machine. It is a thing, instead, of the millennial past. It is a caricature of modern man, not by reason of its primitiveness but, startlingly, because of a modernity outreaching his own. It constitutes, in fact, a mysterious prophecy and warning. For at the very moment in which students of humanity have been sketching their concept of the man of the future, that being has already come, and lived, and passed away.

When one takes a long-term evolutionary look at this problem, it seems that a larger brain size may once have been vitally important in aiding survival—for instance, in a world teeming with wild animals and devoid of harnessed fire. The further development of man would seem to have placed less and less of a premium on the size of his brain. For culture and the benevolence of social life have taken the place of nimble wits as an insurance policy against extinction. Beyond a certain stage in the increase of brain size, we have no evidence that further increase in any way improved man's adaptive abilities.

We have reached a stage in human evolution—and I am convinced we reached it thousands of years ago—where 100 people with smaller brains stand just as good a chance of surviving to child-bearing age as 100 with larger brains, and are likely to leave no fewer children than the others. Brain size is no longer a yardstick to survival as it may once have been. We have used these very brains to create conditions in which mere brain size is of negligible importance; we have used them to develop new mechanisms of adaptation, tools, shelters, clothing, fire, social institutions—and central heating, air-conditioning, refrigeration, disinfection, mink coats, and sunshades. You can have these things with a size 5¾ hat or with a size 8½, just as it makes no difference what size shoe you take.

Within very wide tolerance limits, brain size seems to make no difference as to your ability to avail yourself of the joys and benefits of modern living. So, too, it seems that brain size does not limit your ability to contribute to society, culture, science. Some gifted people, such as Leon Gambetta and Franz Joseph Gall, have had very small brains. Others, also gifted, have had very large brains. Doctor Samuel Johnson had about 2000 gm., while Lord Byron and Oliver Cromwell had about 2350 gm. . . . And some very ordinary persons have had equally large brains.

Small wonder that in 1963 Gerhardt von Bonin could say: "Certainly, the weight of the brain is a very poor indicator of its functional value". . . and, again, "brain size as such is a very poor indicator of mental ability.". . . A century earlier, James Hunt had stood before the Anthropological Society of London and declared, ". . . we now know that it is necessary to be most cautious in accepting the capacity of the cranium as any absolute test of the intellectual power of any race. . . ."

In the parlance of the evolutionary biologist, the selective pressures that once placed a considerable premium on big brains have somewhat relaxed since the late Pleistocene. Whatever advantages a larger cranial capacity must earlier have imported were no longer evident. It was no longer a key to survival to have a large, well-filled braincase.

For Dr. Bronowski, the cave paintings with which cave men adorned the walls of their homes twenty thousand years ago express "the power of anticipation; the forward looking imagination." They were art, and they were in some way magical. These points are explored in more detail by J.Z. Young.

The 'Meaning' of Cave Paintings

J.Z. Young

The [early cave] paintings may have been connected with a system of fertility magic or with a cult of food magic, the representation supposedly serving to increase the probability of finding game. A few of the animals are apparently pierced with spears and this has also suggested that they are enactments of hunting scenes. They could indeed have served as pictorial adjuncts to verbal communication in something of the way suggested . . . for early sculpture. Pictures are a great help to communication even today and would have been even more so if talking itself was a laborious and perhaps boring task, as writing is for young children now. But such explanations are much too simple. For one thing the caves were dark and inaccessible—not at all the place to teach the young to hunt.

Probably by the Upper Palaeolithic period of man's history ideas had already developed on to much more subtle planes. The pictures almost certainly have some kind of magical or mystical purpose, directed to influencing events, perhaps with rituals to pacify spirits, postulated as likely to be inimical if not placated. It is a pity that we have not yet been able to interpret them well enough to understand how men spoke and thought in those days. Probably the motivation to paint was complex then, as it is today. The urge to represent seems strong in creatures like humans who are steeped in symbolism. The annual migrations of masses of animals upon which the men depended must have powerfully aroused this tendency to represent the objects of interest, their multiplicity, their movement, and their death at the hands of hunters. Perhaps it was this very tendency to represent that provided the capacity for communication upon which co-operative hunting depended. It may have spilled over from dance form and ritual into sounds and words, and as these became more abstract, into signs and pictures, conserved deep in the caves and added to by generation after generation of hunters on their regular migrations. The three main theories that have been put forward to 'explain' cave painting may perhaps be summarized as follows.

Art for Art's Sake

The paintings may have been done simply for the pleasure of the painter or to decorate the living space of the group. This motivation cannot be ignored, but neither can it be accepted as the whole answer. Everyday experience tells us that *H[omo] sapiens sapiens* likes drawing and painting. Ethnological studies tell us that food-gathering peoples, such as Australians who decorate caves today, give this as part of their motive. As evidence that the matter is more complex than this is the inconvenience of the placing of many Palaeolithic paintings and the fact that they show only a limited subject-matter—nearly always large food animals, rarely humans, carnivores, or birds, and never plants or inanimate objects.

On the other hand, some paintings are at accessible sites, which were used for occupation; these in fact include more representations of humans than are found in the caves. It may well be that these were done for purposes of decoration or as erotic stimulants, or perhaps just as 'doodles', even by children. Many of the obscure signs may perhaps have no 'meaning' at all; and the superposition of one drawing on another may have no more significance than imitative repetition.

Sympathetic Magic

With all possible allowance for the artistic and casual reasons there remain

strong indications that the paintings have an element of symbolism. The choice of subject suggests that this was concerned with the hunt, serving to make more prey available either by magic or instruction, or by a mixture of both these. There is still disagreement about the interpretation of the marks on the animals that are classified as spears or wounds, and not all carry them. One observer who is himself a hunter has suggested that many of the representations seem to be drawings of *dead* animals. The extent of belly revealed, the perspective of the legs, projecting tongue, and prominent jaw suggest that they were drawn from prostrate corpses, not standing animals. Yet another interpretation is that the large bellies sometimes shown indicate pregnant animals, presumably as part of a fertility cult. Difficulties with all such suggestions are that they generalize from a few instances and take insufficient account of the possibility that the features shown are stylistic conventions. The artist may draw animals as if from the side because that was the accepted way to do it; this was the way animals were thought of and that was how you showed them. The concept may have come originally from seeing dead animals, without the artist meaning to represent dead ones in his picture.

The Pictures as Symbols of Religion or Totemism

The Abbé Breuil, who made a comprehensive study of the paintings, insisted that the caves were in essence 'sanctuaries'. There is certainly much to show that they were indeed connected with some kind of ritual. The very fact of difficulty of access points to some secret ceremonial purpose. An obvious suggestion is that the animal signs were somehow connected with totemism, that is the division of society into a number of groups each connected with an animal. The art of modern hunting communities sometimes has some such connection (but by no means always!). There are some suggestions that the caves may have been used for initiation ceremonies; for example, there may be numerous small foot-marks, said to be those of children (but how big were the Palaeolithic feet anyway?).

Perhaps the strongest evidence of ritual significance is the fact of the varying proportions of the various animals in different parts of the caves. Counts have been made . . . to substantiate this plan, but the identifications of the animals are bound to be difficult and the percentages in different parts of the caves are not very convincing.

Ucko and Rosenfeld[3] conclude that the truth is that we do not understand the paintings. 'It is possible . . . that some representations were the work of children (perhaps some of the floor engravings), that some were used in acts of sympathetic magic (perhaps some of the representations pierced with holes), that some were placed in particular situations in order to please (perhaps some of the open-air low-reliefs), and that some were illustrations of myths and traditions (perhaps those which contain imaginary creatures, anthropomorphs and unexpected combinations of animal species). It is very possible, however, that some and perhaps many Palaeolithic representations were made for reasons which still totally escape the modern observer.'

In this chapter and throughout *The Ascent of Man*, Dr. Bronowski emphasizes man's uniqueness as a major building block in his cultural revolution. It was exemplified in his ability to tailor his environment to his needs, to hunt, to make tools, to create art, and to think ahead. Expanding those tenets, Clifford Geertz suggests that beyond his innate uniqueness, man's behavior within any culture is shaped and programed by a complex of external factors. The ramifications of that viewpoint are expressed in the following excerpt.

The Impact of the Concept of Culture on the Concept of Man

Clifford Geertz

I want to propose two ideas. The first of these is that culture is best seen not as complexes of concrete behavior patterns—customs, usages, traditions, habit clusters—as has, by and large, been the case up to now, but as a set of control mechanisms—plans, recipes, rules, instructions (what computer engineers call "programs")—for the governing of behavior. The second is that man is precisely the animal most desperately dependent upon such extragenetic, outside-the-skin control mechanisms, such cultural programs, for ordering his behavior.

Neither of these ideas is entirely new, but a number of recent developments, both within anthropology and in other sciences (cybernetics, information theory, neurology, molecular genetics) have made them susceptible of more precise statement as well as lending them a degree of empirical support they did not previously have. And out of such reformulations of the concept of culture and of the role of culture in human life comes, in turn, a definition

From Clifford Geertz, "The Impact of the Concept of Culture on the Concept of Man," in Platt (ed.), *New Views of the Nature of Man,* The University of Chicago Press, © 1965.

[3]Ucko and Rosenfeld, *Palaeolithic Cave Art*, Weidenfeld and Nicolson, London, 1967.

of man stressing not so much the empirical commonalities in his behavior, from place to place and time to time, but rather the mechanisms by whose agency the breadth and indeterminateness of his inherent capacities are reduced to the narrowness and specificity of his actual accomplishments. One of the most significant facts about us may finally be that we all begin with the natural equipment to live a thousand kinds of life but end in the end having lived only one.

The "control mechanism" view of culture begins with the assumption that human thought is basically both social and public—that its natural habitat is the house yard, the market place, and the town square. Thinking consists not of "happenings in the head" (though happenings there and elsewhere are necessary for it to occur) but of a traffic in what have been called significant symbols—words for the most part but also gestures, drawings, musical sounds, mechanical devices like clocks, or natural objects like jewels—anything, in fact, that is disengaged from its mere actuality and used to impose meaning upon experience. From the point of view of any particular individual, such symbols are largely given. He finds them already current in the community when he is born, and they remain, with some additions, subtractions, and partial alterations he may or may not have had a hand in, in circulation there after he dies. While he lives he uses them, or some of them, sometimes deliberately and with care, most often spontaneously and with ease, but always with the same end in view: to put a construction upon the events through which he lives, to orient himself within "the ongoing course of experienced things," to adopt a vivid phrase of John Dewey's.

Man is so in need of such symbolic sources of illumination to find his bearings in the world because the non-symbolic sort that are constitutionally ingrained in his body cast so diffused a light. The behavior patterns of lower animals are, at least to a much greater extent, given to them with their physical structure; genetic sources of information order their actions within much narrower ranges of variation, the narrower and more thoroughgoing the lower the animal. For man, what are innately given are extremely general response capacities, which, although they make possible far greater plasticity, complexity, and on the scattered occasions when everything works as it should, effectiveness of behavior, leave it much less precisely regulated. This, then, is the second face of our argument: Undirected by culture patterns—organized systems of significant symbols—man's behavior would be virtually ungovernable, a mere chaos of pointless acts and exploding emotions, his experience virtually shapeless. Culture, the accumulated totality of such patterns, is not just an ornament of human existence but—the principal basis of its specificity—an essential condition for it.

In the modern world of high-speed travel, television, and satellites, the external influences on culture of which Geertz speaks are far greater in number than they were even as recently as a century ago. Thus, anthropologists must wrestle with an ever-increasing rate of cultural evolution and diffusion. The increased rate of change has caused man to look not only to the past for continuity, but also to the future, as discussed in this selection from *Cultural Anthropology Today.*

Change in the Modern World
from *Cultural Anthropology Today* (CRM Books, Publisher)

The emergence of the Western industrial and scientific revolutions opened an era in which man's cultures and social orders have changed more rapidly and extensively than in any other period in human history. And the basic processes of culture change—innovation, diffusion, acculturation, cultural growth, and the development of social institutions in response to changing circumstances and needs—are occurring simultaneously throughout the world at an astounding rate. No human group—not even the most remote settlement of Eskimo in Hudson's Bay, the most backward peasant village in India, the Pygmies in the most inaccessible region of the African rain forest—has been able to avoid change. And most people are confused, bewildered, and overwhelmed by processes of change so swift that each succeeding generation appears a stranger to the one that preceded it.

Although contemporary processes of change are unique in many ways, the basic processes and mechanisms of change are the same as those that have occurred throughout history. The changes affecting Mexican peasants moving from rural villages into urban centers can be compared with the changes affecting rural migrants from Nile villages into Cairo. Contemporary urbanization can be measured against the changes that began with rising civilizations in Mesopotamia, Egypt, India, China, and Peru thousands of years ago that first began to attract villagers from the rural hinterlands. Ultimately, such comparative studies reveal the regular processes of change. It is this knowledge that enables the anthropologist to help plan for the future...

Under the influence of the Western idea of progress, people everywhere have now come to value change and to look for the ideal society in the future instead of in a past "Golden Age." This value orientation encourages and rewards change. Also, in the contemporary world, the development of transportation and communication facilities such as the airplane, radio, television, and the tremendously increased amount of printed material has given contemporary man access to more cultural materials than ever before, and this has increased the number of new raw materials upon which innovation is based.

Diffusion of cultural elements within a population and the passing of elements between cultures have also been intensified by increased travel and contacts between people and, again, by faster and more effective communications. Several thousand years ago, it took centuries for such innovations as the bow and arrow and pottery to spread; today diffusion is almost instantaneous. Paris-dictated dress styles and American popular songs appear almost instantaneously in downtown Tokyo and Hong Kong, and technological advances in German automobiles quickly appear on Japanese automobiles imported into the United States.

2

The Harvest of the Seasons

I turn the lea-green down
Gaily now,
And paint the meadows brown
With my plough.

Patrick Kavanagh

At one time, man and animal moved along their own evolutionary paths, charting their isolated destinies. Then man, through cunning, wisdom, and intellect, gained the upper hand, and animals were put to work for man. F.E. Zeuner summarizes below some of the possible reasons for animal domestication. Two theories of animal domestication that are not discussed by Zeuner are based on the concepts of "imprinting" and "propinquity."

Very early in the development of the young of some species, the newborn fix upon a scent or a moving object. A lasting attachment results, and for a time, the young strive to remain close to the object. The adoption of such an object, sometimes a parent, is called *imprinting.* The duration of this phenomenon is brief and apparently is genetically determined, but man has learned to manipulate the object to be imprinted or the manner in which imprinting takes place. Studies by Konrad Lorenz and others indicate that this process may have influenced the domestication of birds such as ducks and geese and also may have led to the domestication of some large mammals, such as cattle, which can move actively about soon after birth.

The domestication of animals may also have been affected by major climatic changes at the end of the last great glacial period. The drive to find water and warmth may have forced man into close association with wild animals. Close association, called *propinquity,* increases the possibility for domestication of some species. The evidence for propinquity as a factor in animal domestication is strong in North Africa, but it does not appear to have been a factor in other areas where domestication occurred.

Domestication of Animals

F.E. Zeuner

There is no evidence that man had acquired the habit of keeping animals in captivity or domestication before he reached the stage of living in social units of some size. The first animal to pass under his sway was the dog. The hoofed animals that followed were at a certain stage systematically used only as sources of food. Increased security and food supply gave Neolithic man more time, and therefore more opportunity, to improve certain domestic products. Thus domestication is not only a technique introduced by early man but, as so often with the arts, is important for the advancement of other arts.

As to the origin of domestication there are many theories, nearly all based on the conception of purposeful procedure. Man, it was thought, needed certain animals and therefore contrived to tame them. This view includes the theory of a religious origin—the animals being for use as sacrifices—and the theory that domestication was invented to satisfy economic needs for meat and skins. Such views ignore the fact that in Mesolithic times, when domestication began, it would have been far easier to obtain the necessary supplies by the ancient methods of hunting and trapping than to embark on extensive experiments in taming, which could bring their reward only after generations.

It is much more fruitful to consider man as an integral part of his physico-biological environment. It then becomes apparent that both his habits and those of certain animal species make domestication almost inevitable. For the social relation of domestication is not restricted to man and his animal subordinates. He has applied the same practice to his own species, though it is then commonly called slavery. . . .

Man may have tolerated the presence of other species as pets at a very early date. It is conceivable that even Upper Palaeolithic man would have done so, though the form of his economy prevented him from developing such relationship to full domestication. Some Australian tribes illustrate this pet-keeping stage. They have never domesticated animals, but wallabies and opossums, bandicoots and rats, even frogs and young birds, are tied up in the camp, though they are not properly fed or cared for, and most of them soon die. Wallabies and opossums, and occasionally cassowaries, however, learn to fend for themselves within the camp. It is noteworthy that the dingo is similarly treated, being tied up when young and released when it has got used to the human environment. Pets are kept in the same way by certain American Indians.

In view of such facts, it is not surprising that the belief has arisen that the habit of keeping young animals as pets is at the root of domestication in general. This is certainly an overstatement, but it is probable that pet-keeping provided one basis for domestication on an economic scale. The mothering instinct of women may have played a part. Pet-keeping is particularly likely to have been a factor in the domestication of the dog. The scavenging habits of wild dogs brought them into contact with the human social medium, and the pups may have been adopted. . . .

There is another theory of how domestication arose under conditions of amicable relations. It assumes that totem animals, the killing of which is often taboo, would become tame. There is, however, a great difference between the taming of a totem animal (in any case an unusual event) and its economic exploitation under conditions of complete domestication. Moreover, animals which are now domesticated do not appear to have been often chosen as totems. Though totemism is perhaps as old as the Upper Palaeolithic, as witnessed by the bison hunter of Lascaux, it is very improbable that it ever led to domestication, which is based on the motive of exploitation of the properties and abilities of the animals.

Symbioses are not always voluntary. A certain amount of coercion, i.e. transplantation into the social medium of the more intelligent species, or extension of the latter into the social medium of the weaker species, is common. Yet both parties derive certain advantages; thus the animal obtains personal security, while his human master secures an easier and more ample food-supply.

Food-supply appears to have played an important part in the establishment of close associations of animals and man. This is most obvious with the scavengers. Some species enter into a relationship in which one lives regularly on the food-debris or other waste products of the other. If the removal of the waste products is an advantage to the producer, scavenging might closely approach a true symbiosis. But scavengers often have the habit of preying on the host, and especially on its progeny, as opportunity arises. Scavenging thus grades into social parasitism.

Most scavengers are, moreover, not closely host-specific. Thus the jackal can live perfectly well without man, but, where human settlements are available, he will enter into a loose and impermanent, though nevertheless quite regular, relationship with him. The mutual advantage is obvious enough. In other associations of the scavenging type, the relations are permanent. The loose associations of the wild dogs with human society are of great interest, since they illustrate one of the ways in which domestication is likely to have begun. The two social media, that of the dog and that of the man, overlap because man produces offal which dogs will eat. There is no cause for enmity in this, unless one of the species interferes with the habits of the other. This is indeed so in the case of hyenas, for example, which are prone to steal man's food-reserves and are too fierce for the establishment of friendly relations. The smaller species of wild dogs are in a different category. Not only do they fear man, but their association in packs with a recognized leader affords the possibility of transfer of allegiance to man, once he is appreciated as a being of superior strength and cunning.

It is obvious enough that not every scavenger of the dog tribe is a prospective candidate for domestication. The jackals head the list in so far as they are unlikely to attack man. Their social level, however, is lower than that of those wild dog-like creatures which practise active hunting, and which for this reason would be more ready to form with man groups that can be regarded as mixed packs. The smaller races of wolves found in southern countries, which scavenge as well as hunt, would therefore appear to be the most amenable to domestication, but they also tend to regard human settlements as legitimate hunting-grounds, from which, as in India today, they frequently steal goats, domestic dogs, and even children. These propensities are most strongly developed in the large northern races of wolves, and these wolves are thus the least likely to have provided the initial stock from which the domestic dog emerged. . . .

There is one example of domestication of an animal by man which may be regarded as social parasitism. It is that of the reindeer. Both the domesticators and the domesticated have here remained in the state of nomadism. Nevertheless, the activities of man in relation to the reindeer can only be described as those of a social parasite. Like so many guests of ants and termites, the human guest-species has even supplied a delicacy which the animal host-species is eager to obtain. In so far, however, as the exploiting species gains the upper hand in this process and the exploited species degenerates, the case of reindeer and man might equally well be discussed under the head of true domestication; but since it appears to have begun as ordinary social parasitism, it is more usefully treated here.

Of the several theories concerning the domestication of the reindeer, that which derives it from the practice of decoy-hunting is the best founded. There are numerous records of reindeer-hunting with decoy animals. Three methods may be quoted. According to Pallas (an eighteenth-century traveller), the Samoyed hunter, in his time, would select four or five tame hinds and fawns. Under their cover, holding them by ropes, he would approach the herd of wild reindeer against the wind, until near enough to shoot his arrow. A second method now practised by the Tungus is to leave a few tame hinds, during the rutting season, on a feeding-ground of the wild deer. The wild stags associate with the hinds, and are killed when the hunter approaches the herd in a day or two. A third method is adopted by both Tungus and Samoyed. Ropes or thongs are tied around the antlers of a strong, tame, rutting stag. It is set free when a wild herd with a stag is in sight. The tame stag begins to fight the wild one. Their antlers become entangled and give the hunter the chance to approach.

Tame reindeer which have proved their worth would naturally be cared for by the hunter. Since tame hinds will mate with wild stags, any hunter owning decoy reindeer would almost unintentionally become a reindeer-breeder. The ease with which domestication of reindeer can be effected is due largely to the similarity of the social state of man and of deer. Both are nomads, and domestication compels neither to adopt any profound change of habit. To this day, reindeer, both wild and domesticated, remain nomadic as do the tribes who follow the herds, preying on the wild and controlling and exploiting the tame. In doing so, man takes his toll of the species much as wolves do, and the only advantage that accrues to the domesticated reindeer is protection from these and other predators. The reindeer, as a community, have to pay heavily for this protection since, instead of being devoured by wolves and bears, they are eaten by man. Man's numerical inferiority to the reindeer and his adaptation of himself to their habits make him their social parasite. Despite their domestication, hunting of wild reindeer continues. The more elaborate forms of exploitation of the domesticated deer, as for draught or riding, or as a source of milk, are all comparatively modern.

At the outset of the process of domestication, man appears to have availed himself of the eagerness of reindeer for salty matter, notably for human urine. This substance attracts and binds reindeer to human camps. The craving arises from lack of salt in the available water which is derived from melting snow. Even today the human nomads take full advantage of this weakness of the reindeer, so that the supply of a delicacy provides the meeting-ground on which the social media of the two species overlap. This is a most remarkable parallel to the conditions under which societies of social insects are invaded. The reindeer is not the only animal domesticated through salt licks. Another example is the Mithan of Assam, a cross of the gaur with domesticated humped cattle. The gaur is a large wild bovine not closely related to cattle; it is attracted by salt licks placed in suitable localities. . . .

The invention of agriculture brought in its train fresh opportunities for developing the guest-host relation, since cultivated lands are excellent feeding-grounds for several herbivorous and gregarious animals. So long as the land lay fallow, this state of affairs would not differ radically from other cases of scavenging. But when the crop was growing, the same practice of the animals constituted an act of robbery, though it afforded social contacts which may well have led to the domestication of the aurochs and related large bovines, and perhaps of other species. . . .

Domesticated animals proved their worth and played their part in the Neolithic economic revolution. In the urban revolution which followed there were many experiments in domestication, notably in Egypt in Old Kingdom times. Pictorial representations, such as those in the grave of Mereruka at Saqqara, show animals like gazelles, ibex, and addax antelopes with collars round their necks, not to mention monkeys and even carnivores like hyenas. Hyenas appear to have been kept and stuffed like geese. These attempts to tame or fully to domesticate animals, other than the well known standard stock beasts, were abandoned sooner or later. . . .

From the preceding considerations it is possible to deduce an order in which domesticated animals are likely to have appeared. The scavengers were domesticated first. This would have been possible even at a Palaeolithic level. The second group are the nomadic animals, i.e. those which carry out seasonal migrations. For them, a fully fledged Neolithic economy with agriculture would not have been a suitable medium. It is suggested, therefore, that this group was domesticated earlier than the true Neolithic, and at the latest in the Late Mesolithic. These animals enabled man to obtain food by slaughtering in addition to hunting, so that human habitation could be more permanent. If this view be right, the domestication of goat and sheep would have preceded agriculture. The third group includes those animals for the maintenance of which settled life is a prerequisite, at least in the early stages. The main members of this category are cattle. Finally, a fourth category comprises animals which were domesticated primarily as a means of transport, i.e. as beasts of burden, for riding, or for traction. The origin of this group is obscure, though it seems more recent than the third group. Its most curious feature is that its members are not normally slaughtered for food. This suggests that religious inhibitions applied to it that did not exist in the case of other groups (except perhaps the dog). The use of animals for transport was probably developed mainly with cattle. Thus we reach a tentative grouping of some of the early domesticated animals: (i) dog, fox; (ii) reindeer, goat, sheep; (iii) pig (?), cattle; (iv) horse, ass, onager.

This theoretical sequence is to some extent confirmed by the evidence from the Belt cave on the Persian coast of the Caspian Sea. Domestication seems to have begun there in late Mesolithic times, when the cave was occupied by hunters whose chief game consisted of gazelle, wild cattle, sheep, and goat. The proportion of mature to immature specimens suggests that slaughtering of goats was practised already, but, the number of usable specimens being as low as 13, further evidence will be required. In the Early Neolithic level, both goat and sheep are believed to have been domesticated. Again we must await larger numbers of specimens, but it is noteworthy that the same horizon has also yielded domestic fowl.

The later Neolithic, characterized by pottery, weaving, stone axes, and cultivation of cereals, has provided more definite evidence of domestication. Among the goats and sheep the number of immature specimens exceeds that of the mature. Ox and pig may have been domesticated at this time, or a little later. If the tentative view put forward above should be borne out by further evidence, domestication of the sheep and goat would have been practised as early as 6000 B.C. as based on radiocarbon dating; domestication of the goat may well have begun even earlier.

It was "the happy conjunction of natural and human events," according to Dr. Bronowski, that created agriculture in the Fertile Crescent ten thousand years ago. Grass seeds scattered by the wind became wheat, a nutritious crop that only man could cultivate and harvest. Elsewhere, in other cultures, wild maize, rice, and other grains were also domesticated. The history of the domestication of maize, or corn, is detailed in this selection by Paul Manglesdorf.

The Domestication of Maize

Paul C. Manglesdorf

If cultivated maize arose from pod corn, it probably had its beginnings in South America, for there is no doubt that pod corn was known in several parts of America even in recent times. Furthermore, if teosinte is a hybrid of maize and Tripsacum,[1] it may be inferred that the advent of maize in Central America is comparatively recent for, if maize had always been present in a region where Tripsacum is common, the hybridization might have occurred again and again. Finally, there is little doubt that Bolivia and Peru represent the primary center of maize diversity. There are good reasons for suspecting, however, that maize as a wild plant had its habitat, not in the Andes where varietal diversity has reached a peak, but in the adjoining lowlands.

Until recently a systematic search for wild maize in South America had never been undertaken. Now at least a reconnaissance has been made. Dr. Hugh C. Cutler has explored parts of southwestern Brazil, Paraguay and eastern Bolivia. He has not discovered wild maize. He has, however, found a widespread distribution of a group of varieties similar to the maize of the Guarany Indians of Paraguay. This variety, if not primitive, at least exhibits primitive characteristics. . . .

Among cultivated varieties Dr. Cutler has found the maximum diversity on the eastern slopes of the Andes in Bolivia from Sacaba and Cliza on the south to Chulumani and Coroico on the north. The fact that large tributaries of the Amazon, the Rio Beni and the Rio Mamoré, though widely separated in the lowlands, both have their headwaters in this region, may be not without significance.

There is, of course, a possibility that wild corn, if it ever existed in South America, is now extinct. The introduction of cattle and other grazing animals into South America would undoubtedly have led to the decimation of wild maize in many of the places where it occurred. Dr. Cutler reports that large areas of the Matto Grosso region in Brazil are so overgrazed that, if wild

From Paul Manglesdorf, "The Evolution of Maize Under Domestication," *Advances in Genetics*, Vol. 1, Academic Press, 1947.

[1]Tripsacum and teosinte are the names given to wild relatives of maize. Both are native to the Americas.—Ed.

maize had ever grown there, it would no longer be in existence. On the other hand there are still vast areas in South America which have received not even a cursory botanical exploration. There is at least a reasonable possibility that wild maize still exists in South America and that it may eventually be found. At least, there is at present no evidence which is seriously in conflict with the hypothesis that cultivated maize arose from a wild maize indigenous to the lowlands of central South America.

Alternative Possibilities

Anderson, although expressing the opinion that it is much too soon to give serious consideration to the problem of the origin of maize, has listed the following additional possibilities with respect to the method and place of origin of maize:

1. That teosinte is a hybrid of maize and Tripsacum but that the original home of maize was in Central America.
2. That teosinte is not a hybrid of maize and Tripsacum but did cross extensively with a South American maize in producing the cultivated maize of North America.
3. That maize is a South American plant which twice became contaminated with Tripsacum, once at a very early date, in South America, a second time, at a much later date in Central America at which time teosinte originated.
4. That maize originated in Asia, possibly as an amphidiploid hybrid of Coix and Sorghum.

The first three of these items do not warrant extensive additional discussion for they shed no new light on the problem and they have at present little, if any, evidence to support them. The last suggestion, that maize may have had its origin in Asia, demands more detailed treatment, for it revives a question which in recent years has been generally regarded as having been satisfactorily answered.

Many of the earlier botanists were of the opinion that maize had an Old World origin. There has, however, never been any convincing evidence to support such an opinion and, on the contrary, the evidence for an American origin has been almost overwhelming. Later, when it became known that maize was already being extensively cultivated in parts of Asia within less than a century after the discovery of America, it was thought by some students of the problem to have had a pre-Columbian distribution in Asia though its American origin was not denied.

The entire problem of a possible Asiatic origin or pre-Columbian Asiatic distribution of maize received a comprehensive and scholarly treatment at the hands of Laufer many years ago. He concluded that the appearance of maize in China in the sixteenth century is an indication of the rapidity with which it spread, once it was introduced to the Old World, rather than evidence that it had been long established there.

No new evidence has been brought forward to contradict Laufer's well-reasoned and strongly-documented arguments. Several of the items which Anderson mentions as lending some plausibility to the hypothesis of an Asiatic origin prove upon examination to be of little real importance. There is, it is true, a unique race of fowls in Chile which lays blue eggs but there is no evidence that the blue-egg character stemmed from a similar race in Asia in prehistoric times. It is apparently the result of a mutation occurring after the fowl had been introduced into America. Pop corn is grown and utilized as food on the island of Timor but there is no indication that it is an ancient food or the principal one. There is no doubt that maize is extensively cultivated by primitive mountainous tribes in northern India, upper Burma and parts of western China. But this fact loses most of its significance when it is noted that potatoes are also extensively cultivated by some of these tribes and that in others the name of maize is derived from the older name for millet.

Until wild maize is found growing in some part of America or until incontrovertible proof of its previous existence has been established the possibility of an Asiatic origin of maize will remain and cannot be dogmatically dismissed. On the basis of present evidence, however, the odds are very much against it.

The Time of Origin

There is no way of determining from the evidence at hand the time at which the domestication of maize began. Anthropologists have hoped that studies upon the origin of maize might throw light upon the age of prehistoric American cultures and civilizations. It is doubtful whether this will prove to be the case. The most that can be expected from such studies is indirect evidence with respect to the comparative age of different cultures. The data presented here, for example, show that maize probably had its center of diversity as a cultivated plant in the Andean region of Bolivia and Peru and that the first highly developed maize agriculture probably occurred in South America rather than in Middle America. To the extent that the prehistoric civilizations of America were based upon maize agriculture these facts indicate, though they certainly do not prove, that the civilization of South America is earlier than that of Middle America. They furnish no clue regarding the absolute time at which either came into existence.

Anderson [and] Carter . . . have recently utilized races of maize in studying the cultures of the Southwest and in arriving at important conclusions regarding them. Such studies can undoubtedly be very fruitful if the data are interpreted in the light of all other kinds of data which have a bearing on the problem. They can lead to false conclusions if other pertinent facts are ignored or erroneously interpreted.

One more conclusion can be drawn regarding the time of origin of maize. It need not have been as far back in the remote past as some students of the problem have supposed. The diversity of maize in South America represents a relatively long period of domestication but, even here, the evolutionary changes involved have not been great and need not have required many thousands of years. The new diversity resulting from hybridization with Tripsacum was probably of an explosive nature and much of it no doubt was created almost literally overnight.

All of the evidence supports the conclusion that the time which anthropologists allow for the presence of man in America is ample, and more than ample, for all the changes which distinguish maize from its putative wild prototype to have taken place.

Jericho, according to Dr. Bronowski, is a microcosm of all the civilizations that have been lofted into orbit by the agricultural revolution. In developing Jericho's culture, its citizens had become multipliers and harvesters. They had fertile soil and sweet water, they learned to domesticate plants and animals, and they built a protective wall, possibly around the entire settlement. For the archaeologist, multi-leveled "first Jericho" was a treasure trove in every sense of the word, as Kathleen Kenyon describes in the following selection from *Digging Up Jericho*.

The First Jericho

Kathleen M. Kenyon 31

The settlement is quite clearly on the scale, not of a village, but of a town. Its claim to a true civic status is established by the discovery, the first of the exciting finds to be produced by the present excavations, that it possessed a massive defensive wall. This was first found in 1952, at the bottom of a comparatively small sounding in Square G4. It was built of large, undressed stones, and survived to the height of 2.50 m. It looked most impressive as one peered down at it, with its base at what then seemed to us the great depth of 7 m. In this area, we uncovered a length of 20 m., and 25 m. to the north, in Site M I, we subsequently picked up what is very probably its continuation. We cannot as yet say whether it enclosed the whole settlement, as it is of course not impossible that it enclosed only a citadel area. But whether or not it enclosed the whole town, it is clear evidence of community organisation. No one small group or family could have carried out an undertaking of this sort, and someone or some corporate body must have provided the organisation to bring the great boulders from the stream beds at the foot of the mountains, half a mile or so away, someone must have ordained its line and someone have supervised its construction.

The wall was free-standing only on its outer side. On the inner side, the contemporary level was 2.80 m. higher than the external one. This terrace was constructed by cutting away the earlier levels on the outer side, and piling them up on the inside, the wall being built against this pile. The wall must originally have stood at least two metres higher than the surviving top, giving a total minimum height of about five metres, for on the inside it served also as a house wall, and the additional two metres is necessary to give headroom within the house.

It was in this house that the plastered skulls must have been preserved, for they were found discarded in its debris, beneath the floor of the next succeeding building. In the fill beneath the floor of the house, put in when the wall was built, was found the surprising number of burials, from which the skulls must have been derived. It is the association of the great defensive wall with the mass burial that makes it tempting to interpret as a massacre the disaster which the number of burials suggests. If a violent attack on a town at that time undefended had resulted in the death of a large number of the citizens, it would be reasonable that when it was rebuilt, a defensive wall should be added. But this can only be a hypothesis, for the surviving portions of the skeletons do not show any identifiable signs of injuries.

After a period, this wall apparently collapsed. After the first collapse, it was rebuilt on the same line. But after a second collapse, a new wall was built 6.50 m. in advance, against the debris which buried the face of the original one. The new one was on the same principle as the old one, with a free-standing outer face and a higher terrace inside, and the internal house was extended over the new fill, probably up to the new wall, but erosion has destroyed the actual junction. This same erosion makes it impossible to say whether the eight succeeding house levels belonging to the pre-pottery Neolithic period, all with the same type of plan, structure and plastered floors, also run up to the new wall, or to some successor.

As I have just said, the town in the stages preceding the first of these two walls was undefended. But in 1954 we found that they were not the earliest on the site. Beneath the first of them was a deep tipped fill, partly bricky and partly ashy. This dipped down steeply to the west, over the face of a yet earlier wall. As this gradually emerged, it became clear that it was much more impressive than the later one which overlay it. It was not built of such large stones, but it had a considerably more regular face, though the stones were likewise undressed. Moreover, it survived to a height of six metres, with its foundations resting on bedrock at a depth of fifteen metres (nearly fifty feet) from the surface of the mound. Like the later one, it was free-standing only on the outer face, and was built on a slight batter against an interior fill.

In the following two seasons, further details of the system of fortification to which this wall belongs have emerged. On the outer side was a great rock-cut ditch, nine metres wide and three metres deep. The labour involved in excavating this ditch out of the solid rock must have been tremendous. As we have discovered nothing in the way of heavy flint picks, one can only suppose that it was carried out with stone mauls, perhaps helped by splitting by fire and water.

Still more remarkable is the adjunct to the wall on the inner side. This is a massive tower, nine metres in diameter, and solidly built of stone. In conception and construction, this tower would not disgrace one of the more grandiose medieval castles. In its centre is a stone-built staircase with twenty steps, leading down to a horizontal passage. The treads of the stairs are formed by great stone slabs more than .75 m. across, hammer-dressed to a smooth finish, and the roofs of the staircase and the passage are formed of still larger slabs, .95 m. across and up to a metre long, similarly dressed. The whole thing is excellent in both architecture and masonry, and everyone who sees it finds it impossible to believe that it was built eight thousand years or more ago.

We have not yet solved the problem of the purpose of the staircase. The passage at its base most unfortunately leads to just that point in the circumference of the tower which lies beyond the edge of our excavations, and to clear down to its entrance on the outside will mean shifting some fifty feet of filling, as well as twenty feet of our own dump on top. As found, the passage was full of skeletons, twelve in all, jammed in extremely tightly. But this was certainly not the primary purpose, since when they were put in, the fill in the passage had risen to within two feet of the roof. For this same reason, one cannot give the dramatic interpretation that they were the last heroic defenders of the tower. Clearly the passage had ceased to be functional at that stage, and the bodies were put in, after it had gone out of use, simply because it was a convenient hole.

Even the stage of this great wall, ditch and tower is not the beginning of the story of the defences of Jericho. . . . The tower has two outer skins surrounding a stone core. The outermost is not continuous, but is associated with some curiously shaped rooms on the north-west side. These skins proved to represent successive building phases, and it is only to the last of these that the town wall just described belongs, for it is built against the outermost skin, with which it was structurally contemporary. To the north of the tower, the town wall was built against a filling of stone chips, clearly derived from the cutting of the ditch, which is therefore contemporary with it. These chips were piled up against a yet earlier massive stone wall, which survives, free-standing on both faces, to a height of four metres. This wall was cut through by the inner skin surrounding the tower. It is therefore earlier than it, and it almost certainly is the town wall going with the core of the tower, which must represent its earliest phase. The wall contemporary with the intermediate phase, in which the inner skin was added to the core, may be represented by the lower part of the later wall, which is in a different and rather rougher style of building.

We have, therefore, indications that the elaborate system of defences has a very long history. The earlier stages have not yet been fully traced, and it will indeed be difficult to do so without disturbing more than is desirable of the stage at present exposed, which must represent the climax of this remarkable achievement. But the earliest so far traced, with the nucleus of the tower, and a free-standing stone wall must in itself have been sufficiently magnificent, and again and again one returns to the feeling of amazement at the very great antiquity of the remains, and at the evidence they provide for the high degree of organisation and technical ability of their builders.

One question which for long puzzled us was why such fine structures as the town wall and the tower should have been allowed to go out of use. To this question the 1956 season of excavations provided the answer. In the deep excavations towards the north-east end of the *tell* (Site E I, II, V, Square E7), a whole series of superimposed houses with the typical burnished plastered floors, walls built of thumb-impressed bricks . . . had been excavated. In 1956 it became clear that we were penetrating down into a different phase. The most noticeable change was that the floors were of mud and not plaster. Moreover, the walls were curved and the plan of the houses seemed to be round, and when the walls were dissected, the bricks were found to be of a new type. The shape of the bricks was an elongated oval, the base flat and the upper surface curved or hog-backed. As the area was further cleared, another fact emerged. Along the north side of the site the walls and floors were missing, and in their places were deposits of water-laid silt and gravel. This in due course defined itself as a silted-up stream bed. At some stage, therefore, a stream had partially destroyed the houses, cutting into the accumulated walls and floors to a depth of at least 1.50 m. The lines of fill showed that the stream bed had silted up twice and the water had then cut down again, so the stream had three successive periods of activity. Immediately overlying the final silting-up was the floor of the first house with the typical plastered floors. The cutting of the stream therefore neatly divides the phase with the houses built of hog-back bricks from the succeeding phase of the houses with plastered floors. . . . It shows that there was an actual time-gap between them as well as a break in building style. The length of that gap is conjectural. Presumably each fresh cutting down represents a winter's rains, for the stream was almost certainly cut by storm water and not by that of a spring. But rainfall, at any rate locally, in this area is not sufficiently consistent for there to be a strong enough flood to cut the channel out each year to the full depth. The gap may therefore be anything from three years to, say, ten or much more.

It was thus in Site E that the break between two pre-pottery Neolithic phases was sharply defined. But buildings of the earlier phase were reached, in 1956, in two other sites, M and D, and the characteristics of this phase are now becoming familiar. It seems probable that the rooms were of a beehive shape, with the walls circular in plan, and inclining slightly inward in elevation. The floors were sunk below the external level, and in two cases the steps down into the entrance passage are well preserved. In one instance it is clear that the treads of the steps were of wooden planks. Timber was extensively used, for remains of burnt beams are found in all the areas. A curious feature is that in several instances there is a basis of heavy cobbling below the floors, with a deep trough of cobbles along the line of the walls. Not only is the architecture different from that of the later phase, but also most of the articles of equipment are new, the flints of a hitherto unknown type, a remarkably fine and rich bone industry, and greenstone amulets of a different shape. The familiar querns [primitive grain mills] are no longer found, the earlier type having a much less elaborately shaped grinding hollow, and the stone bowls being more roughly shaped.

Though there is so much that is different, the earlier people had one thing in common with the later one. They also had a cult of skulls, and already a considerable number of skulls separated from the bodies have been found. But even here there is a difference, for the skulls are found arranged in groups, in one case closely packed in a circle, all looking inwards, in another arranged in three sets of three, all looking in the same direction. In the third instance there is an unpleasant suggestion of infant sacrifice, for beneath a curious bath-like structure of mud-plaster there is, besides one complete infant burial, a collection of infant skulls with the neck vertebrae attached, showing that the heads were cut off and not merely collected from burials. The ordinary method of disposing of the dead was, as in the later phase, burial beneath the house floors but in deep pits.

It is the break between the hog-back-brick phase and the plaster-floor phase which is responsible for the great town wall and the tower going out of use. The buildings contemporary with the various phases of the tower are in the earlier technique, and those which succeed the layers of debris which cover the face of the wall and tail up over the top of the tower are in the plastered-floor technique. The layers of debris are therefore the equivalent of the stream bed in Site E.

The massive system of defences therefore acquires a new interest, for it belongs to the earliest phase in the history of the town. It might be safer to qualify this statement by adding the phrase 'so far discovered', but we have now reached bedrock in several places without finding any suggestion of a yet earlier phase, though it might be claimed that all our sites are towards the periphery of the mound, and earlier remains might lie beneath the centre.

There is thus abundant evidence that the pre-pottery Neolithic phases at Jericho cover a long period of time. The ruins of the successive buildings raised a mound some forty-five feet in height before the first pottery appeared on the site, a type of pottery which precedes the later type which can . . . be linked with the early village settlements elsewhere in the Near East belonging to the fifth millennium B.C. In the excavation area in the centre of the west side of the *tell*, Sites D and F, nineteen successive building stages belonging to the plastered-floor phase can be identified. In Site E, where the remains of at least two building stages had been removed in the 1955–6 excavations, we have excavated fourteen main stages. Earlier than this whole phase, and separated from it by a gap of unknown duration, is the hog-back-brick phase with all its ramifications, as yet only glimpsed. We are thus carried back a very long way.

To establish any sort of absolute dating, archaeologists have today available what is known as the Carbon-14 method of dating. This is based on the fact that all organic material while living absorbs radio-activity. When it dies, for instance when a tree is felled, it starts to give it up. The rate at which the radio-active matter is lost has been calculated, and measurements can be made of the amount present in suitable specimens. This gives, within a standard deviation, the number of years ago the organism ceased to live. The standard deviation may give a margin of three hundred years or so either way, but when one is dealing with objects several thousand years old, this does not affect the general picture very much.

The material found on ancient sites which is most useful for this purpose is charcoal, and as much charcoal as possible has therefore been collected from all our levels. The process of examination, and the elimination of all possible sources of error, is a complicated one, and so far only preliminary results have been obtained. They concern, moreover, only the plaster-floor stage. From the debris overlying the fourteenth building of this stage from the top in Site F, charcoal fragments have given a date of *c.* 5850 B.C. Below it are five more stages before the debris overlying the hog-back-brick phase is reached. In Site E, a date of *c.* 6250 B.C. has been given for the sixth from the top of the fourteen stages excavated by us. It is thus apparent that a period when the people of the plaster-floor phase had long been established on the site must be dated as long ago as the seventh millennium B.C. Behind it lies all the hog-back-brick phase and the beginning of the plaster-floor phase.*

The finds at Jericho are completely revolutionary for the established ideas of the beginnings of settled life and civilisation. Hitherto, our evidence of the earliest settlements has been of villages of the fifth millennium, primitive, small and unpretentious. Not till towards 3000 B.C. did these develop into anything like towns. Both stratification and the Carbon-14 dating place Neolithic Jericho long before this. We have abundant evidence that, from size, fortifications and general community undertakings, Jericho of the plaster-floor phase must be considered as a town. We do not yet know the full size of the settlement of the hog-back-brick phase, but in all other respects it also must be considered as a town.

Such a community must have been supported by an efficient agriculture, for the number of persons congregated together which is suggested by the size of the settlement could not have been dependent on the wild produce, animal or vegetable, obtainable within convenient reach. In this, the natural resources of the site certainly play a part. The alluvial soil of the Jordan Valley is very fertile, but the very great heat of the district parches vegetation unless there is abundant water. In Jericho this is provided by the spring which emerges at the foot of the ancient site, and this must have been as important in antiquity as it is today. Today, the fertility of the oasis is dependent on an elaborate irrigation system. It is reasonable to assume that fields or gardens, extensive enough to support the considerable population of Neolithic Jericho, could also only have been made available by a system of irrigation, for otherwise the water would not have reached a sufficient area. Our knowledge of the growth of the later towns in Mesopotamia and Egypt has shown what an important part irrigation plays in the development of a community. Community organisation is required for the control of the irrigation channels, and this control provides sanctions which can become the basis of community laws. Thus an agglomeration of families is welded into a unit. Once such community organisation comes into being, other community undertakings can follow. At Jericho we have the evidence of these in the town walls and in the buildings which would appear to be public. The whole process is an interlocking one of cause and effect.

But our most recent evidence shows that Jericho cannot have been unique, though so far it is the only site known. We have suspected this for some time, since for a community to have found it necessary to defend itself by such massive town walls, there must have been some enemies that were feared. The discovery of the two distinct cultures, implying two distinct sets

*Since this was written a Carbon-14 dating of *c.* 6800 B.C. has been obtained for a level comparatively late in the hog-back-brick phase.

of people, provides part of the answer. The defences of the hog-back-brick people were against the plaster-floor people, who eventually overcame them. But it was not a case of townsfolk defending themselves against envious nomads. The plaster-floor people arrived with a fully developed civilisation and a fully developed and even stereotyped architecture, for the houses which immediately succeed the break have all the characteristics which last throughout the long occupation which follows. They therefore come from other towns, and from towns sufficiently near to be an obvious menace to the earlier occupants of Jericho. Somewhere, and moreover somewhere probably in the Jordan Valley, other equally early towns are to be found.

We can therefore deduce that in the sixth and probably seventh millennium B.C. there was a growth of civilisation hitherto completely unsuspected. Any general deductions must await the discovery of other sites, so any further theories must be hypothetical. But since our evidence carries us so very far back, we know that we are getting near to the Pleistocene, the period of the Ice Ages of Europe, which ended about 10,000 B.C. In the Mediterranean and Near East, the counterpart of the Ice Ages was the Pluvials, in which areas now desert must have been fertile. As a hypothesis, it may be suggested that favourable climatic and vegetational conditions may at this time have stimulated a growth of agriculture and therefore of settlement, of which Jericho provides the only evidence so far found. After a long period of flourishing town life, increasing desiccation as conditions approached those of today may have stifled this premature development. The villages of the fifth millennium, which undoubtedly at Jericho represent a period of retrogression, may mark a fresh beginning and a gradual adaptation of settlement to new and less favourable conditions.

We may therefore envisage pre-pottery Neolithic Jericho as a culture with all the attributes of civilisation, except that of a written language. The town must have been almost modern, or at least medieval in appearance, and it must have been surrounded by fertile fields. In the neighbourhood we can assume that there were other similar towns, and since materials such as obsidian, turquoise matrix and cowrie shells must have been obtained from a considerable distance away, we can assume trade with and relations between neighboring districts. This is the revolutionary picture that Jericho has given us of a period nine or ten thousand years ago.

"Joshua fit the battle of Jericho, and the walls come tumblin' down." That is how the Negro spiritual accounts for the destruction of the sturdy walls of Jericho described in the Old Testament. If the walls did in fact fall, Dr. Bronowski would ascribe the cause to the fact that Jericho is earthquake country. It is located at the meeting of two great plates on the earth's surface, where the great stresses produced by their movement are relieved by earthquakes. Geophysicists explain this phenomenon with a modern outgrowth of the theory of continental drift, which asserts that the continents were once joined together in a "supercontinent" and that over a period of more than 200 million years they have drifted further and further apart. The following selection by D.P. McKenzie describes the latest adaptation of the theory of continental drift called *plate tectonics*.

Plate Tectonics
D.P. McKenzie

'The ruins of an older world are visible in the present structure of our planet; and the strata which now compose our continents have been once beneath the sea, and were formed out of the waste of pre-existing continents. The same forces are still destroying, by chemical decomposition or mechanical violence, even the hardest rocks, and transporting the materials to the sea, where they are spread out, and form strata analogous to those of more ancient date. Although loosely deposited along the bottom of the ocean, they become afterwards altered and consolidated by volcanic heat, and then heaved up, fractured, and contorted.' This view of geology, expressed by Hutton in 1788, is still accepted today. . . .

During the past five years the hypothesis of continental drift has been transformed from a highly controversial idea into a theory which is widely accepted by earth scientists. Though there is now overwhelming evidence in favour of large horizontal movements of entire continents, neither Wegener's original hypothesis nor an important later variant of it, known as 'Sea Floor Spreading', proved entirely successful. Recently a new version of these ideas was suggested which retains the satisfactory parts of both earlier theories. The new theory has become known as 'plate tectonics' because it postulates that the outer 50–100 km of the Earth consists of a number of rigid plates in relative motion. Relative motion between plates produces earthquakes, and therefore the location of earthquakes can be used to determine the boundaries between plates. The concept of the rigidity and the importance of relative motion are the principal differences between the new theory and that of continental drift. Drift was often compared with the process of making jam. The Earth probably began as a mixture of its constituent materials without any continents. Convective motions must then have taken place to permit the lighter materials to float to the surface and form continents. The motion and deformation of the continents was believed to be directly related to the convective motions in the mantle beneath them. Similar convective motions in jam-making allow the scum to float to the surface and accumulate into lumps. Jam scum does not, however, form rigid plates, and though jam-making may illustrate the chemical processes involved during the Earth's evolution it is a poor analogy for plate tectonics. A much better one is the behaviour of floating ice, which moves and deforms in a remarkably similar fashion to the outer parts of the earth.

The motion of ice floes illustrates the importance of the rigidity and of relative motion. A man on an ice floe at sea cannot directly observe his motion relative to the ocean floor, and therefore cannot tell if his floe is drifting. The stability of his floe and his peace of mind both depend on the relative motion between his floe and surrounding ones. There are three possible types of motion between any two floes. If they separate, water will well up between them and freeze, forming new ice which is thinner than that of the main floes, and gradually thickens with time. If floes move towards each other one may override the other, or each may thicken to form a pressure

Excerpted from D.P. McKenzie, "Plate Tectonics and Continental Drift." Reproduced with permission from the January 1970 edition of *Endeavour*.

ridge. Both types of deformation are observed in sea ice, the first being common in young thin ice, the second occurring in older thicker pack ice. The third type of motion involves one block of ice sliding past another, and can occur only if both edges are straight or are arcs of a circle. If the margins interlock, sliding cannot take place unless parts of one floe override the other. These three types of relative motion can best be illustrated by tearing a piece of paper and moving the two parts past each other.

The plates on the Earth are much larger than any ice flow, and their motion is much slower. Also many plate boundaries are beneath the sea, and therefore cannot be observed directly. For these reasons the plate boundaries cannot generally be mapped by geologists, but must be determined by indirect methods, of which the most important is the location of earthquakes. Various studies of large shallow earthquakes in many parts of the world have demonstrated that they are the result of sudden movements on large faults. Some of the measured displacements are very large, reaching 10 m in the Great Alaskan Earthquake in 1964. Such displacements are the direct expression of the motion of two plates, separated by the fault which is the boundary between them. Though not all earthquakes produce slip on faults at the Earth's surface, most are caused by surface displacement at the margins between plates which generally takes place by a succession of jumps, each of which is an earthquake, rather than by a steady creeping movement. . . .

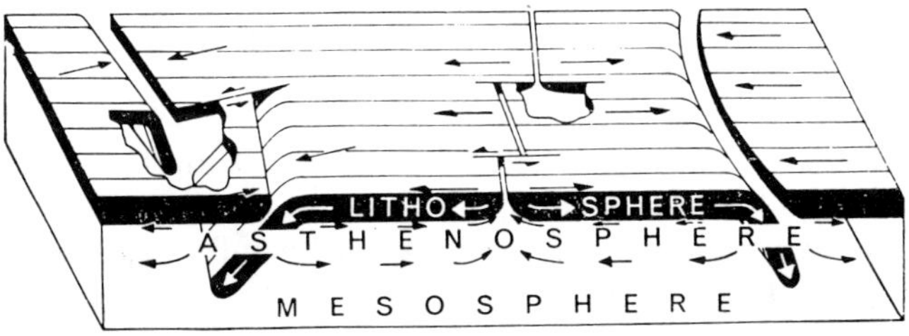

An illustration of the current theories of oceanic tectonics. Plates of lithosphere are formed symmetrically on spreading ridges, move horizontally as rigid blocks and are destroyed by underthrusting beneath island arcs. The arrows on the plates show the direction of relative motions. Arrows showing the flow in the asthenosphere serve merely to indicate that return flow must take place.

From B.L. Isacks, J. Oliver, and L.R. Sykes, *Journal of Geophysical Research*, Volume 73, p. 5855, 1968, copyright by American Geophysical Union.

As in the case of ice floes, there are three possible types of motion between neighbouring plates. . . . Plates of lithosphere are formed symmetrically on spreading ridges, move horizontally as rigid blocks and are destroyed by underthrusting beneath island arcs. . . . If two plates move apart, mantle material from beneath them wells up, cools, hardens and in this way regenerates the plates. Since the upwelling material is hot, it is less dense than the cold plates and therefore wells up above the deep ocean floor to form a linear ridge. There is no longer any reason to believe that each ridge marks the position of the rising limb of a convection cell within the mantle below the plates. Earthquakes occur along the axis of the ridge which is the boundary between the two separating plates. Another type of plate boundary can exist when two plates slide past each other without production or destruction of either, provided that the boundary between them is everywhere parallel to the direction of relative motion. Such boundaries are known as transform faults. . . .

The most seismically active regions of the Earth are areas where two plates are moving towards each other. Here the plates behave as the ice does . . . with one plate overriding the other and driving it down into the mantle. There is, however, one important difference between the tectonics of ice floes and that of plates on the Earth. Since ice is less dense than water it must remain as a floating sheet even after it has been forced below another floe, whereas the cold plates are denser than the surrounding mantle and can therefore sink. As they sink they gradually warm up and their mechanical properties become indistinguishable from those of the surrounding mantle. The trenches and island arcs are the surface features associated with plate destruction, most of which is at present taking place on the Pacific margins. Nothing is yet known about the return flow which carries mantle material from beneath the island arcs to the ridges.

This simple model of the tectonics of the earth has now been tested in a variety of ways, though the most convincing evidence in its favour has come from seismology and paleomagnetism. All quantitative tests of the concepts of plate tectonics depend on a theorem of Euler. He proved that any movement of a rigid spherical cap on the surface of a sphere can be described by rotation about an axis passing through the centre of the sphere. Thus the relative motion between two rigid plates must correspond to a rotation about some axis; if this is not so, the plates cannot be rigid.

E.C. Bullard, J.E. Everett, and A.G. Smith in 1965 were the first to use this theorem to fit together the continents surrounding the Atlantic. The fit of South America and Africa is particularly impressive . . . and shows that both continents have remained undeformed internally throughout the time that the South Atlantic has been opening. Since the present distance of separation of the two continents is 4000 km and has taken more than 80 million years to effect, this striking fit shows that the plates remain rigid for long periods of time while moving enormous horizontal distances.

Man's cultural ascent has been achieved by the creation and use of any number of levers, actual and figurative. Central to the birth and proliferation of the agricultural revolution—and its effect on biological evolution—were the sickle and the plow. The history of the plow is discussed in this selection by E. Cecil Curwin.

The Invention of the Plow

E. Cecil Curwin

The plow was developed as a result of the adaptation either of a digging stick or of a hoe so that it could be dragged continuously through the ground. . . . It is not at all likely that the original invention was made independently at dif-

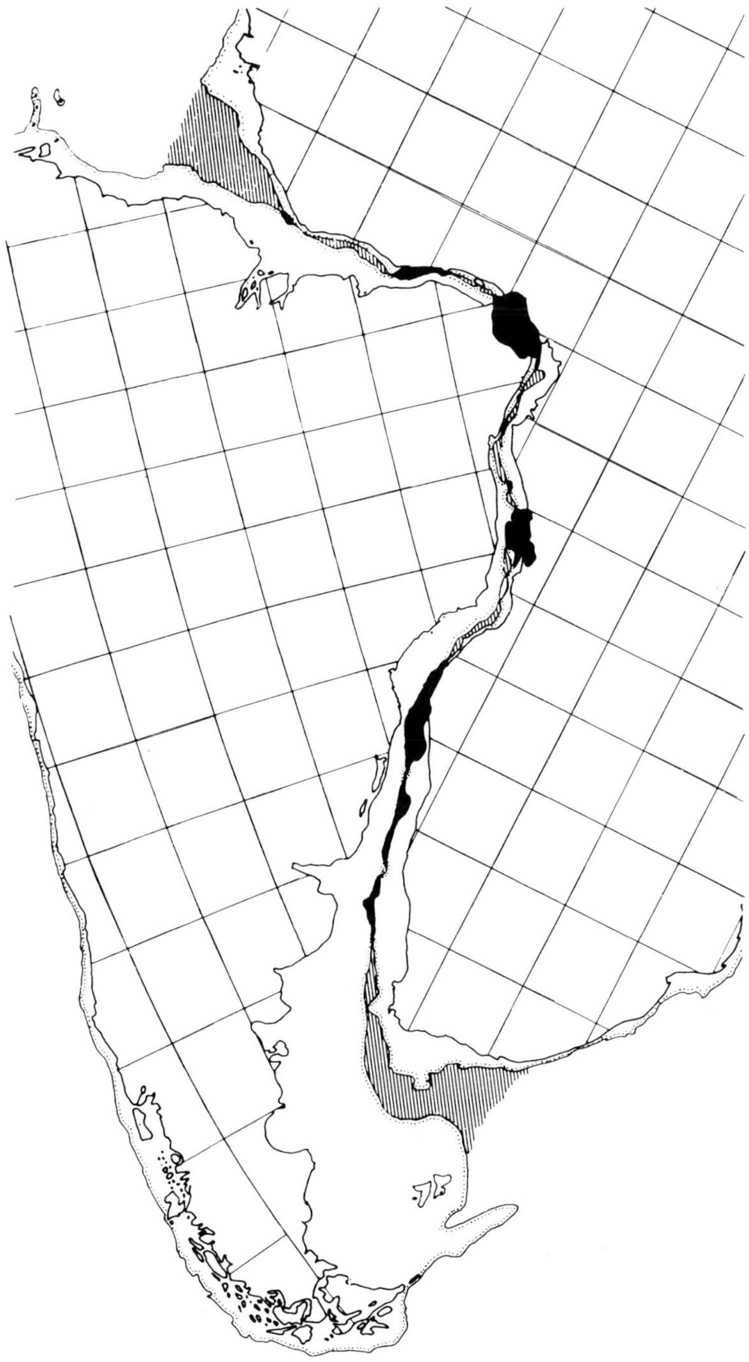

The fit of Africa and South America. ■ *overlap;* |||| *gaps.*

From E.C. Bullard, J.E. Everett, and A.G. Smith, *Philosophical Transactions*, A.250, p. 41. Reprinted by permission of the author and The Royal Society.

ferent times and in different places, but it is quite possible that, once the principle of traction had been established and diffused among neighboring peoples, it was applied by some peoples to digging sticks and by others to hoes, according to the implement which they had been accustomed to using. This would most readily account for the two principal types of primitive plow that have survived down to the present day.

The earliest evidence we have of the plow comes from Mesopotamian cylinder seals and Egyptian paintings going back to rather before 3000 B.C., namely, to the early Dynastic period of both regions. This evidence suggests that the plow, like the wheeled vehicle, may have been a product of the industrial movement of that era. In fact, it is probably not too much to say that the plow made industrial food production and the rise of urban culture possible, because without its help corn could scarcely have been grown on a large enough scale for export. Moreover, whereas hoe culture was women's work, plowing was always done by men; hence the invention of the plow is not likely to have been made by women in order to lighten their labor, for that would have involved persuading the men to take over their work altogether. It must rather have been made by men—probably by the priests, for it was anciently said that the plow was the gift of the gods—as a means of increasing the production of grain for the purposes of taxation and export. Thus the appearance of the plow is an important landmark in the history of agricultural economy.

According to Dr. Carl W. Bishop[2] the plow was first developed from the digging stick in some such way as the following. The lower part of the digging stick may already have been bent at an obtuse angle to facilitate levering up the soil, just as the blade of a spade is set at a slight inclination to the handle. A cord was then tied to the digging stick close to this angle, and while one worker kept the point of the implement in the ground, another pulled on the cord so as to drag the point through the soil, thus producing a furrow. The next step lay with the priests who, for ceremonial purposes concerned with the fertility of the soil, caused a cow or a bull to be tied by the horns to the digging stick and to share with the men the labor of drawing it. Any fertility it may have conferred on the soil will have been by way of manure, and it was later found that the ox was just as effective from this point of view, and a good deal steadier at pulling. In this way the practical value of ox traction overshadowed the theoretical religious advantages, and human traction was rendered unnecessary. It may have been as a result of such experience that ox traction was applied to wheeled vehicles, for the plow and the wagon appeared in Mesopotamia at approximately the same time. Somewhat later, according to the drawings, the cord was replaced by a pole or beam, one end of which was firmly attached to the digging stick and the other to a yoke which was tied to the animal's horns. In the simple plow thus formed the digging stick has become the combined stilt and share beam. A second handle was added in due course. This type of plow, derived from the digging stick, has been called a *spade plow*.

[2]C.W. Bishop, "The Origin and Early Diffusion of the Traction Plow," *Antiquity,* 10:261 (1936).

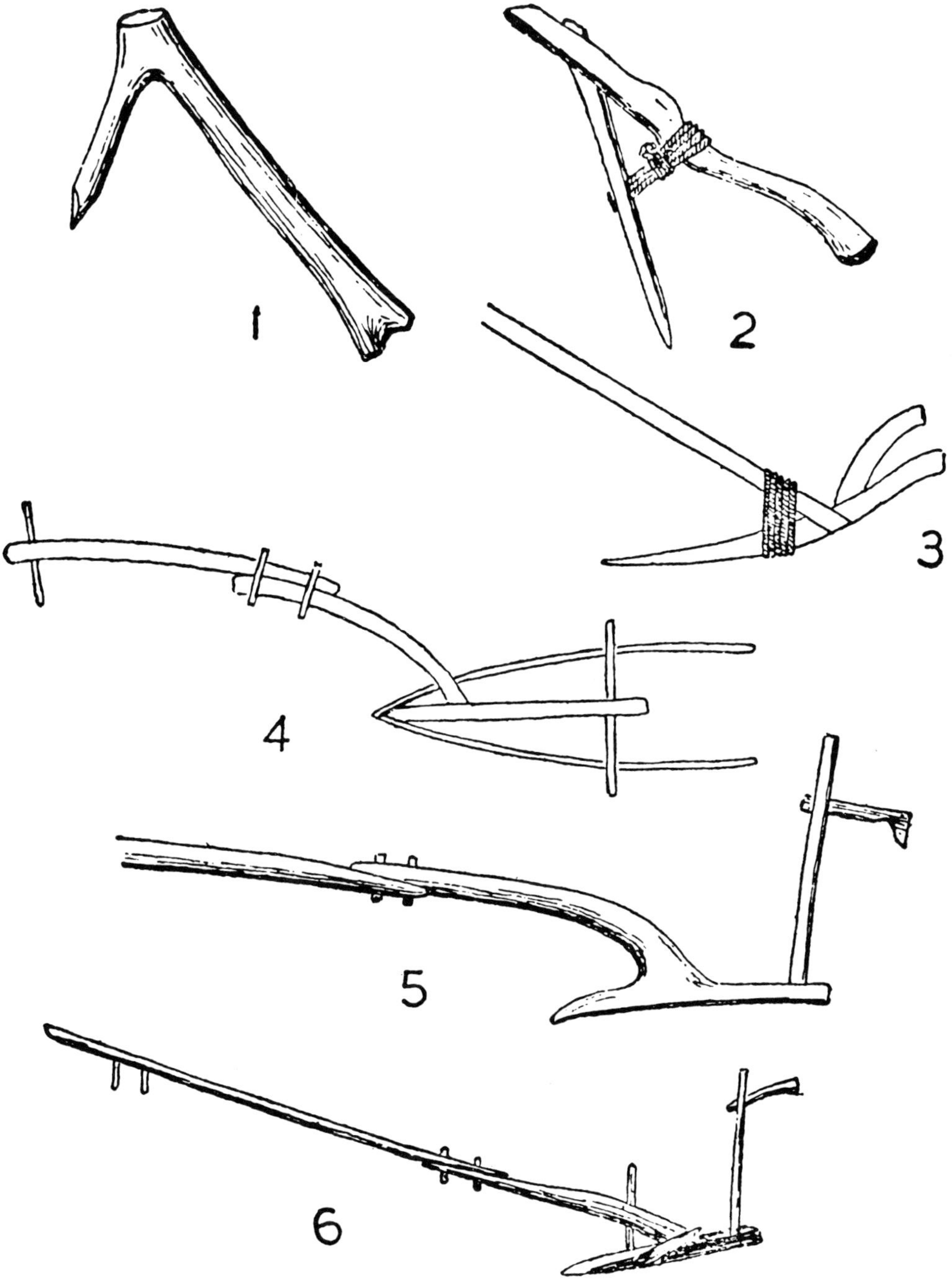

Hoes and Crook Plows. 1, *Most primitive type of hoe, made from forked branch.* 2, *Egyptian hoe of wood.* 3, *Egyptian plow, apparently derived from hoe; note the two handles.* 4, *Sumerian crook plow with "ears"; no stilt visible* 5, *General type of primitive Greek crook plow.* 6, *Modern Cypriote crook plow. Note jointed beams in Nos. 4, 5, and 6.*

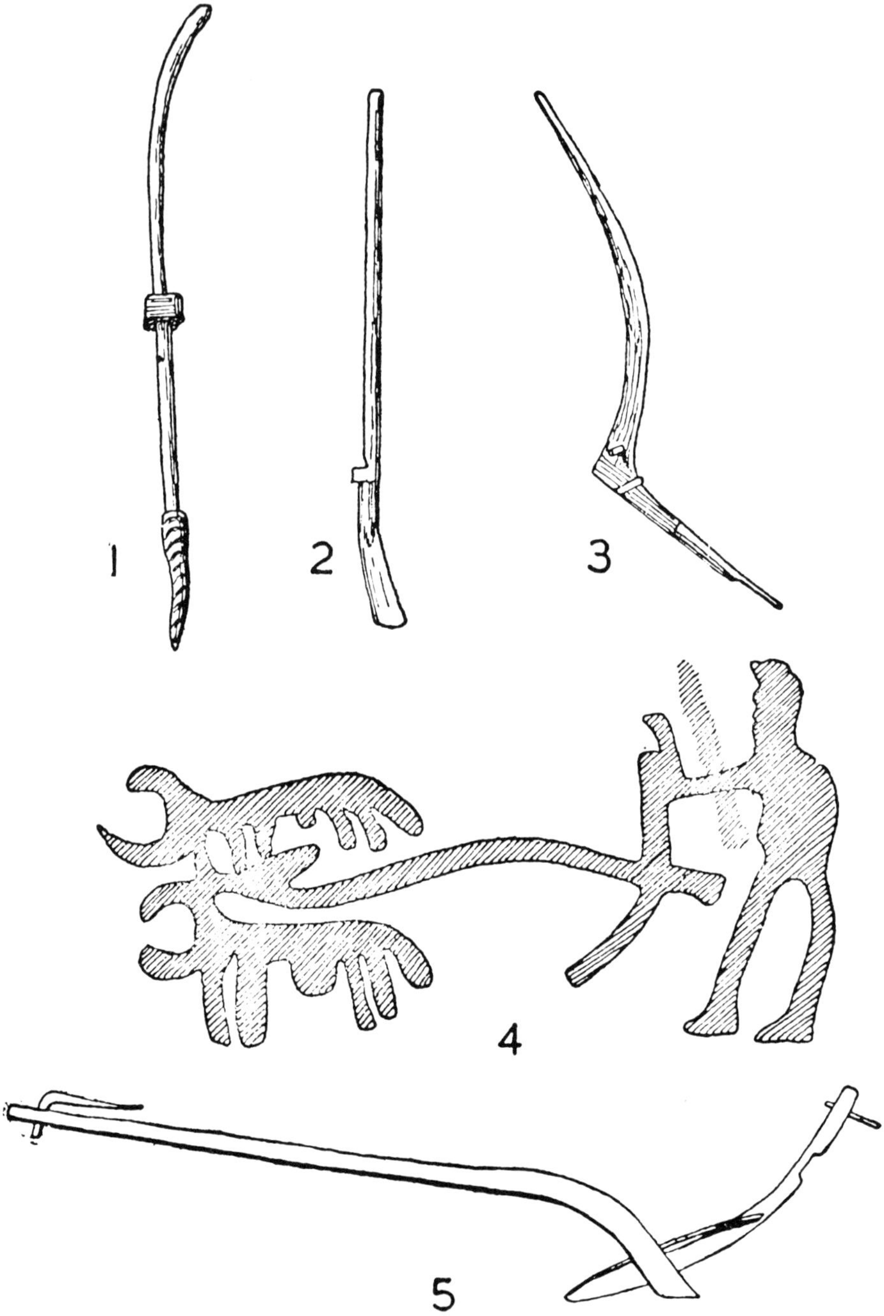

Digging Sticks and Spade Plows. 1, *Bushman digging stick with stone weight and horn point.* 2, *Irish "loy" (angular digging stick).* 3, *Skye "caschrom" (angular digging stick).* 4, *Bronze Age rock drawing of spade plow drawn by two oxen; Bohuslän, Sweden.* 5, *Iron Age spade plow of wood; Døstrup, Denmark. Length of beam, 10 feet.*

But there is another type of plow which has been, and still is, found in widespread use; this had been called the *crook plow* because it has clearly been derived from a forked branch, such as was the primitive hoe, and it may be regarded as a large hoe dragged through the ground by its handle (that is, the pole or beam), a stilt being added at the rear for the purpose of guiding and steadying it. . . . It is evident that no intermediate stage of traction by cords would have been possible in its development, so that it may have been the product of a hoe-using people who had learned the advantages of traction from the users of the spade plow, and who adapted the principle to their own implement.

Whether or not these views are correct, both kinds of plow spread widely over northern Africa, Europe, and Asia, as far as India and China. The original focus is most likely to have been in northern Mesopotamia, but soon after 3000 B.C. the plow is found in Cyprus and Greece to the west, and in India to the east. It had reached the Italian Alps before 1400 B.C., where it is depicted in rock drawings of the Bronze Age. In Denmark and Holland evidence of the use of the plow and cross-plowing, in the form of the actual scratches left by the plowshare in the subsoil, has been found underneath burial mounds of the Bronze Age, dated to approximately 1400 and 1600 B.C., respectively. In Britain, however, there is so far no evidence of the plow before about 1000 B.C., and in the opposite direction it did not reach China till the third century B.C. The crook plow was anciently used in Crete, Greece, and northern Germany, in which last country—at Walle in East Friesland—a specimen has been found preserved in peat, though the early date claimed for it by its discoverer cannot be accepted. On the other hand, the spade plow was the earliest type used in Italy and is depicted in the rock drawings of the Italian Alps as well as in those of Sweden; . . . an actual specimen of one has also been preserved in a peat bog at Dostrup in Jutland, dated to about 400 B.C. . . . Simple plows of one kind or another are still used in Greece and other parts of southern Europe, where the climate is warm and dry.

For the Spanish and the Mongols, the domesticated horse became a means for achieving superiority on the battlefield; for making man greater than he was alone. Wherever and whenever the horse has been used by man, it has left its universal mark on human cultural evolution. This selection reveals how the horse changed the life of the Indians of the American Plains.

The Equestrian Revolution of the Plains Indian

Peter Farb

This culture emerged in the middle of the 18th century as its catalytic agent, the horse, spread northward from the Spanish settlements in New Mexico. Within a few generations the horse was found throughout the central heartland of the continent and Indians from all directions spilled onto the plains.

They all found in the horse a new tool to kill greater numbers of bison than they had ever believed possible. Hunting bison on foot was not productive and could not support large numbers of Indians. Such hunting was practiced largely by the wretched nomads who moved around in small groups and lived off the occasional weakened bison they could kill. Most of the aboriginal cultures on the plains were based on agriculture.

Once the horse arrived on the plain, that way of life changed. The nomadic bison hunters became ascendant over the farmers, who either were driven off their lands or abandoned agriculture to become bison hunters themselves. Indians had never seen the horse until the Spaniards brought it to the New World, for sometime during the great glacial melt it had become extinct in North America.

By the first half of the 18th century, enterprising merchants had already sold the horse to Indians as far north as the Northern Shoshoni of Wyoming and taught them its management. The Shoshoni slowly built up their herds and learned to ride as if they had been born to the saddle. No longer did they have to remain impoverished and secretive inhabitants of the Rocky Mountains, at the mercy of more powerful Indian groups. They scooped down the eastern flanks of the mountains and onto the high plains, where they found a bonanza in bison and a way to even the score with their traditional persecutors, the Blackfoot. From all over, other Indian groups converged on the plains and quickly adapted themselves to an economy based on the bison. The lands of the agriculturalists were usurped and the plains became a maelstrom of varied and often conflicting cultures.

The horse was a new cultural element in the heartland of North America and it changed the entire way of life there. The whole of the plains became peopled by groups of great diversity who had come from all directions. The gross differences in culture among all these peoples had disappeared; it is remarkable that a people from the eastern forests and another from the west, two thousand miles away, should within a few generations have become so nearly identical. Even more remarkable, this homogeneity was achieved with great speed, was not imposed on unwilling people by a more powerful group and was done in the absence of a common tongue.

All this was due to the horse. The herds of bison were pursued on swift horses and the choicest animals killed. The Indians could now afford luxury. Tipis almost twenty-five feet in diameter were filled to overflowing with new found riches. An economic revolution took place. The women no longer toiled in the fields—for gardening was not as profitable as hunting, nor could it be practiced in the presence of nomadic horsemen—and they stopped making pottery because brass kettles were obtained from Whites. Permanent villages disappeared and with them went the elaborate customs and crafts, rules for marriage and residence.

War, the destroyer of men and their civilizations, is, according to Dr. Bronowski, "a highly planned and cooperative form of theft." It is that universal aspect of war that Socrates discussed with Glaucon 2300 years ago.

The Dialogues of Plato—an Excerpt

Plato

The question which you would have me consider is, not only how a State, but how a luxurious State is created; and possibly there is no harm in this, for in such a State we shall be more likely to see how justice and injustice originate. In my opinion the true and healthy constitution of the State is the one which I have described. But if you wish also to see a State at fever-heat, I have no objection. For I suspect that many will not be satisfied with the simpler way of life. They will be for adding sofas, and tables, and other furniture; also dainties, and perfumes, and incense, and courtesans, and cakes, all these not of one sort only, but in every variety; we must go beyond the necessaries of which I was at first speaking, such as houses, and clothes, and shoes: the arts of the painter and the embroiderer will have to be set in motion, and gold and ivory and all sorts of materials must be procured.

True, he said.

Then we must enlarge our borders; for the original healthy State is no longer sufficient. Now will the city have to fill and swell with a multitude of callings which are not required by any natural want; such as the whole tribe of hunters and actors, of whom one large class have to do with forms and colours; another will be the votaries of music—poets and their attendant train of rhapsodists, players, dancers, contractors; also makers of divers kinds of articles, including women's dresses. And we shall want more servants. Will not tutors be also in request, and nurses wet and dry, tirewomen and barbers, as well as confectioners and cooks; and swineherds, too, who were not needed and therefore had no place in the former edition of our State, but are needed now? They must not be forgotten: and there will be animals of many other kinds, if people eat them.

Certainly.

And living in this way we shall have much greater need of physicians than before?

Much greater.

And the country which was enough to support the original inhabitants will be too small now, and not enough?

Quite true.

Then a slice of our neighbours' land will be wanted by us for pasture and tillage, and they will want a slice of ours, if, like ourselves, they exceed the limit of necessity, and give themselves up to the unlimited accumulation of wealth?

That, Socrates, will be inevitable.

And so we shall go to war, Glaucon. Shall we not?

Most certainly, he replied.

Integral to any war is the war machine, a lever for destruction created by man to produce more energy than he may really need: It may be a cannon, a horse, a tank, an atomic bomb. Is the war machine truly an earmark of man's evolving culture? Or, as Dr. Bronowski optimistically inquires, might it be alien? The following selection from *Gulliver's Travels* reflects a similar point of view. Gulliver found in his voyage to Brobdingnag, a kingdom "where the learning of this people is very defective, consisting only in morality, history, poetry and mathematics." Gulliver had great difficulty communicating with the King of that great land.

Gulliver in Brobdingnag

Jonathan Swift

Great allowances should be given to a king who lives wholly secluded from the rest of the world, and must therefore be altogether unacquainted with the manners and customs that most prevail in other nations: the want of which knowledge will ever produce many *prejudices*, and a certain *narrowness of thinking*, from which we, and the politer countries of Europe are wholly exempted. And, it would be hard indeed, if so remote a prince's notions of virtue and vice were to be offered as a standard for all mankind.

To confirm what I have now said, and further to shew the miserable effects of a *confined education*, I shall here insert a passage which will hardly obtain belief. In hopes to ingratiate my self farther into his Majesty's favour, I told him of an invention discovered between three and four hundred years ago, to make certain powder; into an heap of which the smallest spark of fire falling, would kindle the whole in a moment, although it were as big as a mountain; and make it all fly up in the air together, with a noise and agitation greater than thunder. That, a proper quantity of this powder rammed into a hollow tube of brass or iron, according to its bigness, would drive a ball of iron or lead with such violence and speed, as nothing was able to sustain its force. That, the largest balls thus discharged, would not only destroy whole ranks of an army at once; but batter the strongest walls to the ground; sink down ships with a thousand men in each, to the bottom of the sea; and when linked together by a chain, would cut through masts and rigging; divide hundreds of bodies in the middle, and lay all waste before them. That, we often put this powder into large hollow balls of iron, and discharged them by an engine into some city we were besieging; which would rip up the pavement, tear the houses to pieces, burst and throw splinters on every side, dashing out the brains of all who came near. That, I knew the ingredients very well, which were cheap, and common; I understood the manner of compounding them, and could direct his workmen how to make those tubes of a size proportionable to all other things in his Majesty's kingdom; and the largest need not be above two hundred foot long; twenty or thirty of which tubes, charged with the proper quantity of powder and balls, would batter down the walls of the strongest town in his dominions in a few hours; or destroy the whole metropolis, if ever it should pretend to dispute his absolute commands. This, I humbly offered to his Majesty, as a small tribute of acknowledgment, in return of so many marks that I had received of his royal favour and protection.

The king was struck with horror at the description I had given of those terrible engines, and the proposal I had made. He was amazed how so impotent and grovelling an insect as I (these were his expressions) could entertain such inhuman ideas, and in so familiar a manner as to appear wholly unmoved at all the scenes of blood and desolation, which I had painted as the common effects of those destructive machines; whereof he said, some evil genius, enemy to mankind, must have been the first contriver. As for himself, he protested, that although few things delighted him so much as new discoveries in art or in nature; yet he would rather lose half his kingdom, than be privy to such a secret; which he commanded me, as I valued my life, never to mention any more.

Of all the war machines, none is more mighty than the atomic bomb, which unleashes the hidden energy of the sun and stars. It was a war horse of a different color, a force for good that went awry before its positive worth for man could be fully tested. Lt. General Leslie R. Groves was the army officer in charge of the "Manhattan Project," the program that developed the atomic bomb. Here is his justification for the war machine that ended World War II and changed the world overnight.

Reflections on the Manhattan Project

Lt. Gen. Leslie R. Groves

As memories of World War II have faded into history and as the considerations of national survival that determined its course have lost their urgency, many have come forward to question the desirability of our having developed atomic energy. Some people have even gone so far as to assert that the United States was morally corrupt in making the effort. When I consider the sources of these later claims, my first inclination is to ignore them and I would do so if it were not for the fact that if free men are to arrive at a fair appraisal and full understanding of the Manhattan Project, they must be acquainted with the points in its favor, as well as with the views of those who criticized it and of those whose responsibilities were small at the most.

The atomic bombings of Hiroshima and Nagasaki ended World War II. There can be no doubt of that. While they brought death and destruction on a horrifying scale, they averted even greater losses—American, English and Japanese. No man can say what would have been the result if we had not taken the steps that were necessary to achieve this end. Yet, one thing seems certain—atomic energy would have been developed somewhere in the world in the mid-twentieth century. Because of the great costs and difficulties involved and the apparently very small chance of success I do not believe the United States ever would have undertaken it in time of peace. Most probably the first developer would have been a power-hungry nation, which would then have dominated the world completely and immediately. If a peacefully inclined nation had been first, it might have tried initially to use atomic energy for peaceful purposes, but as others exploited its military uses, every country would soon have been forced, as has happened in recent years, to concentrate on the development of weapons. Not until each of the great powers had produced a full atomic arsenal would the threat of one-sided atomic war pass.

Once this state was finally achieved, and I feel that it has been, with sane national leadership, major war is impossible. All that stands in the way of effective international control is the acceptance by all the world's political leaders of this fact. But there is no reason why work on the peaceful uses of atomic energy cannot be pushed ahead vigorously.

The fact that during the years following the end of World War II an unquestioned superiority in atomic weapons rested with the United States has not only averted a catastrophic war without a complete surrender of our principles and our ideals, but it has enabled us to take the first steps toward the realization of a few of the many possible constructive uses of atomic energy. While it is tragic that the forces for destruction that we unleashed are stronger than man's present ability to control them, it is fortunate indeed for humanity that the initiative in this field was gained and kept by the United States. That we were successful was due entirely to the hard work and dedication of the more than 600,000 Americans who comprised and directly supported the Manhattan Project.

Looking back, I think I can see five main factors that made the Manhattan Project a successful operation:

First, we had a clearly defined, unmistakable, specific objective. Although at first there was considerable doubt whether we could attain this objective, there was never any doubt about what it was. Consequently the people in responsible positions were able to tailor their every action to its accomplishment.

Second, each part of the project had a specific task. These tasks were carefully allocated and supervised so that the sum of their parts would result in the accomplishment of our over-all mission. This system of compartmentalization had two principal advantages. The most obvious of these was that it simplified the maintenance of security. But over and above that, it required every member of the project to attend strictly to his own business. The result was an operation whose efficiency was without precedent.

Third, there was positive, clear-cut, unquestioned direction of the project at all levels. Authority was invariably delegated with responsibility, and this delegation was absolute and without reservation. Only in this way could the many apparently autonomous organizations working on the many apparently independent tasks be pulled together to achieve our final objective.

Fourth, the project made a maximum use of already existing agencies, facilities and services—governmental, industrial and academic. Since our objective was finite, we did not design our organization to operate in perpetuity. Consequently, our people were able to devote themselves exclusively to the task at hand, and had no reason to engage in independent empire-building.

Fifth, and finally, we had the full backing of our government, combined with the nearly infinite potential of American science, engineering and industry, and an almost unlimited supply of people endowed with ingenuity and determination.

When I was a boy, I lived with my father at a number of the Army posts that had sprung up during the Indian wars throughout the western United States. There I came to know many of the old soldiers and scouts who had devoted their active lives to winning the West. And as I listened to the stories of their deeds, I grew somewhat dismayed, wondering what was left for me to do now that the West was won. I am sure that many others of my generation shared this feeling.

Yet those of us who saw the dawn of the Atomic Age that early morning at Alamogordo will never hold such doubts again. We know now that when man is willing to make the effort, he is capable of accomplishing virtually anything.

In answer to the question, "Was the development of the atomic bomb by the United States necessary?" I reply unequivocally, "Yes." To the question, "Is atomic energy a force for good or for evil?" I can only say, "As mankind wills it."

3
The Grain in the Stone

. . . all around the self of this
being are growing stones . . .

e.e. cummings

Man, in the prime of his cultural evolution, has been at once scientist, explorer, and artist. With an intellect sharpened by the need for survival and his own betterment, he worked with clay and stone, discovering anew the patterns and structures that nature had already put there, reshaping and refining them. In his architecture he can go as far as the grain in the wood and the stone permits, but often his creations seem to explode beyond those limitations into graceful arches and sweeping buttresses. The selections that follow further elaborate on the blending of mind and matter and on Dr. Bronowski's contention that "the tool that extends the human hand is also an instrument of vision." They explore similarities and differences among the Inca stonemasons who built Machu Pichu and the freemasons who built the cathedrals of Europe, the kinships between the man who made the Watts Towers and the men who pried flamboyant images out of nature's quarries. The chapter spans centuries, continents, cultures, and an enormous breadth of ideas. Some of those ideas came to North America by way of Siberia, passing, when the Glaciers allowed, across the land bridge that once spanned the Bering Sea. The people that brought them may have arrived in Ayacucho Valley of Highland Peru more than 14,000 years ago. How and when this migration occurred is discussed in the following article.

Some Prehistoric Connections Between Siberia and America

James B. Griffin

When western Europeans discovered the Americas they not only took on the task of exploring and colonizing the Western Hemisphere, but they also faced the question of the origin of the peoples who inhabited the New World. The name "Indians" was first applied under the misapprehension that the discovery had been made of the outer regions of India. Among the explanations for the peopling of the New World, one of the most common was a connection to the Lost Tribes of Israel.

One of the major difficulties in obtaining a satisfactory answer to the question of the New World population was the limitation of knowledge available during the 16th and 17th centuries about the age of the earth and about the age of human beings on the earth. With the exploration of the north Pacific area and the discovery of Bering Strait, a great many people concerned with the question of the origin of the Indians, recognizing the resemblance of the American Indian to the Mongoloid people, suggested that Bering Strait was the nearest or the easiest way by which people could have passed from Asia into North America. During the 18th and the early part of the 19th centuries, not only the Bering Strait crossing was suggested as a means by which America was peopled after 2000 B.C., but there were also theories of movements across the Atlantic from the Mediterranean area and of movements of people across the Pacific to populate the southern parts of the New World. Some of these quaint ideas still persist.

During the past 100 years, with the extraordinary growth of knowledge in various scientific fields, it has been possible to provide much more satisfactory explanations of Indian origins. The study of the geological features of the earth, particularly in Europe during the early 1800's proved the great antiquity of the earth. The development of Pleistocene or Ice Age studies in Europe and America indicated very clearly that implements of human manufacture were associated with this last major stage in the shaping of the earth's surface features. The discoveries and study of fossil men in the Old World from *Pithecanthropus* in Java to Neanderthal in Europe, and the recognition of Cro-Magnon in Europe associated with the late phase of the Old World stone industries gave man an antiquity in Eurasia which was not dreamed of in the centuries preceding. The studies of European prehistory certified the existence of a long period of development of human culture and that by the time of the late Pleistocene in Europe man had reached a fairly advanced stage of hunting and gathering economy with a distinctive and spectacular art style. Mural painting begins with the Upper Paleolithic.

Along with these developments in prehistory, the study of the culture of various non-European groups around the world, including groups in the Americas, indicated that there were a number of major groupings or stages of human culture. There were simple hunting and gathering peoples, more advanced food collectors, small-scale food producers, and finally, the more advanced non-European cultures in the New and Old World which had developed complex and well integrated civilizations. In the Americas particularly, it could be seen that the cultures of the primitive groups on the marginal areas of South America and of North America contrasted sharply with the agricultural civilization of Middle and South America, or even with the agricultural groups in the American Southeast and Southwest. Physical anthropologists have indicated that there is considerable variation among the various American Indian tribes and that the Indians did not belong to a single human physical type. There were also indications that the physical type of some of the marginal peoples of the New World corresponded to the physical type of the older American prehistoric skeletal material. Studies of American Indian languages have shown that there are a great many linguistic groups in the New World. The linguistic complexity was one of the main cultural features which indicated a considerable antiquity for man in the New World. For example, it has been said that there was more linguistic diversity within the present area of the state of California than there is in Western Europe. All of the relevant data gathered by scientists on the American Indians implied that the American Indian was indeed derived from Northeast Asia and that he had probably come at a period when he was in a hunting-gathering stage of cultural development. The prevailing opinion is that the major complex agricultural civilizations, which were the wonder of the European adventurers, were developed in the New World.

The discovery of fossil man in the Old World and of ancient Old Stone Age implements was an incentive for scholars and other people of curiosity in the Americas to search for similar materials. During the late 1800's and early 1900's there were claims of fossil men comparable to European forms and also claims of very ancient stone implement finds. There were even claims that the Indians had developed in the New World without connections to Eurasia. Critical studies of almost all of these purported fossil men and ancient implements proved that they were not of the same order of great antiquity as those of the Old World.

For some time archeologists in the New World did not think that man had been here for more than a few thousand years before Christ. In the last 30 years, however, archeological work has been able to produce sound evidence of a considerable antiquity for human cultures not only in the Americas . . . but also in eastern and northeastern Asia. . . . Excavations in China have indicated that modern man was there in an Upper Paleolithic stage of development during the closing phases of the last glaciation. A similar situation is found in south central Siberia where man was present during the closing stages of the Upper Paleolithic and in association with extinct animal forms of the closing phases of the ice age. In North and Middle America a considerable number of finds have human implements in association with extinct American fauna and in association with geological formations attributed to the closing phases of the Wisconsin glaciation in North America.

The Paleo-Indian in North America

On the American side of the Bering Strait the Wisconsin glaciation covered most of the northern United States and practically all of Canada. Glacial ice moved down the eastern slopes of the American and Canadian Rockies and a short distance out onto the plains. In western Canada this ice advance is in contact with the moraines and other glacial debris from the large central Canadian Laurentian ice sheet which moved west. It is not known for certain whether the maximum advance of the continental sheet and the mountain ice occurred simultaneously, or whether there was an ice-free corridor from north to south along the east side of the Canadian Rockies all through the Wisconsin glaciation. The Wisconsin ice advance is thought by some Pleistocene students to have begun about 50,000 B.C., followed by a warmer period corresponding to the Würm interstadial in Europe. This may have provided an ice-free corridor east of the Rockies some 30,000 years ago. The furthest south ice advance which produced the glacial stages in the Great Lakes area during the Wisconsin reached as far as central Illinois about 23,000 B.C. and may have closed the postulated gap in western Canada. It is highly probable that by at least 12,000 to 14,000 B.C. there was a corridor between the mountain and continental glacier in western Canada which would have allowed access from northern Alaska and northwestern Canada to the plains area to the south.

There are a number of finds which support the argument that man may have been in the New World, particularly in the western part of North America, some 30,000 to 20,000 years ago. The evidence for man at this time is, however, not completely convincing, and the majority of American archeologists are hesitant about accepting it. The reason for hesitation varies from find to find. If a clearly defined cultural complex is radiocarbon dated and is in agreement with the correct geological formations and faunal associations, then certainly the majority of American archeologists will be willing to accept such evidence. If early man did come in and occupy the North American continent 20,000 or 30,000 years ago, it would mean that he must have come in during the interstadial between the early and later Wisconsin glaciations or in pre-Wisconsin times, or both. This would imply that the physical type of man associated with such early finds could be very close to Neanderthal in appearance. It would further imply that the cultural type associated with man at this stage would be one resembling the Mousterian of Europe, or the chopper-chopping tool industry of eastern Asia. Neither man nor cultural material has been found in sites of the required age in southern and eastern Siberia.

We are certain that the earliest American Indians were in the United States area by between 10,000 and 12,000 B.C. This is based on radiocarbon dates of some 11,000 years ago from the Rocky Mountain areas in the west and the implied evidence that man was in the southeast and eastern part of the country at approximately this same time. Furthermore, the degree of cultural diversity between finds in the west and in the east are an indication that no small time must be allowed for such a differentiation of culture types in the New World. While there are no skeletal remains clearly associated with the most ancient definite cultures in the New World, there are crania with an antiquity of around 8000 B.C. which are of the same general physical type as that thought to be associated with the Advanced Paleolithic cultures in Siberia.

Our best evidence of man in the period from 10,000 to 12,000 years ago is in the general High Plains area where the economy was that of a hunting-gathering people with heavy emphasis upon the use of game animals for food and for clothing. The general way of life in America was very much the same as in the early Advanced Paleolithic culture of Siberia. Many of the beasts which were hunted are similar, for the mastodon and the mammoth, the bison, horse, camel, and other such herbivores are found associated with remains of early man. The early American hunters of the east and west had about the same inventory of flint artifacts, such as the scraper, the perforator, and the graver, but not the burin, and probably very similar bone implements. These Paleo-Indian hunters did not make their projectile points from carefully prepared blades which were struck off from carefully shaped flint cores, but instead produced their projectile points from large flakes struck off from flint blocks. They also did not make human figurines. A significant new development is the production of bifacially flaked projectiles which have long flakes removed from both faces of the point by means of very careful and excellent chipping techniques from the bases of the projectile. . . . This carefully controlled technique of bifacial flaking is one which does not appear in the Siberian area until close to the Neolithic and is one of the reasons why some of the prehistorians were loathe to accept a high antiquity for these finds in North America. It must be assumed, on the basis of present evidence, that the culture trait is primarily an American development, along with the distinctive fluting which removed flakes from both faces of the projectile.

Mary Haas has recently presented the view that the Muskhogean and Algonkian languages which almost blanketed the area east of the Mississippi are descendants of an ancient common language of some 8000 or more years ago. This could mean that the Paleo-Indian hunters of the east spoke related dialects of a common language and that the marked linguistic diversity of modern times was initiated during the long Archaic period.

We have seen that at the close of the Pleistocene in southern Siberia an Advanced Paleolithic culture was established with a Europoid physical type, and with the amelioration of climatic conditions there was a tendency for the cultural type to deteriorate. It may be that as the climatic conditions were modified, people hunting the big game animals followed the animals northward into the tundra ecological zone and on to a much wider Siberian coastal plain which would have been in existence at that time, because of the lowering of the sea level. From there they could have moved eastward across the Bering Strait, where a land bridge would have been in existence some 10,000 to 15,000 years ago, and thence moved to the valley of the MacKenzie. The northern section of Alaska north of the Brooks Range and a significant area of the Alaskan coastal shelf were tundra vegetation at this time, and unglaciated. This is also true of the Seward Peninsula area and the Chukchi Peninsula. They would still have been in a tundra ecological zone in their spread up the MacKenzie until they came into the prairie grassland areas of the eastern slopes of the American Rockies. It is also possible for man to have moved from the Seward Peninsula area into the boreal forest of the Yukon, up the Yukon Valley to the east and southeast, and finally to have come into the MacKenzie Valley and northern plains in northern Alberta and northeastern British Columbia.

The patterns of migration tracked in the previous article relied heavily on the archaeological dating of tools, skeletal remains, and other artifacts of the past. However, Dr. Bronowski stresses that traces of those treks across the Bering land bridge can be found today in the blood streams of the Indian populations of America. He contends that two distinct migrations brought two major Indian populations to America, one predominantly of blood group O, the other of blood group A. What blood groups mean and how they are passed from one generation to the next are discussed in the following selection from the American Red Cross.

What Are Blood Groups?

The American Red Cross

Blood groups are inherited patterns into which human blood may be divided for scientific purposes. There are four major blood groups: A, B, AB, and O. Groups A, B, and O were discovered by Dr. Karl Landsteiner in 1900, and a fourth, AB, was described by another scientist in 1902. Although the percentage of people belonging to each group tends to vary in different countries (as do other inherited characteristics, such as color of eyes and hair), the following distribution represents the general spread in this country: 45 percent–Group O; 40 percent–group A; 10 to 12 percent–group B; 3 to 5 percent–Group AB.

About the turn of the century, there was great interest in learning how to use blood transfusions as part of the practice of medicine; but doctors were

The American Red Cross.

held back by a forbidding obstacle: while about half the transfusions given to treat illness had a remarkable therapeutic effect, the other half seemed to make the patient much worse and sometimes resulted in death.

In seeking the clue to this riddle, Landsteiner performed a series of tests in the course of which he placed the red cells of one person in suspension in the serum of another person. From this crucial experiment, he noted that in a number of the combinations of red cells and sera, the red cells remained undamaged and freely suspended; in other combinations, the red cells were clumped together in sticky clusters—with resultant damage to the cells. From this evidence, he deduced that certain combinations of cells and sera were incompatible and that it was these unfortunate combinations that were causing illness and death in some patients following transfusion (transfusion reaction).

Further investigation of the factors causing clumping, or agglutination, of the red cells when exposed to incompatible sera revealed that there are two main antigenic substances (agglutinogens) in the red cells and two main antibodies (agglutinins) in the plasma. The antigens in the red cells were named A and B, and the antibodies in the plasma were given names to correspond with the group of red cells that they would agglutinate. Therefore, the plasma antibody causing group A cells to clump was called anti-A; the plasma antibody causing Group B cells to clump was called anti-B. . . . The anti-A and anti-B substances are contained in the globulin protein of the plasma. These basic facts enable us to understand the four simple combinations that are the bases for the major blood groups:

1. *Group A blood* has A substance in the red cells, and anti-B substance in the plasma.
2. *Group B blood* has B substance in the red cells, and anti-A substance in the plasma.
3. *Group AB blood* has both A and B substance in the red cells, and neither anti-A nor anti-B substances in the plasma.
4. *Group O blood* has neither A nor B substance in the red cells, and has both anti-A and anti-B substances in the plasma.

It will be noted that each blood group is named for the type of antigenic substance that is an inherited characteristic of the red cell.

How Are Blood Groups Inherited?

Blood groups are inherited as dominant characteristics according to the Mendelian law: for every inherited characteristic, including blood group, there is a pair of genes, one contributed by each parent. The blood group that a child may inherit is therefore dependent on the combinations resulting from genes contributed by both parents. Owing to intermarriage of people with different blood groups, a variety of combinations can result; but if the parents' blood groups are known, those possible for the children can be predicted.

The Canyon de Chelly is for Dr. Bronowski a place where man learned to use the underlying pattern in the sandstone cliffs of Arizona to create a miniature city with terraced dwellings. It was a communal culture where people used their hands and minds to create pots and baskets, distant echoes of the natural structure underneath. Of the prehistoric Indians who inhabited the Canyon de Chelly, the Anasazi are the best known. Their culture, together with their relationship to their neighbors, the Mogollon and the Hohokam, is described in this selection.

The Anasazi of Arizona

Faculty of the University of Arizona

The Mogollon Indian was of medium build and stature, with a round head, not greatly different from many modern Indians. He wore few, if any, ornaments, though he may have painted his body to offset his lack. At death, he was tightly folded and placed in a shallow pit and was rarely accorded the luxury of accompanying food containers or tools.

Since the northern frontier of the Mogollon touched on the Anasazi, and beyond the western boundary was the Hohokam domain, cultural blending was inevitable, especially because the Mogollon were a culturally impoverished lot and their neighbors were not. The net result of this was a submergency of the old simple Mogollon way of life, and the rise of a more complex society basically Anasazi, also tinged with Hohokam, in which a few old Mogollon elements were retained. Archaeologists are generally interested in a cultural transformation of this kind because it reveals some of the processes of change. Although the Mogollon left an undistinguished record, having given little to their neighbors but having received much more, they added special and challenging problems to the archaeological past. Their modern descendants, if any, have not been identified.

On the plateau to the north of the Mogollon was the homeland of the Anasazi. These people are the best known of all of the prehistoric Southwesterners because archaeologists have studied their ruins for nearly a century and because, until recent years, most expeditions focused their attention to this tribe. Furthermore, the Anasazi have received prominence through the designation of many of their abandoned homes as national monuments.

Much of the plateau is arid, and vast stretches have neither arable land nor readily accessible surface water. These two factors determined the deployment of the population, and because of these, the population density was spotty over the plateau, varying from great centers such as Black Mesa and the Hopi country, the White Mountains, Mesa Verde, the Chaco Canyon, to the sparsely inhabited Painted Desert. The diversified environment of the plateau required the development of specialized techniques before the land could be made to produce foodstuffs in sufficient quantities to support towns. The Anasazi rose to this challenge with an amazing degree of skill. The compelling requisites of water and soil determined the extent to which a community could expand, and expansion was the tribal pattern, demanded

Excerpted from *Arizona, Its People and Resources*, Revised Second Edition, by Members of the Faculty of the University of Arizona, University of Arizona Press, 1972.

by defense, group activities in farming, and proper conduct of religious ceremonies.

The Four Corners country, where Arizona, New Mexico, Colorado, and Utah come together, is generally regarded as the heartland of the Anasazi, and it is here that their oldest remains are found. These date from about the beginning of the Christian Era. Prior to this time, the Anasazi were probably a nomadic people, but the earliest archaeological evidence is from the time when they were just adopting agriculture. Within a few centuries, their living pattern was established. Their permanent dwellings in or out of caves, resembled those of the Hohokam and Mogollon; these were often clustered in loose village arrangements. The bow and arrow replaced the spear-thrower or atl-atl, and pottery-making and the bean were added to the culture. Skills of the artisan were developed in basketry, and in the creation of woven sandals and sashes. Even music made its appearance with the development of flutes. The dog was domesticated. All this represents great progress by the Anasazi people in adaptation to and control over the environment.

Then started a long upward climb, the molding of a new way of life a thousand years in the making. These steps are traceable in the thousands of ruins that dot the plateau. The climax of Anasazi achievements was from about A.D. 1000 to 1400. This was the time when the homes were often built in caves widely scattered through the canyons creasing the plateau. . . . But infinitely more pueblos, as Anasazi towns are known, were built in the open because there simply were not enough caves to supply natural shelter for all. It is incorrect, therefore, to think of Anasazi people as cliff dwellers only. They established their residence wherever land and water permitted, and if a cave was available, so much the better.

Due to a method of dating which was developed on the University of Arizona campus, archaeologists have been able to speak of the age of Anasazi remains with a definiteness not possible for any other New World ruins. Dendrochronology, or the study of tree rings, as applied to the ruins of the Southwest, was the brainchild of Dr. A.E. Douglass. The timbers that went into the Anasazi residences are the main sources for the raw data upon which this system depends. When the archaeologist is able to refer to past events in terms of dated decades, quarter or half centuries, rather than in the vagueness of a relative chronology lacking calendar years as we know them, half of the battle to interpret prehistory is won. The Anasazi story has been enormously enriched because of this method of studying the past.

With the mastery of agricultural techniques, food surpluses followed. The ability to store excess food in jars, baskets, and bins (even to the point of tiding the people over lean years) provided the stability that resulted in an expanding population, large communities, specialization in the arts, and inferentially, increasing complexity in the social, religious and political systems.

The evolution of pueblo architecture from simple beginnings was one of the finest accomplishments of the Anasazi. The joining of room to room whether of stone or adobe, and the stacking of rooms to a height of four or five stories was the Anasazi method for packing people into the least possible space. The reason for this crowding is not clear since there was no shortage of land. While most towns grew by accreation in response to the expanding population, a few of them show planning as though following an architect's design. The monotony of solid room blocks was broken by plazas, providing easier access to the rooms and good space for outdoor activities. The plazas of our oldest Western towns are surely traceable to this old Indian custom.

Another noteworthy architectural feature was the underground kiva with specialized furnishings. As in the modern pueblo Indian villages where kivas are in use today, the prehistoric kivas were probably owned by the male members of the clans, and were used both as club houses and for secret religious rites. The great kivas, as much as seventy feet in diameter and strongly roofed, were the most impressive buildings ever erected in the Southwest until the Conquest. They served the larger communities as a place where rites vital to the group as a whole, such as rain-producing ceremonies, were performed.

A familiar sight on all Anasazi ruins of this climax period are the numerous fragments of pottery. Much of this has black patterns on white backgrounds, the most distinctive of all Anasazi hall marks. . . . Many other kinds are also present, such as multi-colored pottery and the rough-surfaced cooking and storing pottery known as corrugated. All of these have special significance to the archaeologist because the shapes, colors and designs show differences from place to place and time to time.

Other arts also reached a high level of perfection. Textiles of cotton, . . . robes of feather cloth, sandals in a great variety of weaves, baskets, personal ornaments, ritual finery—all of these have been found by archaeologists.

The Anasazi dead were normally buried in the refuse heaps. From the graves come many of the objects that help archaeologists characterize the society. The skeletons reveal that these Indians have living descendants among such tribes as the Hopi, Zuñi, and the Rio Grande pueblos of New Mexico.

By A.D. 1400 some of the larger centers, thriving only one to two centuries before, had already been abandoned. The frontiers were shifting, mainly to the south out of the Four Corners country. The reasons for this still puzzle archaeologists. No satisfactory answer has yet been discovered, but it is certain that the causes were complex, interlocking and that they influenced most of the Southwest.

This movement of people resulted in centers even larger than before, but far fewer in number. These enjoyed a brief spurt of cultural energy. By 1500, however, only the Hopi, Zuñi, and Rio Grande centers remained. Chroniclers of the Spanish Conquest, which began in 1540, record only seventy pueblo towns, and by then most of the old glory was lost. The number today stands at about twenty-six. Perhaps a factor in the concentration of the Anasazi during the fifteenth and sixteenth centuries was the arrival of wandering hunters, the Navajo and Apache Indians. Upsetting as their presence must have been, these nomads did no more than speed up the Anasazi decline; they did not initiate it.

Anasazi history, like that of the Hohokam, matches the worldwide pattern of the rise and fall of ancient societies. Though obscured by time, their achievements have lent drama to Arizona's past.

The Anasazi with their pots and baskets and red sandstone dwellings certainly left no doubt that man in North America was developing a unique culture that was duplicated in some respects elsewhere, whenever man reached the same plateau in his cultural ascent. On the other hand, thousands of miles south in Peru and elsewhere in Latin America, a culture that dazzled the Spanish conquistadors literally pushed man's ascent to a higher plateau. One of its most exemplary cities, high in the Andes, was the highly organized Inca Kingdom of Machu Picchu. Here stone was worked in a new way. "In all Spain I have seen nothing that can compare with these walls and the laying of their stones," an early Spanish intruder remarked. What follows is an elaboration of the Inca's stone-laying techniques.

The Masonry of the Incas

John Hemming

The Inca's skill as masons is their most impressive artistic legacy—in other spheres of the arts they were overshadowed by earlier Andean civilisations. They succeeded in cutting and polishing their stones with dazzling virtuosity. Adjoining blocks fit tightly together without visible mortar. Even when the blocks interlock in complicated polygonal patterns, their joints are so precise that the crevices look like thin scratches on the surface of the wall. And when earthquakes have brought flimsier later walls crashing down, the Inca ashlars [hewn or squared stone] have remained proudly unaffected, with each block still smoothly fixed between its neighbours. . . .

The surfaces of the stones were smoothly polished, but each individual block was bevelled inwards at the edges of its outer face. As a result the joins between blocks were indented and the wall as a whole was crisply rusticated. There was no structural purpose for this bevelling. It was purely decorative, to break an otherwise smooth wall with contrasts of shadow, to give full weight to each individual block, and to draw attention to the brilliant precision of the joining cracks. It proved a successful aesthetic device: the curved surfaces of the stones give a fluid grace to the walls of Cuzco.

Inca masons adopted two styles for their walls. In some the stones interlock in haphazard shapes, with no two stones identical and the joints between them undulating like an elaborate jigsaw puzzle. This style is known as polygonal. In the other style, the stones were cut into rectangles and laid in regular courses, generally with each successive course slightly smaller than the one below. This symmetrical style is known as coursed. The Incas themselves clearly preferred the tidiness of coursed masonry and used it in the walls of important buildings. But to the casual modern observer the polygonal walls are more baffling and impressive. It is almost alarming to see gigantic rocks fitting snugly together like great lumps of putty. . . .

It used to be thought that the polygonal walls were more ancient than the more familiar coursed walls, but recent archaeology has shown that both

Excerpted from *The Conquest of the Incas* by John Hemming, Harcourt Brace Jovanovich, Inc.

styles were common during the Inca empire in the late fifteenth century. There is a plausible explanation for the two styles of stonework. Polygonal masonry was used only for terraces or retaining enclosure walls, where strength was required. The rocks were left in uneven shapes to waste as little as possible of large boulders, and in imitation of the rough fieldstone walls that are common for terracing throughout the Andes. Coursed masonry was used for free-standing house walls. The style may have been an imitation of the turf buildings found in the Cuzco area. The turfs were cut in rectangular pieces and laid with the grass downwards. As they dried the tops and bottoms contracted and the outer sides bulged. This would provide a precedent for the ornamental countersinking of the joints on Inca masonry.

The Incas had a unique system of record-keeping known as the *quipu*. It looks like nothing more than a tangled mass of multi-colored twine, but for the Incas it was an effective mnemonic device, as described here by J. Alden Mason.

The Quipu

J. Alden Mason 63

The Inca had a highly developed mnemonic device known by the native name of *quipu*. Probably in simple form it had a long and widespread history in this region, but, like many other devices, had been perfected by the Inca for their special needs. All the extant examples of the quipu are from the dry coastal graves, but most if not all are of the Inca period. . . . As it was in constant use among the Inca, many of the chroniclers have left us descriptions and explanations of it.

The quipu consists basically of a series of strings in which knots are tied. The great variation possible in the colour and position of the strings, and the nature, number, and position of the knots permits its use for numerical records and mnemonic purposes. All known quipus are different and vary greatly in size and complexity; relatively few of those extant are complete, a necessity for the correct interpretation of any mathematical record.

The main cord, which was held in a horizontal position, is generally of larger size, from a few inches or centimetres to over a yard or metre in length. To this are attached from one to over one hundred pendent strings, of various colours, twists, and other modifications. They may be fastened to the main cord in groups, and subsidiary strings may be attached to them. Knots of various types and positions were tied in these pendent cords. The knots certainly have numerical values; the colours and other qualities of the strings probably signify the nature of the objects thus counted. Several of the chroniclers give interpretations for some of the colours, but the disagreement is so great that no deductions can be drawn.

Studies of quipus by several specialists in this field have demonstrated clearly that the numerical records are given in a decimal arithmetical system very much like our own, with place-value. This was to be expected, since the Quechua numeral system was and is decimal, and the social system was organized on a decimal basis. A simple knot represents 'one'; digits from two to nine are denoted by longer knots in which the cord was wound or looped a given number of times before it was pulled tight. The concept of zero was understood but required no symbol; the absence of any knot in the expected position denoted zero. Place-value was indicated by distance from the main cord; the unit digits were at the farther or lower end of the string, the higher multiples—tens, hundreds, and thousands—closer to the main cord. In the known extant quipus calculations in thousands are rare, and apparently only one instance of ten thousand is known. Generally the long knots of many loops were employed only for the unit digits; multiples of the higher orders were represented by the proper number of single knots close together.

The quipu was a recording and mnemonic, not a calculating device. Its principal purpose was doubtless that of statistical record; this is obvious from the statements of the chroniclers who saw them in use. Probably the majority were censuses of the population by age-classes in given districts, as well as records of domestic animals, quantities of agricultural products, and such statistics. Probably the figures could be read by any Inca *quipucamayoc* or professional quipu-interpreter, who doubtless also knew the meaning—or various meanings—of the string colours, but certainly in many or most instances some verbal information of interpretive value had to accompany a quipu.

However, the quipu could be, and certainly was, employed as a mnemonic device for the recitation of traditional material such as historical ballads and genealogical records. Thus, according to some of the chroniclers, the life and activities of each emperor were recorded on a quipu. Such quipus could, of course, be interpreted only by the maker or by someone fully familiar with the data. . . .

It is rather certain that an abacus was employed in making the calculations that were later recorded on the quipus. No example has yet been discovered archaeologically, but some chroniclers have described its use and even given a picture of one. . . . This seems to have been a rectangular block with twenty (5 by 4) compartments in which from one to five kernels of corn or other small objects were placed and manipulated.

Tracing man's architectural history, it becomes clear that there is more to building than squarely placing stone upon stone. To quote Dr. Bronowski, "the physical limitations of stone could not be overcome without a new invention." There are better ways to fit stone into and around space, better entryways than rectangular doors and windows. The Romans overcame the physical strains inherent in stone beams and columns by building arches, which were not only stronger, but also more esthetically pleasing. Centuries later, the pointed Gothic arch was one of the crowning achievements of medieval stone masons. It was, as William Fleming writes, "a struggling, striving, dynamic urge that reaches upward to embrace infinity." The Roman arch is discussed here in sections from Fleming's *Arts and Ideas*.

The Roman Architectural Contribution

William Fleming

The Roman contribution to architecture was fourfold: (1) building for use,

(2) development of the arch and vault as a structural principle, (3) emphasis on verticality, and (4) design of significant interiors. In the first case, Roman architecture was marked by a shift in emphasis from religious buildings to the civil engineering projects that had such an important bearing on the solution of the practical problems of the day. This did not mean that the Romans neglected their shrines and temples or that they lacked religious feeling. As in the 19th and 20th centuries, however, the main architectural expression was to be found in secular rather than religious structures. In this category come the basilicas, aqueducts, roads, bridges, even the sewer systems, which so admirably served the utilitarian purposes of the Romans.

Second—and perhaps most important of all—was the Roman exploitation of the possibilities inherent in the arch as a building principle to implement the above objectives. The construction of a true arch by means of the wedge-shaped blocks known as *voussoirs*, as well as some of the implications of the system, can be seen more easily . . . than explained in words. When such arches are placed side by side in a series, the resulting arcade can be used for such structures as aqueducts and bridges, as seen in the Pont du Gard. . . .

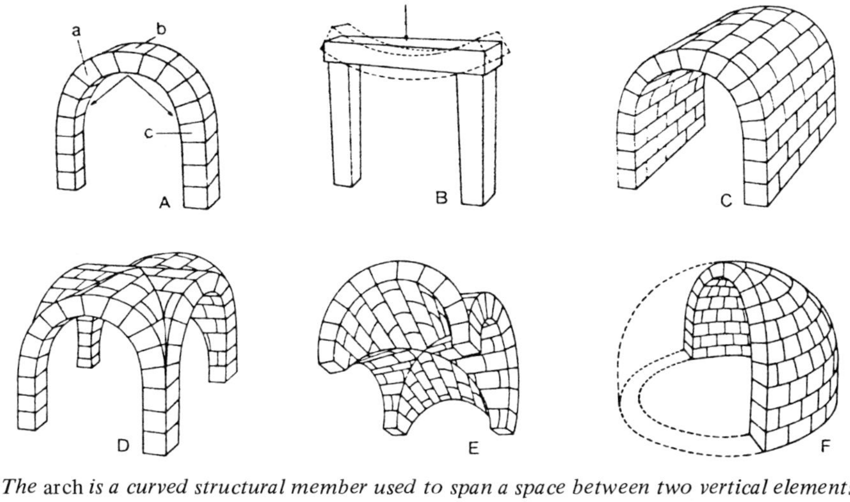

The arch *is a curved structural member used to span a space between two vertical elements, such as posts (piers) or walls (A). Wedge-shaped blocks, called* voussoirs *(a), give the arch its form and stability by virtue of their compressed relationship to each other and to the* keystone *(b) at the apex of the semicircle. The curve of the arch rises from the* springing *on either side of the opening (c), and the* compression *of its stones is better able than the* tensile *strength of a* lintel *on posts (B) to resist the downward thrust of the load above. It does this by directing the weight laterally and by transmitting the load to the ground through the line of the arch and the vertical elements supporting it. Placed one after the other, a series of arches make a* tunnel vault *(C). Two tunnel vaults intersecting each other at right angles (D) create a* cross vault *also known as a* groin vault *(so-called for the groin line [E] along which the vaults join). Intersecting each other around a central axis, a series of arches would form a* dome *(F).*

When placed in a series, from front to back, the result is a *barrel vault* . . . , which can be seen in the Arch of Trajan at Benevento . . . and which was useful for roofing interiors. When a series of arches span a given space by intersecting each other around a central axis, the result is a dome, as exemplified in the Pantheon. . . . In greatly oversimplified form, these constitute the technical principles that underlie the Roman architectural achievement.

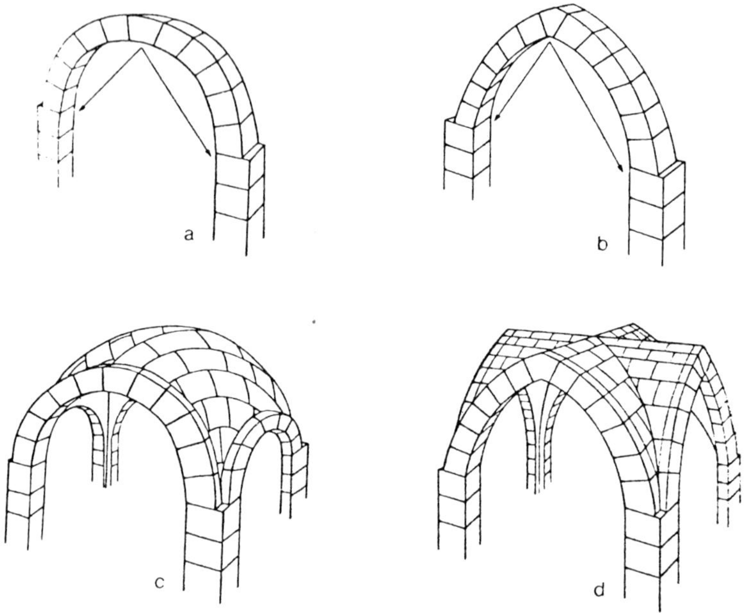

The pointed arch and ribbed groin vault are fundamental to Gothic architecture, making it a light and flexible building system that permits generous openings in walls for large, high windows, the result of which are well-illuminated interior spaces. Being more vertical, pointed arches thrust their load more directly toward the ground (b), whereas the lower round arches spread the load in a lateral direction (a), which makes them less stable than Gothic arches. Too, pointed arches can rise to any height, whereas the height of semicircular arches is governed by the space they span. In (c) and (d) the space, or bay, *that has been vaulted is rectangular in shape, rather than square. In (c) the round arches create a domical vault whose shapes and openings are irregular and restricted. In (d) the pointed arches rise to a uniform height and form a Gothic vault with ample openings.*

Third, by their technical advances, they were able to increase the height of their buildings in proportion to the growing size of their large structures. The six-story mercantile buildings of Trajan's Forum were an impressive demonstration of the practical advantages of such verticality, which allowed the combination of many small shops into a single structure in a crowded city location. The multifamily apartment houses in Ostia and Rome were also cases in point. The trend was to be seen in the great height that imparted such pleasing spatial proportions to the halls in the Baths of Trajan and Caracalla as well as to the Pantheon—all made possible by cross vaulting and the dome.

Last, the enclosing of large units of interior space was made necessary by the expansion of the city's population. The direction of architectural thought in meeting this need can easily be seen by contrasting a Greek agora with the Forum of Trajan, a Hellenistic theater with the Colosseum, or the Parthenon with the Pantheon. Special attention to space composition and the problems of lighting are in evidence in the planning of such interiors as those of the Basilica Ulpia, the Pantheon, and the halls of the great baths. In all instances, the increasing Roman awareness of the value, tangibility, and reality of the spatial medium is discernible.

In his exposition of the "monuments of unaging intellect" that make up *The Ascent of Man*, Dr. Bronowski used what he calls "a herringbone effect," that is, cultural, scientific and artistic innovations—past and present—merging in time and space, overlapping and reinforcing each other. This is another illustration of the uniqueness of the whole mankind and man the individual. A culture can be created anywhere if the conditions are right. For example, while the Inca masons were building their stone cities in Peru, Europeans who carved new shapes from stone in the Middle Ages were becoming a freewheeling intellectual and artistic elite. They were "freemasons" in every sense of the word, and the following selection reveals how they got there.

The Rise of Freemasonry

Jean Gimpel

The Latin terms that designated workers who cut stone in the Middle Ages do not generally permit a distinction to be made between those who simply quarried the stone and those who cut the vault stones, window tracery, and monumental portal sculpture. The sculptors were lost among the general mass of stonecutters. This is really rather extraordinary to us, because an enormous difference seems to exist between those who perform a seemingly mechanical task, such as cutting blocks of stone, and those who sculpt, *with their very soul*, the magnificent statues in the cathedrals. The truth is that for the great majority of men in the Middle Ages there was between a good *work* and a *masterpiece* only a difference of degree, not a difference of kind. The idea that there is an unbridgeable gulf between a worker and an artist (in the modern sense of the word) did not really occur until the Renaissance when it was expressed by intellectuals who judged, classified, and evaluated manual labor which was very foreign to them.

It was the Renaissance writers who, for the first time in history, extolled the personal merits of painters and sculptors, resulting in an excessive deification of both, the consequences of which are still felt. The Renaissance fabricated the notion of "the artist." The medieval thinker hardly ever expressed himself in his writings on actual questions of aesthetics, and when he did concern himself with what we call "art," it was in terms of theology or philosophy. Why? Did he more or less consciously hold manual labor in disdain and relegate both sculptor and mason to the same ill-favored category? Was he insensitive to the beauty of forms? Or was it because he intuitively perceived that a beautiful statue or a beautiful window were complete works in themselves, sufficient in themselves and thus requiring no comment? (In this connection it should be noted that the glorification of the "artist" since the Renaissance, even though it would seem to be a promotion, has been nothing more than a placement of art under the guardianship of literature.

Art henceforth was always to await its justification from literature.)

To avoid this problem of terminology, the word "artist" is deliberately not used here, since it adds nothing to the greatness of the cathedral builders and because its current meaning is essentially foreign to the spirit of the Middle Ages. Only in the dictionary of the French Academy of 1762 was "artist" mentioned in the sense that we now understand it.

In England, however, the terms that designated stonecutters permit a certain distinction to be made between laborers who did the heavy work and those who executed the more refined work. The distinction is based on the characteristics of the stone worked. For example, those who worked the particularly hard stone found in Kent were known as "hard hewers." They were distinguished from "freestone masons" who worked the beautiful calcareous stone found in great abundance from Dorset to the Yorkshire coast and which was perfect for the delicate work of sculpture-carving. "Freestone masons" were also distinguished from "rough masons" who simply rough-hewed the stone.

The expression "freestone mason" was replaced after a while by the simplified "freemason," a word which evidently had to do with the quality of stone and not with any franchise which would have benefited builders. When honorary freemasonry was introduced into France from England about 1725, it was naturally translated as *franc-maçonnerie*, an expression unknown in the Middle Ages. On the other hand, it is known that in London in 1351 a *maître maçon de franche peer* was more or less the Anglo-French equivalent of two Latin expressions: *sculptores lapidum liberorum* (London, 1212) and *magister lathomus liberarum petrarum* (Oxford, 1391). The modern English and French translations of this second term would be: *a master mason of freestone* and *un maître tailleur de franche pierre*.

Application of the words *franc* or *franche* ("free") to stone in France has survived uninterruptedly to the present. The beautiful building-stone found in the region about Paris has long been spoken of as "free liais," a term which is still applied to veins of excellent quality. When I visited the underground quarries of Paris, an engineer in charge of supervising the subterranean part of the capital showed me the inner side of a tunnel and told me, much to my surprise, that to designate veins of stone of the very best quality one would say "free veins."

Stone used for sculpture did not often have the same grain as that used for building the walls to which the statues were attached. A rapid glance at the *Portail Royal* at Chartres is sufficient to show that the stone of these famous statue-columns was not taken from the quarry at Berchères. And at Vézelay the capitals are not of the same stone used in buildings the rest of the church. . . .

Becoming a sculptor, the stonecutter rose to the world of the spirit. He was closer to the theologians and learned more through this contact; he had the wonderful opportunity to thumb through precious manuscripts in abbeys; he learned to look, observe, think. His intellectual horizon was expanding, which permitted him to participate spiritually as well as physically in his creations. Thanks to the manuscript minatures leafed through and admired in various abbeys, the sculptor could humbly suggest slight variations in the subjects proposed by the Fathers. Since the sculptor and the theologian worked toward the same goal, the sculptor could consider himself free, for in this association he did not submit to any constraint. In this sense it could be said that the difference between the artist of today and the medieval sculptor is that the latter was not an individualist since he did not claim the rights of personal inspiration. However, a legitimate pride did overcome the spirit of these workers with humble backgrounds, and sculptors did not hesitate, especially in the twelfth century, to sign their works. Gislebert signed the famous tympanum of Autun: "*Gislebertus fecit hoc opus*." Giraud signed the portal of Saint-Ursin in Bourges; Umbert, a capital of the porch at Saint-Benoît-sur-Loire; Rettibitus, a capital of Notre-Dame du Port in Clermont. . . .

Intoxicated by his freedom and his extraordinary success in the spiritual and physical order of things, he dreamed only of putting sculpture everywhere. He wanted literally to cover churches with his creations and he suffocated them. Compositions became confused. Some of the twelve hundred sculptures at Notre-Dame de Paris were misplaced on the building. At Reims it was necessary to number the cathedral's three thousand sculptures in much the same way that prefabricated parts are numbered today. At Tournai sculptors established themselves and produced works to order. In 1272 the abbot of Cambron bargained over prices with two sculptors to cut and ship cut-to-specification window mullions to Brussels.

Having broken with the architect, the sculptor next dissociated himself from the theologian. Did he already consider himself better than the mortals around him? Was this the dawn of the Renaissance? He constantly had to be brought back into line. The period of Christian expansion was over. In 1306 the sculptor Tideman executed a Christ for a London church which did not conform to the accepted conventions. The bishop actively intervened, had the Christ removed, and ordered Tideman to return his fee.

Such freedom for the sculptor from all tradition, something that would have been impossible a century earlier, coincided with a general weakening of the Faith. The powerful and rich who previously had given their riches to the cathedral now devoted their fortunes to increasing their own comfort and satisfying their taste for life. They built private chapels and city houses. Activity in the great workshops ground to a halt as the better sculptors and builders were hired away and hired to build and ornament the mansions and chapels of the nobility. The situation was similar to that in Athens in the fifth and fourth centuries B.C.: the progressive era for Athens ended with the Peloponnesian Wars. Throughout the fifth century sculptors worked on such projects as the Parthenon, but Athens was ruined in the fourth century and rich individuals hired sculptors—Praxiteles, for example—to appease their egotistic appetite for beauty.

But the French sculptor of the fourteenth century for a long time could not profit from working for private individuals because the Hundred Years' War, beginning in 1337, disrupted both his life and tranquility. It ended prosperity and travel for the sculptor. National revenue fell sharply. Misery and poverty were felt everywhere, and plagues, famine, and battle all but decimated the population. Rare indeed were the privileged who could still engage sculptors. Life was of and for the war, a situation in which sculpture was only a luxury. Workshops either closed or slowed down, most sculptors were impressed into military service and then forced to build military structures such as prisons and fortified castles. Needless to say, delicate skill was not a primary concern. Local stone was used since it was impossible to haul better stone from great distances during the war. With the impossibility of travel, new generations of workers forgot where the famous quarries which had provided stone for monumental sculpture were located. It is difficult to recognize in these sculptors—compelled as they were to carve stone cannonballs—the descendants of those who built Chartres and Mont-Saint-Michel.

When the Hundred Years' War ended in the midfifteenth century, and stonecutters once again began to sculpt, they could not return to the great tradition of the cathedral crusade regardless of their efforts. The world had changed, and in its turn stone sculpture had ceased to be a living means of expression. . . .

"Brain and hand unite." That has been obvious since evolving man was endowed with the fortuitous arrangement of thumb and fingers that distinguished him from the ape. It allows him to release underlying forms in nature, unleashing his creative spirit in new and unexpected ways. The following section reveals the closeness to nature that is essential to the creations of Henry Moore, Michelangelo, Gordon Parks, and Eskimo carvers.

On Creativity

"Landscape has been for me one of the sources of this energy. It is generally thought that no sculptor is much interested in landscape, but is only concerned with the solid, immediate form of the human figure or animals," writes Henry Moore. "For myself, I have always been very interested in landscape. (I can never read on a train—I have to look out of the window in case I miss something). As well as landscape views and cloud formations, I find that all natural forms are a source of unending interest—tree trunks, the growth of branches from the trunk, each finding its individual air space, the texture and variety of grasses, the shape of shells, of pebbles, etc. The whole of Nature is an endless demonstration of shape and form, and it surprises me when artists try to escape from this. Not to look at and to use Nature in one's work is unnatural to me."

For Michelangelo Buanarroti there was something uplifting about the act of unlocking hidden images from a stone, as is expressed in this selection from *Art Through the Ages.*

One of the best-known observations by Michelangelo is that the artist must proceed by finding the Idea, the image, locked in the stone, as it were, so that by removing the excess stone he extricates the Idea, like Pygmalion bringing forth the living form:

The best artist has no concept which some single marble does not enclose within its mass, but only the hand which obeys the intelligence can accomplish that. . . . Taking away . . . brings out a living figure in alpine and hard stone, which . . . grows the more as the stone is chipped away. . . .

The artist, Michelangelo felt, works through many years at this unceasing process of revelation and "arrives late at lofty and unusual things and remains little time thereafter".

Michelangelo did indeed arrive "at lofty and unusual things", for he broke sharply from the lessons of his predecessors and contemporaries in one important respect: He mistrusted the application of mathematical methods as guarantees of beauty in proportion. Measure and proportion, he believed, should be "kept in the eyes". Vasari quotes Michelangelo as declaring that "it was necessary to keep one's compass in one' eyes and not in the hand, for the hands execute, but the eye judges". Thus, he would set aside Vitruvius, Alberti, Leonardo, Albrecht Dürer, and others who tirelessly sought the perfect measure, being convinced that the inspired judgment could find other pleasing proportions, and that the artist must not be bound except by the demands made by the realization of the Idea. This insistence upon the artist's own authority is typical of Michelangelo and anticipates the modern concept of the right of talent to a self-expression limited only by its own judgment.

Something of the same creative zest is expressed by Gordon Parks, photographer, film maker, writer, and composer. Although he has worked in different artistic media than Michelangelo and Henry Moore, he, too, feels the need to release the underlying essence of nature. Here he relates how, as a child, he became literally possessed of his love for music.

I was about seven when a peculiar fantasy kindled my love for music. The Kansas day was hot, and I was hunting june bugs in our cornfield when I heard a murmuring in the cornstalks. The murmuring grew into music, and I stood there, my mouth full of mulberries, puzzled, looking up at the slow-drifting clouds—wondering if they were the music's source. The violins, horns and drums were as true to me as the sunlight, and I had a feeling that the music was trapped inside my head, that it would be there even if I had no ears. I covered them with my hands, and the sounds were still there and they continued until all the clouds moved away and there was nothing but pale sky. Then it was gone as mysteriously as it had come, and I ran toward the house a little frightened, a little joyful. Then in a frenzy I started banging on our old kimball upright, trying to reproduce the sounds I had heard. My father came in from another part of the field and stood for a moment watching me with astonishment. Later that night he told my mother that he thought her youngest child was going batty.

From Horst de la Croix and Richard G. Tansey, *Gardner's Art Through the Ages* (5th ed.), Harcourt Brace Jovanovich, Inc., 1970. Reprinted by permission of the publisher.

Excerpted from *Creativity: A Discussion at the Nobel Conference*, North-Holland Publishing Company, Amsterdam, 1970.

The same creative enthusiasm for discovering form in nature is exemplified by Eskimos who fashion art from ivory and whale bone, as discussed in *Eskimo Realities* by Edmund Carpenter.

Carving, like singing, isn't a thing. When you feel a song within you, you sing it; when you sense a form emerging from ivory, you release it. As the carver holds the unworked ivory lightly in his hand, turning it this way and that, he whispers, "Who are you! Who hides there!" And then: "Ah, Seal!" He rarely sets out to carve, say a seal, but picks up the ivory, examines it to find its hidden form and, if that's not immediately apparent, carves aimlessly until he sees it, humming or chanting as he works. Then he brings it out: Seal, hidden, emerges. It was always there: he did not create it, he released it; he helped it step forth.

Along with the construction of something new and different from nature's building blocks is the simple pleasure that the artist finds in his skill. That feeling must have inspired the men who covered the walls of their Ice Age caves with deer and bison, and it obviously was part of Simon (Sam) Rodia as he fabricated his eccentric but magnificent Watts Towers.

Simon Rodia
Robert Bryan

With Alice it was a simple matter of stepping through the looking glass, while Dorothy, of Kansas, journeyed to the Land of Oz on the wings of a midwestern tornado. All of which proves that there are various roads to wonderland.

Each year some 25,000 visitors, including numerous ones of an age in which the sense of wonder is very much alive, journey to the Simon Rodia Towers in Watts, California. Most of them come via a rolling roadway known as the Harbor Freeway, which rides high above a settlement of frame and stucco houses bordered by palm trees.

At Century Boulevard or at Imperial Highway, the visitors turn east, in search of something that would never have been except for the obstinate, persistent obsession of an immigrant Italian tile laborer who may have been baptized Sabatino, was called Simon, but came to be known to his friends as Sam. Sam gave thirty-three years of his life, from 1921 to 1954, to accomplishing the impossible.

In Los Angeles almost everything that lifts to the sky must have a clearly discernible purpose. The high voltage pylons one sees in Watts bring electrical benefit to Long Beach, while the telephone poles that rise above the one-story houses carry lines that transport messages of all sorts. But then as one turns into East 107th Street he is confronted by a fantastic group of structures with no purpose at all but the giving of pleasure and the instilling of wonder.

James Barrett, the man at the gate who welcomes visitors to the unique structures, is a treasure house of information about them. His affection for the Watts Towers, as they have come to be known, is evident in every word he speaks, especially as he recalls the times when he and his childhood friends rode by them on their bicycles. He will tell you that they considered them some strange sort of cathedral, for these were the years when entrance to the Towers was denied most people; when the builder, grown weary of the taunts of the neighborhood children as well as the endless questions asked by the curiosity-seekers, had secluded himself in his house. What a change from the Sam Rodia of the twenties and thirties who was willing to have his magical gardens with their fanciful water basins used for baptisms of the total immersion type. The Sam of those years was an unattached gay blade who squired the girls of the neighborhood and drove his 1927 Hudson touring car about at a tremendous speed. At other times, when working in the Towers, Sam would regale the entire neighborhood with his full-throated singing of Italian songs or arias from the operas.

Today's visitors to the Towers find the children of Watts all about them, perhaps skipping rope in front of that entranceway where years ago Sam had truckloads of sand and gravel dumped, perhaps playing softball along 107th Street. To the children who live with the Towers always about them, the unusual and unique have become almost accepted. And yet not entirely. One recent morning a child walking down the tracks of the Pacific Electric right-of-way, thinking himself unobserved, was seen gazing up at the towers for a full minute or two as he passed by. Did he expect to see Sam—"El Italiano" as he was once called in the neighborhood—aloft on one of the Towers with his window washer's belt, his pail of cement and his burlap pouch filled with broken tile and sea shells? Twenty-five years ago he might well have.

If there are children among today's visitors to the Towers the effect on them is immediate. Although their elders will stand about with that sense of awe and reverence that most of us assume in churches or museums, the children seem to know immediately where they are and what fun awaits them. In no time at all they will be playing about the Ship of Marco Polo at the very end of the gardens, or exclaiming at the tiny china animals imbedded in Sam's Hall of Mirrors before the front door of what was once his house. They will pull their parents with them to show the corncobs and the old boots immortalized in cement or the place in the Cactus Garden where Sam left his thumb prints. Perhaps that was the day he temporarily ran out of sea shells and broken glass. . . .

What, then, was it like to be a child in Watts during the roaring twenties and the starving thirties? What did it mean to live down the street from an altogether extraordinary artist during those years when he raised his monument to his adopted country, to his heroes, and to himself?

From all accounts it was a great day to be alive. Watts was a village with a mayor and police force of its own. Its connection with the distant city of Los Angeles was via the Big Red Cars whose right-of-way ran alongside the northern line of the property that Sam Rodia purchased in the early twenties at the end of 107th Street. It may well have been that he purchased the property in order that his future work might be seen by passing travelers.

Every artist, after all, likes recognition.

Louis Sauceda was a child during the years when Sam was building. He recalls a 107th Street that is difficult to picture today; a dirt road with few houses but a good supply of chickens and goats. He tells of a community of open fields and a swamp where Jordan High School now stands. East of Wilmington everything was truck gardening, where a kid in season could earn fifty cents a day pulling onions. And in the open fields one could hunt small game.

Along the street one had to watch for the horse-drawn buggies. Later there was the 1922 Chevy which belonged to the local bootlegger, and then Sam Rodia's famous 1927 Hudson touring car, which one day its owner would decide to bury in his backyard.

Everyone agrees that there was a special relationship between Sam Rodia and the Sauceda family, his dear friends. "El Italiano," as Sam was called along the street then, was a widower and certainly must have welcomed the concern and friendship accorded him. It might include everything from sewing on a button to an evening meal with the family.

Louis Sauceda, the child, just naturally gravitated to the house at the end of the street, he states, where already the lacework of steel rods, chicken wire and cement with tile and sea shells was rising. He was privileged to watch the raising, rung by rung, of the Towers, to be a witness to the gradual emergence from Sam's fertile imagination of all sorts of wondrous things. Would he work today on the first fountain which bears resemblance to a crown? Or would he, perhaps, choose tile to be placed at the base of the Santa Maria Tower? What child could ask for anything more?

There were days when Sam was in a good mood, Mr. Sauceda affirms, and then a welcome would be accorded to all. Children from the neighborhood would bring bits of tile and broken bottles. Sam's payment for such cast-off scraps was usually in cookies or pennies. The price he paid for broken plates, which were used throughout the walls surrounding the Towers, was one cent each—a handsome price. In fact, so good a price that the children hit on a way to increase the supply. They would take a family plate, break it and then take it to Sam who would, unwittingly, pay them the going rate. It was a good business until one day a mother counted her plates, and that was the end of that.

There were other days when Sam was absent from his work, perhaps having caught one of the Big Red Cars for Wilmington Beach. There he hunted for sea shells of a size and shape to fit in some fastidious design he had in mind. He would walk back along the Pacific Electric right-of-way all the way home. By the time he returned, his cloth cement sack would be filled with broken crockery, bathroom tile, or bits of china dogs and squirrels from some forgotten world's fair. Someone had thrown them all away as of no further use, but Sam found a use for most of them. What he did not want he buried in the backyard.

"Everybody said he was crazy," Mr. Sauceda states, "but I remember him up in the Towers singing away."

Like any builder, Sam was obsessed with time running out on him. "I'm worried to death to get this work done before I leave," he once said, "but I'm a happy man."

And he must have been very proud, too, especially during the series of earthquakes that shook the Los Angeles area in the early thirties. "We ran over to him one day when the earth began to shake," a resident of 107th street recalled. "That crazy guy was up there in the Towers and the earth was shaking and we told him to come right down. But he just laughed. 'It won't fall,' he said, and he was right. By gosh, he was right!"

4

The Hidden Structure

Nature operates by unseen bodies.

Lucretius

In Chapter 3 it was told how man, with his intellect and his hands, had learned to take advantage of the underlying forms in nature to create monumental works of art, domiciles and cathedrals, cities, and lesser artifacts that he made simply because they pleased him. But within those natural forms is the hidden structure that Dr. Bronowski explores in Chapter 4, the atomic framework of which all the gases, liquids, and solids that make up the world around us are composed. Dr. Bronowski places the beginning of the search for the hidden structures with the alchemists, who believed that with fire they could transform lesser substances into gold, a substance that Roger Bacon said is "able to remove the corruptions of the human body to such an extent that it would prolong life for many ages." Bacon merely borrowed a supposition from Chinese alchemists of thirteen centuries earlier who had maintained that "if you make sacrifice to the furnace, you will be able to transmute cinnabar into gold. When the gold shall have been produced, you may make use of it utensils for eating and drinking. Through using them your life will be prolonged." (Although Bacon and the Chinese alchemists had little basis for their conviction that gold prolonged life, there is some justification for their belief in the fact that gold is used to treat the crippling debilitations associated with rheumatoid arthritis.) In any event, the alchemists and their reverence for fire did lay the foundations for modern science, even if they did not realize that their notable achievement was the chemical transformation of one substance to another.

The Chinese alchemical recipe for *Scarlet Snow and Flowing Pearl Elixir* began with realgar, an orange-red mineral that we now recognize as a chemical compound of arsenic and sulfur. In the following reading, Nathan Sivin recounts his attempt to duplicate that Chinese recipe.

A Chinese Alchemical Recipe

Nathan Sivin

One *chin* of realgar is pounded and sifted through silk gauze. It is then mixed with wine vinegar to a creamy consistency and is dried in the sun. When dry the mixing is repeated, the process being carried out a total of ten times. The material is mixed with powdered white salt and [placed in a two-part reaction vessel] with a bed and cover of salt. The vessel is luted and after a day and a night is warmed with a low flame to render the six-one lute bone dry. The fire is gradually increased—there is no need [at this stage] for a roaring fire—for a day and a night longer, and then a roaring fire is applied. The lower part of the combustion chamber must be kept the same color as the fire the whole day through; the intensity of the fire should not be allowed to slacken even momentarily. The heating is continued in this way for three days and three nights. The reaction vessel is then cooled for twenty-four hours. Open it and gather the medicinal essence in the upper section. It is further ground slightly. The dregs in the lower section are also pounded, mixed with the essence, and then with cooked rice until moist. The mixture is placed in the reaction vessel as before and heated first by a gentle and then by a strong flame just as was done previously. When the medicine is done it will have a brilliant radiance. It will be of the shape of a pendant string of pearls or of colored silken threads. Again, its configuration will be that of stretched knotted netting. Its fresh brilliance dazzles the eye. Those who see it will, unawares, feel a shock. But it is well to be calm. For those suffering from an illness following a sudden fainting attack, those on the point of death, and those who have already expired, grind the elixir fine [and make pills] the size of three or four hempseeds. They are washed down with raw egg yolk and a little wine. Put the medicine into the patient's mouth and hold his head up. In a short while he will recover. If his mouth is firmly shut and will not receive the medicine, his upper teeth are to be prized up and the medicine washed into his mouth. It is pushed down on its way to the stomach with the fingers. The patient is shaken so that the *ch'i* [= activity] of the medicine will be disseminated. In a moment the patient will come to life again. The elixir also cures diseases of malevolent possession and minor autumnal intermittent fevers as soon as it is taken. This medicine has wonderful efficacy, which cannot be fully described. My regret is that in these unsettled times there is no one who will understand its preparation and use.

Commentary

I carried out the first part of this preparation—up to the adding of the rice—at the University of Singapore in 1962. The realgar was chemically pure, but the vinegar was a household rice wine vinegar of local manufacture. The reaction vessel was two laboratory crucibles placed mouth to mouth and luted with "six-one lute" made according to Sun's instructions from red bole and alum. Very careful drying was required to keep the lute from cracking. The combustion was carried out at 900°C in an electric furnace. This is, roughly speaking, the temperature at which iron glows the same bright red as a charcoal fire. The sublimed "essence" proved to be metallic arsenic, which badly corroded the nickel crucibles I was forced to use. Reference to a phase diagram indicates that this elixir would have been equally destructive to an iron vessel. The chemical identity of the elixir which results from the second sublimation cannot, of course, be determined until the apparent textual corruption is dealt with. If cooked rice (or any other reducing agent) is in fact not added, it is possible that the final product contains crystalline or fibrous arsenic trioxide, which could more or less tally with the description. The lute is sufficiently porous—and sufficiently likely to crack—to allow the diffusion of air into the reaction vessel during long heating, particularly if slight variations in the temperature of a charcoal fire cause the container to "breathe."

There are those who argue that alchemy, with its reliance on incantation and ritual actually impeded the progress of the field that we call chemistry. The alchemists did actually use fire to break down chemical compounds into their constituents, but they were less interested in that accomplishment than the magic and mystique that surrounded it. This was especially true of the Greeks, who, as early as 400 B.C., had established a workable theory of the atomic structure of matter. Yet it was also the Greeks who argued that our universe was made up of four elements: fire, water, air, and earth. Greek alchemists became bogged down in scientifically pointless experiments designed primarily to "illustrate preconceived theories rather than to discover new truths," as Charles A. Browne elaborates in the following reading.

Greek Alchemy: Pluses and Minuses

Charles A. Browne

Since the conversion of the four elements into one another was taught by Plato (427–347 B.C.) and Aristotle, it was argued that the transmutation of copper or lead into silver and gold was a possible accomplishment. All fusible substances such as wax, glass, and metals were held by Plato to consist of earth with a certain amount of water to give them fluidity. The rust or corrosion of the baser metals was regarded as an exudation of this earthy ingredient. Of all the metals, gold was considered to contain the least amount of earth, this residual impurity being of so fine a quality that there was no exudation of rust or tarnish. The removal of the excess of the supposed earthy constituent of base metals by a process of derusting was thus the first step of transmutation.

The belief in transmutation was enhanced by observing the gold- and silver-colored alloys that the skilled Egyptian technologist prepared from base metals—alloys which in many cases were used for counterfeiting. A number of their workshop receipts are contained in old Greek papyri in Leyden and Upsala; their fraudulent character is clearly indicated by the following translations:

Clean white soft tin four times; melt six parts of the same with one mina of white Galatian copper. It becomes prime silver that will deceive even skilled workmen who will not suppose it to be made by such a treatment.

Add six parts of purified tin and seven parts of Galatian copper to four parts of silver and the resulting product will pass unnoticed for silver bullion.

These quotations are good examples of the chemical receipts of the Greco-Egyptian period before they were obscured by the allegorical and mystical interpretations of the later alchemists, who, under the influence of philosophical, gnostic, astrological, and religious ideas, believed a simulation of the colors of silver and gold by means of counterfeit alloys to be an actual transmutation of the base into precious metals.

Alchemy, as a pseudoscience, began to attract the attention of Greek writers during the third century A.D. Poets, priests, philosophers, and physicians, as well as laymen, became interested in this so-called "sacred art." In the five centuries between 300 and 800 A.D. there was a vast output of manuscripts upon the subject. The oldest collection of these remains of Greek alchemy is the famous Codex 299, in the Library of St. Mark at Venice, written in a hand of the eleventh century. Some of these manuscripts bear

Reprinted from Joseph Jastrow (ed.), *The Story of Human Error*, Books for Libraries Press, Inc., 1967.

the names of mythical authors, as Hermes and Isis; others are pseudographs falsely attributed to Moses, Democritus, and other ancient sages; others are genuine productions of Zosimus (fifth century A.D.), Synesius (about 370–430 A.D.), Olympiodorus (fifth century A.D.), Stephanus (seventh century A.D.), and other writers of the later Alexandrian and Byzantine periods.

The literary remains of the Greek alchemists form a mixed collection of philosophical treatises, dialogues, poems, enigmas, prayers, invocations, oaths of secrecy, allegories, visions, and magical incantations, together with numerous laboratory receipts. Astrology, mythology, philosophy, and religious ideas of Egyptian, Chaldean, Greek, Jewish, and Christian origin are all hopelessly intermingled, in language which is deliberately obscure, in a vain effort to explain some theory or process of transmutation.

But even in this period of decadence we note the same desire of the early Greek philosophers to arrive at a broad, unifying conception of the universe that would include man and all the works of Nature. In the same way that man, the microcosm, was viewed as a small counterpart of the great universe or macrocosm, the Greek alchemists regarded the metals as having a similar cosmic relation to man upon the one hand and to the universe upon the other. Each of the seven alchemistic metals was supposed to be generated in the earth by emanations from the seven heavenly bodies—gold by emanations from the sun, silver from the moon, quicksilver from Mercury, copper from Venus, iron from Mars, tin from Jupiter, and lead from Saturn. Remains of these old cosmic ideas in Greek alchemy still linger in such chemical terms as "microcosmic salt" and "lunar caustic."

The ancient technologists imparted the color of gold to base metals by means of yellow-colored oxides and other mineral substances. In this way originated the alchemical idea of the philosopher's stone, or powder of projection—a miraculous tinctorial substance of infinite potency, of which, it was said, a single grain could convert oceans of base metal into gold. This transmuting substance was held to be a universal medicine, an elixir which would cure disease, prolong life, and restore youth. The quest of this fabulous stone or elixir became eventually the obsession of all classes of men, from slaves to emperors.

No better illustration of the fatuity of alchemy can be found than that during its revival at the time of the Renaissance the writers of the West could only reëcho what the Greeks had written ten centuries before. As in the alchemy of the Greeks, so in that of a thousand years later we find the same symbols, allegories, prayers, oaths, invocations, poems, and obscurities of language. The Western writers could only retrace the landmarks of the ancient Greeks.

We may summarize the accomplishments of the Greeks in chemistry by saying that their work was distinguished on the one hand by many brilliant conceptions, some of which had a great influence on the future history of the science, and that it was characterized on the other by a great lack of experimental ability, such as was needed for subjecting their speculations to practical tests. They failed to make proper use of such experimental knowledge as they already possessed, for had they applied the densimetric method of Archimedes (287–212 B.C.) to testing the gold and silver that they thought to obtain by the transmutation of base metals, they would have demonstrated the fallacy of alchemy at the outset and thus saved the world from one of the greatest illusions of the human mind.

Nicephorus Blemmydes, one of the later Greek writers of the thirteenth century, once stated that the four requirements for successful operations in chemistry were "a balance, a furnace, a supply of fuel, and a mind that was subtle and unlimited." The subtle and unlimited mind the Greeks always possessed, but their recognition of the value of the balance came too late. The opportunity of drawing fundamental deductions from experimental work with this instrument was reserved for future time to the younger civilizations of the West.

Those early attempts at transforming matter, and unknowingly rearranging its inner, hidden structure, were not completely pointless, as Dr. Bronowski points out. They gave us bronze nearly six thousand years ago and steel some three thousand years ago. The making of bronze and steel in fact marked the beginning of the field of metallurgy, as described here by Cyril Stanley Smith.

The Beginning of Metallurgy

Cyril Stanley Smith

Melting is the basis of a primary means of shaping metals, for they solidify essentially unchanged. When solid, metals have the property of resisting stress only up to a certain point, beyond which they deform without breaking. Shaped by locally high stresses under the hammer, metals will resist the less severe stresses of use indefinitely. This property rests very intimately on the structure of the metal and on the state of its electrons, but the theory of deformation was not necessary for the discovery that metals are deformable, any more than the theory of alloys was needed for the discovery of bronze. The practical-cookbook method of handling materials revealed nearly all of the different types of behavior of metals and alloys long before man had any ideas concerning their atomic and electronic basis. It was enough to have found out that properties were reproducible. They were simply attributed to the different natures of materials which could be identified and put to use.

There are some metals (notably copper and gold) which occur in nature in conspicuous malleable lumps, and man could have found these in his early searches for toolmaking stones. The oldest known artificially shaped metal objects are some copper beads found in northern Iraq and dating from the beginning of the 9th millennium B.C. Two Anatolian finds date from the 8th millennium. (These have not yet been examined metallographically, but there is no reason to doubt that the objects are native copper that has been hammered, supposedly by means of a stone hammer and anvil, and perhaps abraded to shape.) The Indians of the Great Lakes region of America were using native cooper [*sic*] about 4000 years ago. Most metals, however, occur in nature in abundance only in the form of mineral compounds (oxides, carbonates, sulfides, or others), which, though they possess attention-catching color and density, have not the malleability of a metal. At some point in time—not well established but probably shortly after 5000 B.C., and in the mountains that form the northern boundary of the Fertile Crescent—it was found that heating certain greenish or bluish minerals in the proper kind of fire would produce metal—in other words, smelting had been discovered. This was a chemical event second in importance only to the initial discovery of the effectiveness of fire, and it vastly increased the amount of metal that could be made available. Separation of metal from ore requires a higher temperature than is commonly available in open fires, and many archeologists think it may have occurred first in a pottery, for clay pots were being fired in kilns at moderately high temperatures and oxide colors were probably being experimented with before copper was smelted. Or smelting may have been discovered by another route. Hammering of native copper may have been followed by the discovery that copper could be more extensively hammered if it were repeatedly heated (annealed) in a fire. Once the critical temperature had been reached, the melting and casting of native copper would have followed, making it possible to utilize previously useless scrap metal and small nuggets. A final step might have been the discovery that the green matter accompanying native copper could be transmuted to produce more metal when it was strongly heated with wood or charcoal. But this need not have been an entirely accidental discovery. Once the change-producing properties of a hot fire were known, the fundamental human instinct of curiosity could have led someone to experiment empirically, for no other reason than that he enjoyed the sight of unexpected things happening when colored and heavy stones were heated, alone and in mixture. A plot, against time, of the highest temperature regularly produced for any purpose by a given culture would serve as a good graph of the waxing and waning of technological skill in general.

Smelting, like most phase transformations or social innovations, had invisible beginnings and can only be recognized clearly after a certain stage of growth had occurred. However, regardless of the uncertain origin of smelting, it is well established that before the end of the 3rd millennium B.C. nearly all the metals that can be reduced by carbon from a common distinctive mineral had been discovered and put to some use; most of the useful alloys that can be made by mixing such metals had been discovered, and definite proportions had been selected; the effects of work-hardening and annealing (heat-softening) were known and used; and most of today's methods for shaping metal by casting, deformation, and joining were being used, except for those that require power or high-precision tools. Even iron was then known, though its wide utilization was delayed for a millennium. Steel was made with some degree of control by 1200 B.C.; cast iron was made by 500 B.C. (though only in China; it was not appreciated in Europe until 2000 years later). Zinc has an interesting history for it was used in an alloy (brass) long before it had been seen as a metal.

Aluminum cannot be reduced by carbon, and therefore, despite the vast abundance of its ores, the metal was not made until after electrolysis had revealed the electropositive alkali metals which were needed for its reduction. Many other metals, such as titanium and niobium, appeared on the scene relatively late, partly because they cannot be reduced with carbon except at very high temperatures, but principally because they are contaminated to the point of uselessness by carbon and the atmospheric gases, which the primitive metallurgist could not control (and which not even the advanced chemist could control until the present century).

Properties of Metals and Alloys

The principal methods of hardening metals that were employed in antiquity . . . are (i) work hardening; (ii) hardening by the formation of solid solutions, as in the copper-tin alloys (bronzes); and (iii) hardening by partially restrained allotropic transformation, until recently known only in the case of the quenching of steel from a suitable high temperature, which produces the most spectacular change of properties. The three methods of hardening are now known to be simply three different ways of reducing imperfections in the regularity of the stacking of atoms in the metal crystals. Work hardening breaks up the crystals mechanically and introduces "dislocations". Solid solution replaces some atoms of the main component in the crystal by others of a different size. The quench hardening of steel produces both of these effects and in addition produces extremely small crystals which oppose the deformation of each other. Tempered steel owes its hardness partly to a dispersion of tiny particles of iron carbide. (A somewhat similar dispersion, though produced without a major transformation in crystal form, is the basis of precipitation hardening, the only hardening technique not used by the ancients. It was discovered in an aluminum alloy in 1906, though hints of it could have been found previously in the behavior of sterling silver and certain dental alloys.) But theory was in no way necessary for discovery.

Other critically useful modifications of metal properties that were used well before the origins of the classical civilizations were the changes of melting point produced by alloying, useful in soldering; diffusion in solids; and various chemical changes (particularly those involving oxides, sulfides, and silicates) which gave immiscible liquids useful in separating metals from earthy matter and from each other, as well as in refining.

A close relative of almost every one of the alloy compositions and methods of hardening that were employed for any purpose up to the very end of the 19th century had been found and put to use over 4000 years earlier! The engineering achievements of the Romans, the marvels of Byzantine architecture and ecclesiastical art, even the developments of the Industrial Revolution and 19th-century engineering were, with few exceptions, all based on materials of types, if not exact compositions, that had long been known. For most of this long period the history of metallurgy is mostly that of the geographic diffusion of techniques from the Middle East, where they originated and that of an extending scale. Metallurgists were innovators only to the extent that they produced their materials in ever-increasing quantities at ever-decreasing cost. Craftsmen and engineers used their age-old materials with skill and with increasing reliability, but there was only trivial change in the range of characteristics available. For millennia, sophisticated alloys were associated with rather primitive mechanical technology. The empirical search and trials of the early metallurgists of the Middle East had indeed left little to be discovered. It seems that there was a personal immediacy in man's sensual contact with materials, in contrast to his more slowly developed, more intellectual, approach to mechanics. As R.J. Forbes has remarked, there is an important difference between the *discovery* of the properties of materials and the *invention* of things to do with them.

A more fanciful description of the discovery of some noteworthy metallic elements is provided by Lucretius (Titus Lucretius Carus), a Roman poet who lived from about 98 to 55 B.C.

Origin of the Metals

Lucretius

Copper and gold and iron were discovered and at the same time weighty silver and the substance of lead, when fire with its heat and burnt up vast forests on the great hills, either by a discharge of heaven's lightning, or else because men waging with one another a forest-war had carried fire among the enemy in order to strike terror, or because drawn on by the goodness of the soil they would wish to clear rich fields and bring the country into pasture, or else to destroy wild beasts and enrich themselves with the booty; for hunting with the pitfall and with fire came into use before the practice of enclosing the lawn with toils and stirring it with dogs. Whatever the fact is, from whatever cause the heat of flame had swallowed up the forests with a frightful crackling from their very roots and had thoroughly baked the earth with fire, there would run from the boiling veins and collect into the hollows of the ground a stream of silver and gold, as well as of copper and lead. And when they saw these afterwards cool into lumps and glitter on the earth with a brilliant gleam, they would lift them up attracted by the bright and polished lustre, and they would see them to be moulded in a shape the same as the outline of the cavities in which each lay. Then it would strike them that these might be melted by heat and cast in any form or shape soever, and might by hammering out be brought to tapering points of any degree of sharpness and fineness, so as to furnish them with tools and enable them to cut the forests and hew timber and plane smooth the planks, and also to drill and pierce and bore. And they would set about these works just as much with silver and gold at first as with the overpowering strength of stout copper, but in vain, since their force would fail and give way and not be able like copper to stand the severe strain. At that time copper was in higher esteem and gold would lie neglected on account of its uselessness, with its dull blunted edge: now copper lies neglected, gold has mounted up to the highest place of honour. Thus time as it goes round changes the seasons of things. That which was in esteem, falls at length into utter disrepute; and then another thing mounts up and issues out of its degraded state and every day is more and more coveted and blossoms forth high in honour when discovered and is in marvellous repute with men.

If anyone broke the alchemists' reliance on magic incantations it was Aureolus Phillipus Theophrastus Bombastus von Hohenheim (1493–1541), more commonly known as Paracelsus. It was that outrageous and irreverent gentleman who announced to the dismay of other alchemists that man suffered from specific diseases which could be treated by specific remedies. His colleagues believed that all diseases could be ascribed to one bad all pervading substance, or humor. The following selection treats of Paracelsus at a pivotal high point in his somewhat Rabelaisian career.

Paracelsus

Henry M. Pachter

At Basle, at the age of thirty-three, Paracelsus was at the height of his career, and had an opportunity, that was never to occur again, to teach and to carry out the reform of medicine in an official capacity. But he wasted it. With a sovereign disdain for tact, he neglected to make the courtesy calls on his colleagues which academic custom required and which any newcomer would wish to make. Nor did he conform with the necessary formalities of University registration. He did not ask the Dean to enter his name in the register; he failed to present his diploma to the Faculty; he did not invite his colleagues to a disputation. Instead of discussing his intended lecture with the department, he published a program for the students over the heads of the Faculty.

This "intimation," given here in full, contains his famous formula, "*experimentum et ratiocinium.*"

Theophrastus Bombast of Hohenheim, doctor of both medicines and professor, greetings to the students of medicine. Of all disciplines medicine alone, through the grace of God and according to the opinion of authors divine and profane, is recognized as a sacred art. Yet, few doctors today practice it with success and therefore the time has come to bring it back to its former dignity, to cleanse it from the leaven of the barbarians, and to purge their errors. We shall do so not by strictly adhering to the rules of the ancients, but exclusively by studying nature and using the experience which we have gained in long years of practice. Who does not know that most contemporary doctors fail because they slavishly abide by the precepts of Avicenna, Galen, and Hippocrates, as though these were Apollo's oracles from which it is not allowed to digress by a finger's breadth. If it pleases God, this way may lead to splendid titles, but does not make a true doctor. What a doctor needs is not eloquence or knowledge of language and books, illustrious though they be, but profound knowledge of Nature and her works. The tasks of a rhetorician is to bring the judge over to his opinion. The doctor must know the causes and symptoms of the disease and use his judgment to prescribe the right medicine.

Thanks to the liberal allowance the gentlemen of Basle have granted for that purpose, I shall explain the textbooks which I have written on surgery and pathology, every day for two hours, for the greatest benefit of the audience, as an introduction to my healing methods. I do not compile them from excerpts of Hippocrates or Galen. In ceaseless toil I created them anew upon the foundation of experience, the supreme teacher of all things. If I want to prove anything I shall not do so by quoting authorities, but by experiment and by reasoning thereupon. *If, therefore, dear reader, you should feel the impulse to enter into those divine mysteries, if within a brief lapse of time you should want to fathom the depths of medicine, then come to me at Basle and you will find much more than I can say here in a few words.*

To express myself more plainly, let me say here, by way of example, that I do not believe in the ancient doctrine of complexions and humors which has been falsely supposed to account for all diseases. It is because of these doctrines that so few physicians have correct views of disease, its origins and its course. I bid you, do not pass a premature judgment on Theophrastus until you have heard him. Farewell, and come with a good will to study our attempt to reform medicine.

Basle, June 5, 1527.

The Faculty was outraged. It solemnly advised the University that Paracelsus' students would not be admitted to take examination and could not be graduated. This refusal had a semblance of legal justification. Paracelsus had invited barbers, alchemists, and others who lacked academic background to attend his lecture. Privileges were in danger of being trodden upon should the self-taught be admitted to equal rank with accredited students.

The Dean turned the tables on the city. If the University had no jurisdiction over the course, it need not make its facilities available. Paracelsus received no lecture room on the campus.

Actor, Glutton, Teacher, and Scholar

Thirty students (twenty-two more than had registered in the preceding year) assembled in the off-campus hall where Paracelsus was forced to conduct his course. The opening session was as theatrical as the mountebank could hope to make it. Paracelsus appeared at the lectern in a professor's robe, paused, and tore the robe off. With a grandiose gesture, he flung it out into the room and proceeded to lecture in the sooty apron of the alchemist, to the applause of his friends, the barber-surgeons and itinerant *scholares*.

After this piece of showmanship, however, he set to work in earnest. He lectured throughout the summer term and the winter term 1527–1528. His subjects were pathology and therapeutics, pharmacology, diagnostic through the examination of the pulse and the urine, purgation, venesection, injuries and disorders requiring surgery. He delivered the medical lectures in a Latin version prepared by his secretary Oporinus, but in his surgical lectures he used German. The only subject he treated in the traditional way, by explaining a text, were the aphorisms of Hippocrates.

In addition to teaching, he saw patients, inspected pharmacies, gave personal consultations to his students, wrote essays and pamphlets, experimented in his chemical kitchen, and drank with his friends. How he spent his day is reported in a letter by Oporinus:

. . . As to Paracelsus, he has been dead for a long time and I should hate to speak against the spirit of his death (as the saying goes). While he was living I knew him so well that I should not desire again to live with such a man.

Apart from his miraculous and fortunate cures in all kinds of sickness, I have noticed in him neither scholarship nor piety of any kind. It makes me wonder to see all the publications which, they say, were written by him or left by him but which I would not have dreamt of ascribing to him. The two years I passed in his company he spent in drinking and gluttony, day and night. He could not be found sober, an hour or two together, in particular after his departure from Basle. In Alsace, noblemen, peasants and their womenfolk adulated him like a second Aesculapius. Nevertheless, when he was most drunk and came home to dictate to me, he was so consistent and logical that a sober man could not have improved upon his manuscripts. I had to translate them into Latin and there are several books which I and others thus translated. All night, as long as I stayed with him, he never undressed, which I attributed to his drunkenness. Often he would come home tipsy, after midnight, throw himself on his bed in his clothes wearing his sword which he said he had obtained from a hangman. He had hardly time to fall asleep when he rose, drew his sword like a madman, threw it on the ground or against the wall, so that sometimes I was afraid he would kill me. I would need many days to tell what I had to put up with. . . .

. . . He was a spendthrift, so that sometimes he had not a penny left, yet the next day would show me a full purse. I often wondered where he got it. Every month he had a new coat made for him, and gave away his old one to the first comer; but usually it was so dirty that I never wanted one. . . .

I never heard him pray or inquire after the Evangelical doctrine which then was practiced in our town. He not only despised our good preacher, but threatened that one day, as he had done to Hippocrates and Galen, he would set Luther's and the Pope's heads right. He also said that none so far who had written about the Holy Scriptures had grasped their right meaning. I don't know how much more nonsense he said which I would be ashamed to relate here. . . .

He did not care for women and I believe he never had doings with any. In the beginning he was very modest, so that up to his twenty-fifth year, I believe he never touched wine. Later on he learned now to drink and even challenged an inn full of peasants to drink with him and drank them under the table, now and then putting his finger in his mouth like a swine.

Dr. Johan Weyer (Wierus), who had requested this information on Paracelsus, inserted it in his work on *The Magic of Demons*. Wierus' approach to the problem of witchcraft has earned him the fame of being the first to publish a book on psychiatry. However, he was well aware that Paracelsus had preceded him in this field and appeared pleased to have a rival disparaged. Posterity, however, has discerned, in this damaging letter, involuntary tributes to Paracelsus' genius, his fanatical zeal, and the unrelenting energy which made him an exacting master. His secretaries could not keep up with the stream of his lectures and manuscripts. Paracelsus' obsession with his work is confirmed by other witnesses: "It is true that he enjoyed drinking. But, on the other hand, when he had undertaken anything he scarcely ate or drank until he had finished his work, and then, when he felt free, he became a mightily merry." The testimony of Paracelsus' bulter Franz . . . contradicts Oporinus. According to Franz, Paracelsus' conduct was exemplary, and he praised his professional skill, his generosity, and philanthropic practices. Franz reports:

He gave a florin to a cantor in Basle who was his patient, to buy some food and wine. I also know that he cured lepers, dropsicals, epileptics, syphilitics, and podagra [gout] patients, besides innumerable other diseases. The Galenic doctors could not do likewise, and envied the honors he earned.

Oporinus, himself, admired Paracelsus as a doctor and teacher. When his master had to leave Basle, he followed him, and so great was his loyalty that Paracelsus mentioned him in his book on syphilis as his "faithful Oporinus" who stood by him when others betrayed him. Oporinus little deserved the honor. When he wrote his letter, in 1555, he believed that Paracelsus' prestige was buried with him. A few years later, however, the Paracelsus revival was well under way, and Oporinus, who in the meantime had fallen out with Wierus, wrote to Paracelsus' disciple Toxites that his unfavorable remarks had been published without his intention and knowledge. As an excuse he said that he had not realized Paracelsus' importance. Oporinus was to win fame as publisher of Vesalius' *Fabrica*; later he became a professor in Basle.

Avicenna at the Stake

On the 24th day of June, 1527, Paracelsus staged the symbolical act which marked the zenith of his career and which has gone down in history under his signature. It was St. John's Day, traditionally given over to commencement celebrations and the students' Midsummer Night's antics. On such occasions peaceable citizens stayed behind shutters. Into the St. John's Fire went rubbish and bric-a-brac, whatever students no longer needed or did not like. And many a prince, cleric or unpopular professor was burned in effigy by the hooting students. Three hundred years later, German students threw into the St. John's Fire the uniform of a Prussian policeman, a pigtail, and other paraphernalia of the *ancien régime* in Germany.

On this St. John's Day of 1527, they built a beautiful fire. Theophrastus came to join in the merry-making. Behind him, his students dragged a heavy volume, the *Summa* of the books on medical science, the venerable *Canon* of Avicenna. At his command they hurled it into the fire.

A brief controversial note is in place here, too. Sudhoff, ever anxious to preserve the

Master's respectability, has tried to make the famous incident appear innocuous. The gist of his argument is that the Canon was too heavy to be carried to the fire. He assumes that Paracelsus had a much smaller book, the *Summa* Jacobi de Partibus (published in 1521), under his arm, and when he saw the students hooting around the fire, a whim of the moment impelled him to throw the book into the flames. This assumption almost refutes itself. If the incident had happened as Sudhoff construes it, nobody would have noticed or found it worth while to report or to dwell on it.

Nor was the act a whim of the moment. Everything we know about Paracelsus' character and the elated feelings which possessed him at the height of his career supports the assumption that he deliberately chose the heaviest volume for the symbolic burning. His own report asks for no extenuation: "I threw into the St. John's Fire the Summa of the books, so that all the misfortune might go up in the air with the smoke. Thus the realm of medicine has been purged."

As soon as chemists began to realize that the world was composed of something more than the four elements espoused by the Greek philosophers, they found that their universe was at once more simple and more complex. New elements were discovered. Salt was broken down into its component elements—sodium and chlorine. And the air itself was also found to be divisible into smaller fractions, each with its unique properties. Thus, Joseph Priestley (1733-1804), a dissenter in the religious and scientific sense, found that the atmosphere was composed of different types of "air": "nitrous air," which did not support life or combustion, and "dephlogisticated air," which did. The latter was actually oxygen, and in discovering it, Priestley had simply repeated experiments that the alchemists had done a thousand years earlier. The difference was that he interpreted the results of his experiments with cool objective insight and careful observation, two important facets of what has come to be known as the scientific method. For tools, Priestley used whatever was close at hand and cheapest, such as old beer glasses and a steel gun barrel. This biographical note by F.W. Gibbs expands on Priestley's great discovery and the method he used to achieve it.

The Discovery of Oxygen

F. W. Gibbs

Priestley obviously enjoyed telling about the way he came to make one of his greatest discoveries—dephlogisticated air. He kept his readers in suspense while he pointed some morals from his experience.

When I . . . compare my last discoveries relating to the constitution of the atmosphere with the first, I see the closest and easiest connexion in the world between them, so as to wonder that I should not have led immediately from the one to the other. That this was not the case, I attribute to the force of prejudice, which, unknown to ourselves, biasses not only our judgments, *properly so called, but even the perceptions of our senses: for we may take a maxim so strongly for granted, that the plainest evidence of*

Excerpted from F.W. Gibbs, *Joseph Priestley: Adventurer in Science and Champion of Truth*, Thomas Nelson & Sons Ltd., 1965.

sense will not entirely change, and often hardly modify our persuasions; and the more ingenious a man is, the more effectually he is entangled in his errors; his ingenuity only helping him to deceive himself, by evading the force of truth.

Few maxims had taken so firm a hold on the human mind as that air was a simple, elementary substance, indestructible and unalterable. All his experiments, however, had led him to believe that it was a composition of some kind. He next asked what this composition was—'what is the thing that we breathe'—and how could it be made from its constituents. He thought this could be done by heating various substances in air over mercury, suspending them or resting them on a small support inside a tube. He found he could not do this effectively with the aid of his mirror, and decided that he needed a strong burning lens.

In June 1774 he obtained a lens of 12-inch diameter and 20-inch focal length, and immediately set about "with great alacrity to examine . . . what kind of air a great variety of substances, natural and factitious, would yield."

It was not until 1 August, however, that he tried to obtain an air from *mercurius calcinatus per se*, that is, the substance produced when mercury is heated alone in the air. He readily obtained an "air" that did not dissolve when water was admitted to the tube. He was extremely surprised to find also that a candle burned in this air remarkably well, much the same as it did in the air formed when iron was left in nitrous air. He knew that no nitrous material had been used in the preparation, and so he was utterly at a loss to account for it. The candle, indeed burned with more splendour and heat than in the "modified nitrous air." Red-hot wood sparkled in it—like paper that had been dipped in a solution of nitre and dried—and it consumed fast.

At the same time he obtained the same air from "the red precipitate" (the salt obtained from mercury and spirit of nitre), and in this case he thought the air was formed by the acid. Clearly, the mercury calx had obtained something from the air that was also common to nitre. This appeared more extraordinary than it ought to have done, and he suspected the purity of the mercury calx; had he been anything of a practical chemist, he admitted, he could not have entertained any such suspicion. He therefore obtained a pure specimen of the substance from Warltire, who happened to be staying with him at Calne at the time, and from that specimen he got even more air than before.

This experiment might have satisfied any moderate sceptic: but, however, being at Paris in the October following, and knowing that there were several eminent chymists in that place, I did not omit the opportunity, by means of my friend Mr. Magellan, to get an ounce of mercurius calcinatus prepared by Mr. Cadet, of the genuineness of which there could not possibly be any suspicion; and at the same time, I frequently mentioned my surprize at the kind of air which I had got from this preparation to Mr. Lavoisier, Mr. le Roy, and several other philosophers, who honoured me with their notice in that city . . .

Right at the start, too, he had got the same air from red lead, a yellow product being left behind. Though it contained some fixed air, as he had found before, a candle burned in this specimen very strongly and with a crackling noise. He argued that, in both cases, since the calces were prepared in the same way, they must have got "this property of being able to yield this particular kind of air" from the atmosphere itself.

As I never make the least secret of any thing that I observe, I mentioned this experiment also, as well as those with the mercurius calcinatus, and the red precipitate, to all my philosophical acquaintance at Paris, and elsewhere; having no idea, at that time, to what these remarkable facts would lead.

Immediately on his return to London in November 1774 he had got to work on Cadet's specimen, and prepared some air in which a candle burned with a vivid flame. He them submitted it to various tests to see if it resembled nitrous air. He grew more and more surprised as he found it behaved quite differently. He still thought of it as having something in common with nitrous air and nitrous acid. Now nitrous air diminished common air. Would this new air do the same? Nothing happened. The result surprised him.

He was much more surprised to find that, after the new air had stood over water for two days, a candle still burned in it as well as in common air. After similar treatment, the nitrous airs he had prepared would extinguish a candle. He was therefore forced to conclude that there was a "very material" difference between them and his new air.

From November of that year to the end of February 1775 he was busy with other work, including his experiments on vitriolic acid air [sulfur dioxide]. It was at this time that he burned one of his hands badly, and he also suffered from very painful boils, which began in December and recurred at intervals until the following spring; they were still troubling him when he took up his work again at the point where he had left it.

Meanwhile he had been meditating on the problem of the new air, and by 1 March made the decisive step; he tested it to see if it could be diminished *with* nitrous air. This gave results; the new air was diminished "quite as much as common air," and now, for the first time, he realized that it was more like common air than nitrous air. He therefore thought that it would be fit for breathing.

But he did not then take any notice of what in reality was the most surprising thing of all, namely that the diminution was *greater* than it would have been with common air; for he was "so fully possessed by the notion of there being no air better than common air"—a reasonable enough assumption. The next day he was more surprised than ever to find that, after the mixture of nitrous air and the new air had stood all night over water, a candle still burned in the remaining air "and even better than in common air." Despite this evidence, he thought he had found a way of obtaining common air, and told this to his friends, including Price. He realized that respiration tests were necessary to complete the proof of its being common air.

It took him until 8 March to obtain a mouse, which he put into two ounce-measures of the air, expecting it to live about a quarter of an hour. But it lived a full half-hour, and, though it appeared dead when he pulled it out, he was able to revive it by holding it near the fire. This made him suspect, but still with a little doubt, that this air was indeed better than common air. On 9 March he tested the air the mouse had breathed, expecting it to be noxious; but it was still better than common air. This so puzzled him that he lay awake that night thinking about it. In the morning he decided to add more nitrous air to it, and, to his "utter astonishment," he found it further reduced to about half its volume. As he still had some of the air the mouse had breathed, he put it in again, and it lived anouther half-hour; and this time it came out lively and vigorous. By the next day the air had become reduced in volume from one and a half to two-thirds of an ounce-measure, but even then it was still almost as good as common air. He now decided to measure the "goodness" of this new air more accurately, and found it four to five times as good as common air.

Turning to the question of why red lead had given a somewhat different result, he obtained the new air from it and from massicot (fused lead monoxide), but found that it always contained a quantity of fixed air. This time he concluded that the presence of fixed air was due to some circumstance connected with their preparation. He therefore asked Dr. Higgins to make him some fresh red lead. This was not a good specimen, for it had a yellowish cast and some "pieces" were entirely yellow. But though it contained little true red lead, the specimen did give a little air, which this time was pure. He then tried to convert the specimen to the form that gave much air, by adding mineral acids, drying, pulverizing, and heating in a gun-barrel. The specimen prepared with spirit of nitre became exceedingly red and gave air as soon as it was heated, though this time it was mixed with some fixed air. On moistening the red substance with nitric acid (thus removing any carbonate) he obtained air five times as pure as common air. He thought about a name for the new air, and decided to call it dephlogisticated air (air eminently suited for combustion, that is, to receive phlogiston from combustible substances).

He now tried the effect of nitric acid on all kinds of earths, and in every case obtained dephlogisticated air on heating. He repeated his former experiment with saltpetre, and found that this also gave dephlogisticated air on heating. He thought that inflammable air might explode with more violence in this air than in common air, but the result far exceeded his expectations; the report, he considered, was about fifty times as loud.

He then breathed the air himself, drawing it into his lungs through a syphon until he had reduced it to the standard of common air, as judged by the nitrous air test.

The feeling of it to my lungs was not sensibly different from that of common air; but I fancied that my breast felt peculiarly light and easy for some time afterwards. Who can tell but that, in time, this pure air may become a fashionable article in luxury. Hitherto only two mice and myself have had the privilege of breathing it.

This also prompted the thought that "pure" air might be very useful in medicine, but he advocated caution, for, as a candle burns out much faster in it, so a man might "live out too fast."

While Priestley was experimenting with air, a young contemporary named John Dalton was finding experimental sense in the "atomic theory" first proposed by the Greek philosopher Democritus (born about 460 B.C.). Democritus and some of his colleagues alleged that the physical realm is made up of atoms, which are very small, indivisible, and unchangeable. They also maintained (with no experimental evidence) that different substances are composed of different types of constantly moving atoms which had different weights. What Dalton did was to adapt many of those tenets to show that the weight of different atoms could be determined by measuring the proportions in which they combined.

As a child, Dalton was considered a mathematical genius and it is fortunate that the laboratory of his mind—where he devised his modern atomic theory—was neater and more precise than the laboratory where he did his supportive experiment. The entire laboratory cost Dalton about thirty-three dollars, and a contemporary wrote that his experiments often failed to work and that "the simplicity of his apparatus and the awkwardness with which he used it were not calculated to rivet the attention of his audience." Yet Dalton's experiments are a true monument in the scientific ascent of man. His key findings are reported here in his own words.

Atomic Theory
John Dalton

When any body exists in the elastic state, its ultimate particles are separated from each other to a much greater distance than in any other state; each particle occupies the centre of a comparatively large sphere, and supports its dignity by keeping all the rest, which by their gravity, or otherwise are disposed to encroach up it, at a respectful distance. When we attempt to conceive the *number* of particles in an atmosphere, it is somewhat like attempting to conceive the number of stars in the universe; we are confounded with the thought. But if we limit the subject, by taking a given volume of any gas, we seem persuaded that, let the divisions be ever so minute, the number of particles must be finite; just as in a given space of the universe, the number of stars and planets cannot be infinite.

Chemical analysis and synthesis go no farther than to the separation of particles one from another, and to their reunion. No new creation or destruction of matter is within the reach of chemical agency. We might as well attempt to introduce a new planet into the solar system, or to annihilate one already in existence, as to create or destroy a particle of hydrogen. All the changes we can produce, consist in separating particles that are in a state of cohesion or combination, and joining those that were previously at a distance.

In all chemical investigations, it has justly been considered an important object to ascertain the relative *weights* of the simples which constitute a compound. But unfortunately the enquiry has terminated here; whereas from the relative weights in the mass, the relative weights of the ultimate particles or atoms of the bodies might have been inferred, from which their number and weight in various other compounds would appear, in order to assist and to guide future investigations, and to correct their results. Now it is one great object of this work, to shew the importance and advantage of ascertaining *the relative weights of the ultimate particles, both of simple and compound bodies, the number of simple elementary particles which constitute one compound particle, and the number of less compound particles which enter into the formation of one more compound particle.*

If there are two bodies, A and B, which are disposed to combine, the following is the order in which the combinations may take place, beginning with the most simple: namely,

1 atom of A + 1 atom of B = 1 atom of C, binary.

1 atom of A + 2 atoms of B = 1 atom of D, ternary.

2 atoms of A + 1 atom of B = 1 atom of E, ternary.

1 atom of A + 3 atoms of B = 1 atom of F, quaternary.

3 atoms of A + 1 atom of B = 1 atom of G, quaternary.
etc. etc.

The following general rules may be adopted as guides in all our investigations respecting chemical synthesis.

1st. When only one combination of two bodies can be obtained, it must be presumed to be a *binary* one, unless some cause appear to the contrary.

2d. When two combinations are observed, they must be presumed to be a *binary* and a *ternary*.

3d. When three combinations are obtained, we may expect one to be a *binary*, and the other two *ternary*.

4th. When four combinations are observed, we should expect one *binary*, two *ternary*, and one *quaternary*, etc.

5th. A *binary* compound should always be specifically heavier than the mere mixture of its two ingredients.

6th. A *ternary* compound should be specifically heavier than the

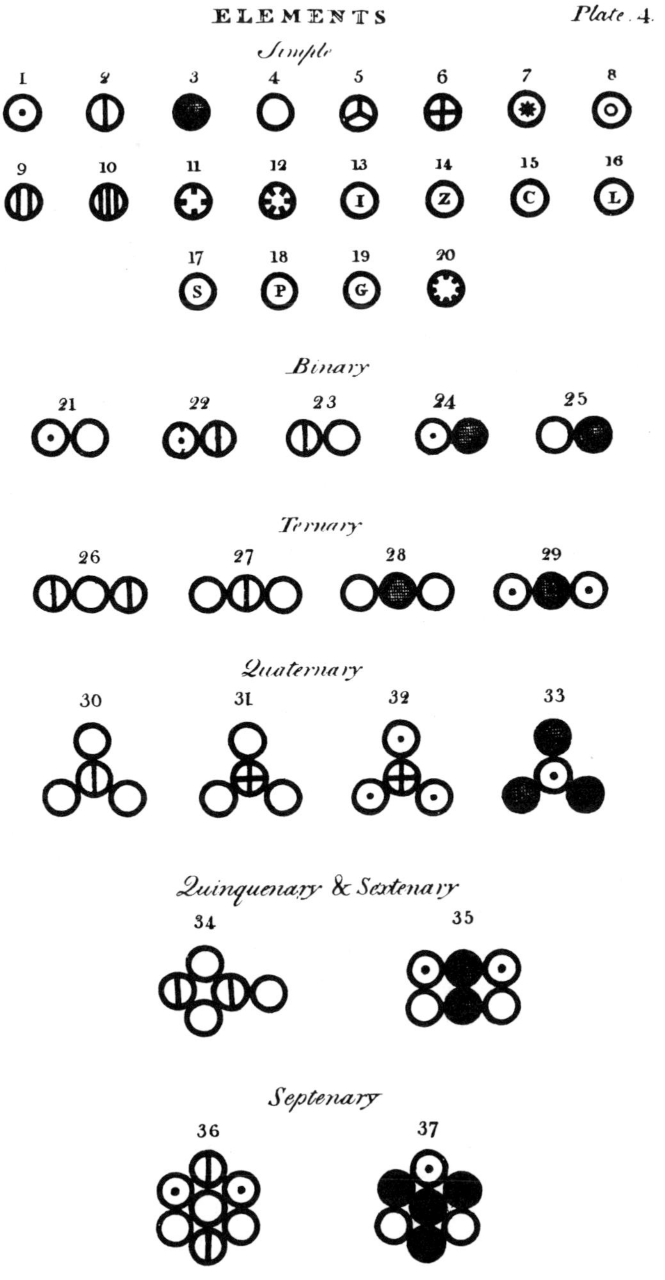

From the novelty as well as importance of the ideas suggested in this chapter, it is deemed expedient to give plates, exhibiting the mode of combination in some of the more simple cases. A specimen of these accompanies this first part. The elements or atoms of such bodies as are conceived at present to be simple, are denoted by a small circle, with some distinctive mark; and the combinations consist in the juxta-position of two or more of these; when three or more particles of elastic fluids are combined together in one, it is to be supposed that the particles of the same kind repel each other, and therefore take their stations accordingly.

mixture of a binary and a simple, which would, if combined, constitute it; etc.

7th. The above rules and observations equally apply, when two bodies, such as C and D, D and E, etc. are combined.

From the application of these rules, to the chemical facts already well ascertained, we deduce the following conclusions; 1st. That water is a binary compound of hydrogen and oxygen, and the relative weights of the two elementary atoms are 1:7, nearly;[1] 2d. That ammonia is a binary compound of hydrogen and azote [nitrogen], and the relative weights of the two atoms are as 1:5, nearly; 3d. That nitrous gas is a binary compound of azote and oxygen, the atoms of which weigh 5 and 7 respectively; that nitric acid is a binary or ternary compound according as it is derived, and consists of one atom of azote and two of oxygen, togethering weighing 19; that nitrous oxide is a compound similar to nitric acid, and consists of one atom of oxygen and two of azote, weighing 17; that nitrous acid is a binary compound of nitric acid and nitrous gas, weighing 31; that oxynitric acid is a binary compound of nitric acid and oxygen, weighing 26; 4th. That carbonic oxide is a binary compound, consisting of one atom of charcoal, and one of oxygen, together weighing nearly 12; that carbonic acid is a ternary compound, (but sometimes binary) consisting of one atom of charcoal, and two of oxygen, weighing 19; etc. etc. In all these cases the weights are expressed in atoms of hydrogen, each of which is denoted by unity.

The atomic theory was a great leap from the abstract imaginings of Democritus and other Greek philosophers to the cluttered laboratory of John Dalton. Since then it has been refined and re-refined, as later chapters will show, and, in fact, what was a revolutionary theory in Dalton's time is now part of every schoolchild's lessons. Interpretations still vary, however, as Flann O'Brien points out in this excerpt from his novel *The Third Policeman*, in which an Irish police sergeant mulls over his own free-form interpretation of atomic theory with an incredulous friend.

The Police Constable's Atomic Theory

Flann O'Brien

"Did you ever discover or hear tell of the Atomic Theory?" he inquired.

"No," I answered.

He leaned his mouth confidentially over to my ear.

"Would it surprise you to be told," he said darkly, "that the Atomic Theory is at work in this parish?"

"It would indeed."

"It is doing untold destruction," he continued, "the half of the people

[1] Dalton was wrong on two counts here. He failed to realize that *two* atoms of hydrogen and one atom of oxygen combine in weight ratios of 1 to 8 (not 7) in a molecule of water. Thus according to his system, H_20 would actually be a ternary compound.

are suffering from it, it is worse than the smallpox."

I thought it better to say *something*.

"Would it be advisable," I said, "that it should be taken in hand by the Dispensary Doctor or by the National Teachers or do you think it is a matter for the head of the family?"

"The lock stock and barrel of it all," said the Sergeant, "is the County Council."

He walked on looking worried and preoccupied as if what he was examining in his head was unpleasant in a very intricate way.

"The Atomic Theory," I sallied, "is a thing that is not clear to me at all."

"Michael Gilhaney," said the Sergeant, "is an example of a man that is nearly banjaxed from the principle of the Atomic Theory. Would it astonish you to hear that he is nearly half a bicycle?"

"It would surprise me unconditionally," I said.

"Michael Gilhaney," said the Sergeant, "is nearly sixty years of age by plain computation and if he is itself, he has spent no less than thirty-five years riding his bicycle over the rocky roadsteads and up and down the hills and into the deep ditches when the road goes astray in the strain of the winter. He is always going to a particular destination or other on his bicycle at every hour of the day or coming back from there at every other hour. If it wasn't that his bicycle was stolen every Monday he would be sure to be more than half-way now."

"Half-way to where?"

"Half-way to being a bicycle himself," said the Sergeant.

"Your talk," I said, "is surely the handwork of wisdom because not one word of it do I understand."

"Did you never study atomics when you were a lad?" asked the Sergeant, giving me a look of great inquiry and surprise.

"No," I answered.

"That is a very serious defalcation," he said, "but all the same I will tell you the size of it. Everything is composed of small particles of itself and they are flying around in concentric circles and arcs and segments and innumerable other geometrical figures too numerous to mention collectively, never standing still or resting but spinning away and darting hither and thither and back again, all the time on the go. These diminutive gentlement are called atoms. Do you follow me intelligently?"

"Yes."

"They are lively as twenty leprechauns doing a jig on top of a tombstone."

"*A very pretty figure*," Joe murmured.

"Now take a sheep," the Sergeant said. "What is a sheep only millions of little bits of sheepness whirling around and doing intricate convolutions inside the sheep? What else is it but that?"

"That would be bound to make the beast dizzy," I observed, "especially if the whirling was going on inside the head as well."

The Sergeant gave me a look which I am sure he himself would describe as one of *non-possum* and *noli-me-tangere*.

"That remark is what may well be called buncombe," he said sharply, "because the nerve-strings and the sheep's head itself are whirling into the same bargain and you can cancel out one whirl against the other and there you are—like simplifying a division sum when you have fives above and below the bar."

"To say the truth I did not think of that," I said.

"Atomics is a very intricate theorem and can be worked out with algebra but you would want to take it by degrees because you might spend the whole night proving a bit of it with rulers and cosines and similar other instruments and then at the wind-up not believe what you had proved at all. If that happened you would have to go back over it till you got a place where you could believe your own facts and figures as delineated from Hall and Knight's Algebra and then go on again from that particular place till you had the whole thing properly believed and not have bits of it half-believed or a doubt in your head hurting you like when you lose the stud of your shirt in bed."

"Very true," I said.

"Consecutively and consequentially," he continued, "you can safely infer that you are made of atoms yourself and so is your fob pocket and the tail of your shirt and the instrument you use for taking the leavings out of the crook of your hollow tooth. Do you happen to know what takes place when you strike a bar of iron with a good coal hammer or with a blunt instrument?"

"What?"

"When the wallop falls, the atoms are bashed away down to the bottom of the bar and compressed and crowded there like eggs under a good clucker. After a while in the course of time they swim around and get back at last to where they were. But if you keep hitting the bar long enough and hard enough they do not get a chance to do this and what happens then?"

"That is a hard question."

"Ask a blacksmith for the true answer and he will tell you that the bar will dissipate itself away by degrees if you persevere with the hard wallops. Some of the atoms of the bar will go into the hammer and the other half into the table or the stone or the particular article that is underneath the bottom of the bar."

"That is well-known," I agreed.

"The gross and net result of it is that people who spent most of their natural lives riding iron bicycles over the rocky roadsteads of this parish get their personalities mixed up with the personalities of their bicycle as a result of the interchanging of the atoms of each of them and you would be surprised at the number of people in these parts who nearly are half people and half bicycles."

I let go a gasp of astonishment that made a sound in the air like a bad puncture.

"And you would be flabbergasted at the number of bicycles that are half-human almost half-man, half-partaking of humanity. . . ."

"Are you certain about the humanity of the bicycle?" I inquired of him. "Is the Atomic Theory as dangerous as you say?"

"It is between twice and three times as dangerous as it might be," he replied gloomily. "Early in the morning I often think it is four times, and what is more, if you lived here for a few days and gave full play to your observation and inspection, you would know how certain the sureness of certainty is."

5
The Music of the Spheres

Now is the air made of chiming balls.

Richard Eberhart

The "music of the spheres," a harmony in nature, triggered the evolution of modern mathematics in the ascending mind of man. The followers of Pythagoras took the phrase literally and believed that the whirling planets made their own music. But for the Greek philosophers, in general, the rational view of nature that they developed over centuries of thought went well beyond that celestial chorus, defining space and time within a framework of mathematical design. They were the spiritual ancestors of Newton and Einstein, the inspiration for a new, numerical view of the universe. This point is explored further in this selection from Morris Kline's *Mathematical Thought from Ancient to Modern Times*.

The Greek Rationalization of Nature

Morris Kline

The civilizations that preceded the Greek or were contemporary with it regarded nature as chaotic, mysterious, capricious, and terrifying. The happenings in nature were manipulated by gods. Prayers and magic might induce the gods to be kind and even to perform miracles but the life and fate of man were entirely subject to their will.

From the time our knowledge of Greek civilization and culture begins to be reasonably definite and specific, that is, from about 600 B.C., we find among the intellectuals a totally new attitude toward nature: rational, critical, and secular. Mythology was discarded, as was the belief that the gods manipulate man and the physical world according to their whims. The new doctrine holds that nature is orderly and functions invariably according to a plan. Moreover, the conviction is manifest that the human mind is powerful and even supreme; not only can the ways of nature be learned by man, but he can even predict the occurrences.

It is true that the rational approach was entertained only by the intellectuals, a small group in both the classical and Alexandrian periods. Whereas these men opposed the attribution of events to gods and demons and defied the mysteries and terrors of nature, people in general were deeply religious and believed that the gods controlled all events. They accepted mystical doctrines and superstitions as credulously as did the Egyptians and the Babylonians. In fact Greek mythology was vast and highly developed.

The Ionians began the task of determining the nature of reality. We shall not describe the qualitative theories of Thales, Anaxagoras, and their colleagues, each of whom fixed on a single substance persisting through all apparent change. The underlying identity of this prime substance is conserved but all forms of matter can be explained in terms of it. This natural philosophy of the Ionians was a series of bold speculations, shrewd guesses, and brilliant intuitions rather than the outcome of extensive and careful scientific investigations. They were perhaps a little too eager to see the whole picture and so naïvely jumped to broad conclusions. But they did substitute material and objective explanations of the structure and design of the universe for the older mythical stories. They offered a reasoned approach in place of the fanciful and uncritical accounts of the poets and they defended their contentions by reason. At least these men dared to tackle the universe with their minds and refused to rely upon gods, spirits, ghosts, devils, angels, and other mythical agents.

The Development of the Belief in Mathematical Design

The decisive step in removing the mystery, mysticism, and arbitrariness from the workings of nature and in reducing the seeming chaos to an understandable ordered pattern was the application of mathematics. The first major group to offer a rational and mathematical philosophy of nature were the Pythagoreans. They did draw some inspiration from the mystical side of Greek religion; their religious doctrines centered about the purification of the soul and its redemption from the taint and prison of the body. The members lived simply and devoted themselves to the study of philosophy, science, and mathematics. New members were pledged to secrecy at least as to religious beliefs and required to join up for life. Membership in the community was open to men and women.

The Pythagoreans' religious thinking was undoubtedly mystical, but their natural philosophy was decidedly rational. They were struck by the fact that phenomena that are most diverse from a qualitative point of view exhibit identical mathematical properties. Hence, mathematical properties must be the essence of these phenomena. More specifically, the Pythagoreans found this essence in number and in numerical relationships. Number was their first principle in the explanation of nature. All objects were made up of points or "units of existence" in combinations corresponding to the various geometrical figures. Since they thought of numbers both as points and as elementary particles of matter, number was the matter and form of the universe and the cause of every phenomenon. Hence the Pythagorean doctrine "All things are numbers." Says Philolaus, a famous fifth-century Pythagorean, "Were it not for number and its nature, nothing that exists would be clear to anybody either in itself or in its relation to other things. . . . You can observe the power of number exercising itself not only in the affairs of demons and gods but in all the acts and the thoughts of men, in all handicrafts and music."

The reduction of music, for example, to simple relationships among numbers became possible for the Pythagoreans when they discovered two facts: first, that the sound caused by a plucked string depends upon the length of the string; and second, that harmonious sounds are given off by equally taut strings whose lengths are to each other as the ratios of whole numbers. For example, a harmonious sound is produced by plucking two equally taut strings, one twice as long as the other. In our language, the intervals between the two notes is an octave. Another harmonious combination is formed by two strings whose lengths are in the ratio 3 to 2; in this case the shorter one gives forth a note called the fifth above that given off by the first string. In fact, the relative lengths in every harmonious combination of plucked strings can be expressed as ratios of whole numbers. The Pythagoreans also developed a famous Greek musical scale. Though we shall not devote space to the music of the Greek period, we note that many Greek mathematicians, including

Euclid and Ptolemy, wrote on the subject, especially on harmonious combinations of sounds and the construction of scales.

The Pythagoreans reduced the motions of the planets to number relations. They believed that bodies moving in space produce sounds; perhaps this was suggested by the swishing of an object whirled on the end of a string. They believed, further, that a rapidly moving body gives forth a higher note than one that moves slowly. Now according to their astronomy the greater the distance of a planet from the earth the more rapidly it moved. Hence the sounds produced by the planets, which we do not hear because we are accustomed to them from birth, varied with their distances from the earth and all harmonized. But since this "music of the spheres," like all harmony, reduced to no more than number relationships, so did the motions of the planets.

The Pythagoreans and probably Pythagoras himself wanted not just to observe and describe the heavenly motions but to find regularity in them. The idea of uniform circular motion, seemingly obvious in the case of the moon and sun, suggested that all the planetary motions were explainable in terms of uniform circular motions. The later Pythagoreans made a more striking break with tradition; they were the first to believe that the earth was spherical. Moreover, because 10 was their ideal number, they decided that the moving bodies in the heavens must be 10 in number. First, there was a central fire around which the heavenly bodies, *including the earth*, moved. They knew five planets in addition to the earth. These six bodies, the sun, the moon, and the sphere to which the stars were attached made only 9 moving bodies. Hence they asserted the existence of a tenth one, called the counter-earth, which also revolved around the central fire. We cannot see this tenth one because it moves at exactly the same speed as the earth on the opposite side of the central fire and also because the inhabited part of earth faces away from the central fire. Here we have the first theory to put the earth in motion. However, the Pythagoreans did not assert the rotation of the earth; rather, the sphere of fixed stars revolves about the center of the universe.

The belief that the celestial bodies are eternal, divine, perfect, and unchangeable and that the sublunar bodies, that is the earth and (according to the Greeks) the comets, are subject to change, decomposition, decay, and death may also have come from the Pythagoreans. The doctrine of uniform circular motion and the distinction between celestial and sublunar bodies became embedded in Greek thought.

Other features of nature also "reduced" to number. The numbers 1, 2, 3, and 4, the *tetractys*, were especially valued because they added up to 10. In fact the Pythagorean oath is reported to have been: "I swear in the name of the Tetractys which has been bestowed on our soul. The source and roots of the everflowing nature are contained in it." The Pythagoreans asserted that nature was composed of fournesses; for example, point, line, surface, and solid, and the four elements, earth, air, fire, and water. The four elements were also central in Plato's natural philosophy. Because 10 was ideal, 10 represented the universe. The ideality of 10 required that the whole universe be describable in terms of 10 categories of opposites: odd and even, bounded and unbounded, good and evil, right and left, one and many, male and female, straight and curved, square and oblong, light and darkness, rest and motion.

Clearly, Pythagorean philosophy mingled serious thoughts with what we would consider fanciful, useless, and unscientific doctrines. Their obsession with the importance of numbers resulted in a natural philosophy that certainly had little correspondence with nature. But they did stress the understanding of nature, not, like the Ionians, through a single substance, but through the formal structure of number relationships. Moreover they and the Ionians both saw that underlying mere sense data there must be a harmonious account of nature.

We can now see why the discovery of incommensurable lengths was so disastrous to Pythagorean philosophy: a ratio of incommensurable lengths could not be expressed as a ratio of whole numbers. In addition, they had believed that a line is made up of a finite number of points (which they identified with physical particles); but this could not be the case for a length such as $\sqrt{2}$. Their philosophy, based on the primariness of the whole numbers, would have been shattered if they had accepted irrationals as numbers.

Because the Pythagoreans "reduced" astronomy and music to number, these subjects came to be linked to arithmetic and geometry; these four were regarded as the mathematical subjects. They became and remained part of the school curriculum even into medieval times, when they were called, collectively, "the quadrivium." As we have noted, the Pythagorean interest in arithmetic (i.e. the theory of numbers) was due not to the purely aesthetic value of that subject but to a search for the meaning of natural phenomena in numerical terms; and this value caused the emphasis on special proportions and on triangular, square, pentagonal, and higher forms into which numbers could be arranged. Further, it was the Pythagorean natural philosophy centering about number that gave the subject importance with such men as Nichomachus. In fact, modern science adheres to the Pythagorean emphasis on number—though, as we shall see, in a much more sophisticated form—while the purely aesthetic modern theory of numbers derives from Pythagorean arithmetic per se.

The philosophers who came chronologically between the Pythagoreans and Plato were equally concerned with the nature of reality but did not involve mathematics directly. The arguments and views of men such as Parmenides (5th cent. B.C.), Zeno (5th cent. B.C.), Empedocles (*c*. 484–*c*. 424 B.C.), Leucippus (*c*. 440 B.C.), and Democritus (*c*. 460–*c*. 370 B.C.) were, like those of their Ionian predecessors, qualitative. They made broad assertions about reality that were, at best, barely suggested by observation. Nevertheless, each affirmed that nature is intelligible and that reality can be grasped by thought. Each was a link in the chain that led to the mathematical investigation of nature. Leucippus and Democritus are notable because they were the most explicit in affirming the doctrine of atomism. Their common philosophy was that the world is composed of an infinite number of simple eternal atoms. These differ in shape, size, order, and position, but every object is some combination of these atoms. Though geometrical magnitudes are infinitely divisible, the atoms are ultimate indivisible particles. (The word *atom* in Greek means indivisible.) Hardness, shape, and size are physically real properties of the atoms. All other properties, such as taste, heat, and color are not in the atoms but in the perceiver; thus sensuous knowledge is unreliable because it varies with the perceiver. Like the Pythagoreans, the atomists asserted that the reality underlying the constantly changing diversity of the physical world was expressible in terms of mathematics and, moreover, that the happenings in this world were strictly determined by mathematical laws.

Plato, the foremost Pythagorean next to Pythagoras, was the most influential propagator of the doctrine that the reality and intelligibility of the physical world can be comprehended only through mathematics. For him there was no question that the world was mathematically designed, for "God eternally geometrizes." The world perceived by the senses is confused and deceptive and in any case imperfect and impermanent. Physical knowledge is unimportant, because material objects change and decay; thus the direct study of nature and purely physical investigations are worthless. The physical world is but an imperfect copy of the ideal world, the one that mathematicians and philosophers should study. Mathematical laws, eternal and unchanging, are the essence of reality.

Plato went further than the Pythagoreans in wishing not merely to understand nature through mathematics but to substitute mathematics for nature itself. He believed that a few penetrating glances at the physical world would supply some basic truths with which reason could then carry on unaided. From that point there would be no nature, just mathematics, which would substitute for physical investigations as it does in geometry.

Plato's attitude toward astronomy illustrates his position on the knowledge to be sought. This science is not concerned with the movements of the visible heavenly bodies. The arrangement of the stars in the heavens and their apparent movements are indeed wonderful and beautiful to behold, but mere observation and explanation of the motions fall far short of true astronomy. Before we can attain to the latter we "must leave the heavens alone," for true astronomy deals with the laws of motion of true stars in a mathematical heaven of which the visible heaven is but an imperfect expression. Plato encourages devotion to a theoretical astronomy, whose problems please the mind, not the eye, and whose objects are apprehended by the mind, not visually. The varied figures that the sky presents to the eye are to be used only as diagrams to assist the search for the higher truths. The uses of astronomy in navigation, calendar-reckoning, and the measurement of time were alien to Plato.

Plato's views on the role of mathematics in astronomy are an integral part of his philosophy, which held that there is an objective, universally valid reality consisting of forms or ideas. These realities were independent of human beings and were immutable, eternal, and timeless. We become aware of these ideas through recollection or anamnesis; although they are present in the soul, it must be stimulated to recall them or fetch them up from its depths. These ideas are the only reality. Included among them but occupying a lesser rank are mathematical ideas, which are regarded as intermediate between the sensible world and such higher ideas as goodness, truth, justice, and beauty. In this comprehensive philosophy, mathematical ideas played a double role; not only were they part of reality themselves but . . . they helped train the mind to view eternal ideas. As Plato put it in Book VII of *The Republic*, the study of geometry made easier the vision of the idea of goodness: "Geometry will draw the soul toward truth, and create the spirit of philosophy. . . ."

Aristotle, while deriving many ideas from his teacher Plato, had a quite different concept of the study of the real world and of the relation of mathematics to reality. He criticized Plato's otherworldliness and his reduction of science to mathematics. Aristotle was a physicist; he believed in material things as the primary substance and source of reality. Physics and science generally must study the physical world to obtain truths; genuine knowledge is obtained from sense experience by intuition and abstraction. Then reason can be applied to the knowledge so obtained.

Matter alone is not significant. As such it is indeterminate, simply the potentiality of form; matter becomes significant when it is organized into various forms. Form and the changes in matter that give rise to new forms are the interesting features of reality and the real concern of science.

According to Aristotle matter is not, as some earlier Greeks believed, composed of one primitive substance. The matter we see and touch is composed of four basic elements: earth, water, fire and air. Also, each element has its own characteristic qualities. Earth is cold and dry; water is cold and moist; air is hot and moist; and fire is hot and dry. Hence the qualities of any given object depend upon the proportions of the elements that enter into it; and thereby solidity, hardness, coarseness, and other qualities are determined.

The four elements have other qualities. Earth and water have gravity; air and fire have levity. Gravity causes an element to seek to be at rest at the center of the earth; levity causes it to seek the heavens. Thus by knowing the proportions of the elements that enter into a given object, one can also determine its motion.

Aristotle regarded solids, fluids, and gases as three different types of matter, distinguished by the possession of different substantial qualities. The transition from solid to fluid, for example, meant the loss of one quality and the substitution of another. Thus changing mercury into rigid gold involved taking from mercury the substance that possessed fluidity and substituting some other substance.

Science also had to consider the causes of change. For Aristotle there were four types of causes. The first was the material or immanent cause; for a statue made of bronze, bronze is the immanent cause. The second was the formal cause; for the statue it is the design or shape. The formal cause of harmony is the pattern of 2 to 1 in the octave. The third cause was the effective cause, the agent or doer; the artist and his chisel are the effective cause of the statue. The fourth was the final cause, or the purpose that the phenomenon served; the statue serves to please people, to offer beauty. Final cause was the most important of the four because it gave the ultimate reasons for events or phenomena. Everything had a final cause.

Where was mathematics in this scheme of things? The physical sciences were fundamental to the study of nature, and mathematics helped by describing formal properties such as shape and quantity. It also provided explanations of facts observed in material phenomena. Thus geometry provided the reasons for facts provided by optics and astronomy, and arithmetical proportions could give the reasons for harmony. But mathematics was definitely an abstraction from the real world, since mathematical objects are not independent of or prior to experience. They exist in human minds as a class of ideas mediating between the sensible objects themselves and the essence of objects. Because they are abstracted from the physical world, they are applicable to it; but they have no reality apart from visible and tangible things. Mathematics alone can never provide an adequate definition of substance. Qualitative differences, as among colors, cannot be reduced to differences in geometry. Hence in the study of causes, mathematics can provide at best some knowledge of the formal cause—that is, a description. It can describe what happens in the physical world, can correlate concomitant variations, but can say nothing about the efficient and final causes of movement or change. Thus Aristotle distinguished sharply between mathematics and physics and assigned a minor role to mathematics. He was not interested in prediction.

From this survey we may see that all the philosophers who forged and molded the Greek intellectual world stressed the study of nature for comprehension and appreciation of its underlying reality. From the time of the Pythagoreans, practically all asserted that nature was designed mathematically. During the classical period, the doctrine of the mathematical design of nature was established and the search for the mathematical laws instituted. Though this doctrine did not motivate all of the mathematics subsequently created, once established it was accepted and consciously pursued by most of the great mathematicians. During the time that this doctrine held sway, which was until the latter part of the nineteenth century, the search for the mathematical design was identified with the search for truth. Though a few Greeks—for example, Ptolemy—realized that mathematical theories were merely human attempts to provide a coherent account, the belief that mathematical laws were the truth about nature attracted some of the deepest and noblest thinkers to mathematics.

We should also note, in order to appreciate more readily what happened in the seventeenth century, the Greek emphasis on the power of the mind. Because the Greek philosophers believed that the mind was the most powerful agent in comprehending nature, they adopted first principles that appealed to the mind. Thus the belief that circular motion was the basic type, defended by Aristotle on the ground that the circle is complete whereas a rectilinear figure, because it is bounded by many curves (line segments), is incomplete and therefore secondary in importance, appealed to the mind on aesthetic grounds. That the heavenly bodies should move with only constant or uniform velocity, was a conception which appealed to the mind perhaps because it was simpler than nonuniform motion. The combination of uniform and circular motion seemed to befit heavenly bodies. That the sublunar bodies should be different from the planets, sun, and stars seemed reasonable also, because the heavenly bodies preserved a constant appearance whereas change on earth was evident.

"There is no place and no moment in history where I could stand and say 'arithmetic begins here now,' " Dr. Bronowski observed on the mystical isle of Samos. Wherever and whenever mathematics originated, it became a universal language for defining the numerical relationships that limit the world in time and space. The Greeks knew it and so did the Babylonians. And on the other side of the world where civilizations reached their peaks long before Columbus convinced Queen Isabella to finance his exploration, the Mayans had developed a system of mathematics, astronomy, and calendrics amazingly similar to the Babylonian. Their accomplishments are discussed in this reading from *Maya* by Charles Gallenkamp.[1]

Mayan Mathematics
Charles Gallenkamp

108 Above all else the Maya were to accomplish during the upward surge of Classic expansion, it was in the fields of calendrics and mathematics that their ingenuity was most pronounced. As these realms of knowledge were of the utmost importance—so far as their relation to supernatural factors was concerned—the priests expended unlimited effort toward their elucidation. Astronomy became the primary stimulus of their attainments, and the supreme proof of their intellectual abilities is evidenced by their mastery of its principles. Tirelessly they labored to explore the multiple aspects of time, to understand and placate its awesome influence upon the destiny of individuals and empire alike. And from their attempts evolved a calendrical lore which reached back millions of years in scope and encompassed a profoundly complex philosophy.

To the Maya, time was never a purely abstract denominator by which events were arranged into orderly sequences: rather it was an infinite beyond-world inhabited by omnipotent forces of creation and destruction. Its alternating cycles—the days, months, and years—were believed to bring with them the benevolence or evil of gods who bore them along their endlessly recurring cycles. Each bearer was the patron of a sacred number and assumed a form by which it could be portrayed in hieroglyphic inscriptions, and divisions of the calendar were regarded as "burdens" carried on the backs of these divine guardians of time. If a deity of malevolent intent happened to assume the burden of a particular cycle, grievous consequences could be expected until it was relinquished to a more provident bearer at the end of its natural course. Whether a certain month or year held promise of good or bad fortune was a matter predetermined by the temperament of the god on which it was borne.

It was a curious belief, and one which explains in part the far-reaching power of the priesthoods over the populace who must surely have deemed

[1]See also Chapter 6.

survival impossible without learned mediators to interpret the irascible tendencies of their gods. Only the astronomer-priests stood between normal continuation of life and catastrophes brought about by misjudging the inclinations of the divinities. Having recognized the varying aspects of the gods and plotted their restless paths along the highways of time and space, they alone could determine when it was that beneficial gods were in possession of specific periods; or, as was more often the case, when the greatest number of benevolent deities marched in conjunction with less sympathetic ones. Thus the Mayan obsession with time was very largely tantamount to a grand-scale quest for lucky and unlucky periods in the hope that once forewarned of future prospects their destiny could be guided along the most favorable possible course.

The Mayan calendar was the most accurate such invention known to the ancient world. It was extremely complex in structure and interpretation, and consisted of a series of independently revolving "wheels" recording interrelated cycles of time. Two separate year measurements were observed: the *tzolkin* or sacred year made up of two hundred and sixty days and a civil year known as a *haab* composed of eighteen months of twenty days each to which an extra month of five days was added, making a three-hundred-and-sixty-five-day year. The civil year was used in ordinary calendrical reckoning, whereas the shorter *tzolkin* determined matters pertaining to religious observances.

Each day and month was given a proper name and an identifying hieroglyph. There was a so-called "normal form" for each of these glyphs (usually an abstract design used in secular inscriptions) and a "head-variant form"—stylized images derived from humans, animals, and mythological creatures. Both forms functioned as graphic symbols in representing specific periods of time, and the "head-variant forms" were ordinarily reserved for inscriptions of a more formal or occult nature.

The basic unit of the Mayan calendar was the day or *kin*. Twenty *kins* comprised a *uinal* or month. Beginning with the *uinal*, the calendar was divided into cycles as follows:

18 *uinals*	= a *tun* (a year with the addition of five extra days)
20 *tuns*	= a *katun* (7,200 days)
20 *katuns*	= a *baktun* (144,000 days)
20 *baktuns*	= a *pictun* (2,880,000 days)
20 *pictuns*	= a *calabtun* (57,600,000 days)
20 *calabtuns*	= a *kinchiltun* (1,152,000,000 days)
20 *kinchiltuns*	= a *alautun* (23,040,000,000 days)

Each of these cycles revolved simultaneously, one within the other, expanding from *katuns* through *alautuns* by multiples of twenty. Each also had its identifying or accompanying glyph by which inscriptions relating to it can easily be recognized. Numerals accompanying these glyphs denoted the calendric cycles that had revolved since the earliest point at which the

Maya began to record time. A typical date inscription as interpreted by archaeologists might read: 9.11.5.0.0 or nine elapsed *baktuns*, eleven *katuns*, five *tuns*, no *uinals*, and no *kins*—a date equivalent to A.D. 657. These "Long Count" or "Initial Series" inscriptions, as they are called, also contained glyphs denoting the day of the sacred year on which the date terminated; and were augmented by a "Supplementary Series"—glyphs pertaining to the age and position of the moon in relation to the accompanying Long Count. Eventually a simplified system of expressing dates known as the "Short Count" came into widespread use, which eliminated the representation of all but the day endings within a previously determined *katun*.

One of the major problems faced by epigraphers was that of arriving at the point in Mayan history from which time was first recorded. No correlation of the Mayan calendar with our own Gregorian system was possible without bringing the two systems to bear upon each other. The Gregorian calendar is oriented from the birth of Christ, but it was not known at what point this event coincided with Mayan chronology. The two calendrical systems were like halves of a gigantic slide rule; the problem was to bring their calibrations together to form corresponding equations.

In working out the calendric inscriptions it was discovered that every Initial Series sequence was counted from a cryptic date recorded as 13.0.0.0.0 4 *Ahua*, 8 *Cumhu*, which apparently marks the beginning of Mayan chronology. It is referred to as the "Zero Date" of all calendrical reckonings, and denotes a point roughly five thousand years in the past. Spinden interprets it literally as October 14, 3373 B.C. (it is 3113 B.C. according to the Goodman-Hernández-Thompson correlation). Oddly enough, this date was some three millennia *before* the earliest known dates on the Leyden Plate and Stela 9 from Uaxactún; thus archaeologists are agreed that it denotes a hypothetical event rather than one of historical actuality. Perhaps, as has been suggested by Morley, the Maya considered it the day of the world's creation or the birth date of their gods. It was from this point that the astronomer-priests began marking off divisions of lapsed and future time.

While unraveling the mutiple complexities of the Mayan calendar, scholars were awakened as never before to the incredible technical prowess of the astronomer-priests that made possible its formulation. Over a period of many centuries they had meticulously observed and recorded the passage of celestial bodies back and forth across the skies—the "gods marching in never-ending relays." Using towers and raised platforms with pre-fixed lines of sight, their courses had been carefully plotted. Eventually accurate tables were devised for predicting solar and lunar eclipses, and specialized studies were concentrated on the synodical revolution of the planet Venus until the worship of the so-called Venus year became the inspiration of a highly covert cult.

Mayan astronomers had been well aware that seemingly minute errors in dividing time into its recurring cycles would eventually multiply irreconcilable flaws. As a result of cautious observation the margin of error in their computations was incredibly slight, and corrections were made to allow for the discrepancies brought about by leap years and other synodical variations.

And always there remained the difficult problem of synchronizing the various calendrical divisions with one another, as well as to enmesh the three-hundred-and-sixty-five-day civil year with the two-hundred-and-sixty-day sacred year. Without a workable mathematical system the Maya would have been hopelessly at sea in such endeavors.

Like the astronomical glyphs, two methods were employed in recording numerical notations; elaborate head-variant numerals and a more commonly used system of bars and dots. A dot was equivalent to one unit, and a bar was equal to five. Numbers from one to nineteen were written in varying combinations of each:

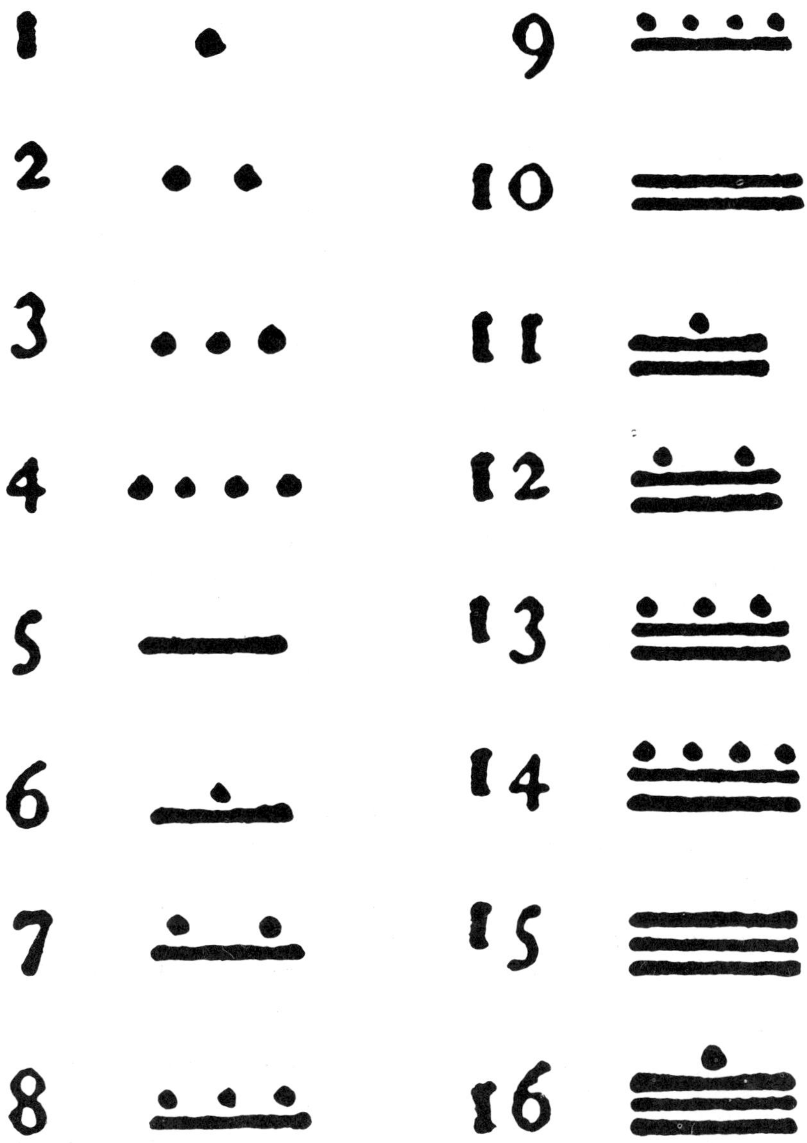

17  19 

18  0 

Numbers above nineteen were noted according to their placement in vertical columns, and each ascending position increased in value by multiples of twenty from bottom to top. Numbers placed opposite a specific position were multiplied by its corresponding multiple of twenty; the column was then added to formulate the sum total.

One of the most notable innovations brought about by the Maya in the field of mathematics was the principle of the zero. This abstract concept, essential to all but the most simplified calculations, had eluded discovery by the most advanced civilizations of the Old World. At only one other place in antiquity—among the ancient Hindus—was its use independently evolved. Subsequently it passed to Arabia and was acquired by the Moors who introduced it into Europe during the Middle Ages, but the Maya were making use of the zero at a date far earlier than even Hindu mathematicians. It was represented in their inscriptions by means of a shell, an open hand, or one of several variant glyphs.

Sometime in the vicinity of 300 B.C. a Greek named Euclid carried the ideas of Pythagoras to Egypt. His book *Elements of Geometry* was mathematically unassailable until the present time and Albert Einstein once recalled that Proposition 47—Euclid's version of the Pythagorean theorem—had obsessed him as an eleven-year-old youngster. It is given below, in translation, essentially as Euclid wrote it more than 2200 years ago.

Proposition 47

Euclid

In right-angled triangles the square on the side subtending the right angle is equal to the squares on the sides containing the right angle.

Let *ABC* be a right-angled triangle having the angle *BAC* right; I say that the square on *BC* is equal to the squares on *BA, AC.*

For let there be described on *BC* the square *BDEC*, and on *BA, AC* the squares *GB, HC*; through *A* let *AL* be drawn parallel to either *BD* or *CE*, and let *AD, FC* be joined.

Then, since each of the angles *BAC, BAG* is right, it follows that with a straight line *BA*, and at the point *A* on it, the two straight lines *AC, AG* not

From *Euclid's Elements*, translated by Sir Thomas L. Heath, published by Cambridge University Press.

lying on the same side make the adjacent angles equal to two right angles; therefore *CA* is in a straight line with *AG*.

For the same reason *BA* is also in a straight line with *AH*. And, since the angle *DBC* is equal to the angle *FBA*: for each is right: let the angle *ABC* be added to each; therefore the whole angle *DBA* is equal to the whole angle *FBC*. And, since *DB* is equal to *BC*, and *FB* to *BA*, the two sides *AB, BD* are equal to the two sides *FB, BC* respectively; and the angle *ABD* is equal to the angle *FBC*; therefore the base *AD* is equal to the base *FC*, and the triangle *ABD* is equal to the triangle *FBC*.

Now the parallelogram *BL* is double of the triangle *ABD*, for they have the same base *BD* and are in the same parallels *BD, AL*.

And the square *GB* is double of the triangle *FBC*, for they again have the same base *FB* and are in the same parallels *FB, GC*.

[But the doubles of equals are equal to one another.]

Therefore the parallelogram *BL* is also equal to the square *GB*.

Similarly, if *AE, BK* be joined, the parallelogram *CL* can also be proved equal to the square *HC*; therefore the whole square *BDEC* is equal to the two squares *GB, HC*.

And the square *BDEC* is described on *BC*, and the squares *GB, HC* on *BA, AC*. Therefore the square on the side *BC* is equal to the squares on the sides *BA, AC*.

Therefore etc. Q.E.D.

As mentioned earlier, Pythagoras and his followers were obsessed with the harmony of nature, and the rotation of the planets, in which they heard the "music of the spheres." It was a romantic thought, and in fact, it is now known that celestial bodies do give off radio signals with their own characteristic frequencies. In the Pythagorean sense, however, the idea of celestial harmonies titillated the fancy of Johannes Kepler (1571-1630), a tavern keeper turned astronomer, whose accomplishments are covered more fully in Chapter 6. In this reading, Kepler wonders, "In the celestial harmonies, which planet sings soprano, which tenor, which alto, and which bass?"

The Celestial Chorus

Johannes Kepler

Although these words are applied to human voices, while voices or sounds do not exist in the heavens, on account of the very great tranquillity of movements, and not even the subjects in which we find the consonances are comprehended under the true genus of movement, since we were considering the movements solely as apparent from the sun, and finally, although there is no such cause in the heavens, as in human singing, for requiring a definite number of voices in order to make consonance (for first there was the number of the six planets revolving around the sun, from the number of the five intervals taken from the regular figures, and then afterwards—in the order of nature, not of time—the congruence of the movements was settled): I do not know why but nevertheless this wonderful congruence with human song has such a strong effect upon me that I am compelled to pursue this part of the

comparison, also, even without any solid natural cause. For those same properties which custom ascribed to the bass and nature gave legal grounds for so doing are somehow possessed by Saturn and Jupiter in the heavens; and we find those of the tenor in Mars, those of the alto are present in the Earth and Venus, and those of the soprano are possessed by Mercury, if not with equality of intervals, at least proportionately. For howsoever in the following chapter the eccentricities of each planet are deduced from their proper causes and through those eccentricities the intervals proper to the movements of each, none the less there comes from that the following wonderful result (I do not know whether it is occasioned by the procurement and mere tempering of necessities): (1) as the bass is opposed to the alto, so there are two planets which have the nature of the alto, two that of the bass, just as in any Mode of song there is one [bass and one alto] on either side, while there are single representatives of the other single voices. (2) As the alto is practically supreme in a very narrow range [*in angustiis*] on account of necessary and natural causes unfolded in Book III, so the almost innermost planets, the Earth and Venus, have the narrowest intervals of movements,
114 the Earth not much more than a semitone, Venus not even a diesis. (3) And as the tenor is free, but none the less progresses with moderation, so Mars alone—with the single exception of Mercury—can make the greatest interval, namely a perfect fifth. (4) And as the bass makes harmonic leaps, so Saturn and Jupiter have intervals which are harmonic, and in relation to one another pass from the octave to the octave and perfect fifth. (5) And as the soprano is the freest, more than all the rest, and likewise the swiftest, so Mercury can traverse more than an octave in the shortest period.

The snowflake, the crystal of table salt, even rock candy growing on a string, all are bound by what Dr. Bronowski describes as "the nature of the space . . . the three dimensions, the flatness within which we live." Whatever makes up a crystal, whether it be the charged "ions" of sodium chloride or the billions of carbon atoms that combine to form the dense packed crystalline network of diamond, the units of a crystal determine its size and shape, within the confines of surrounding space. Much of what we now know about the architecture of crystals comes from a technique called x-ray diffraction, which employs x rays to peer into what chemists define, appropriately enough, as a crystal's space lattice. The concept of the crystal lattice is described here by Michell J. Sienko and Robert A. Plane.

Space Lattices and Packing of Atoms

Michell J. Sienko and Robert A. Plane

The pattern of points which describes the arrangement of molecules or atoms in a crystal is known as a *space lattice*. In Fig. [1] is shown the space lattice of sodium chloride, NaCl. Each of the points corresponds to the position of the center of an ion [Cl^- or Na^+]. The open circles locate the centers of the positive sodium ions, and the filled circles locate the positions of the negative

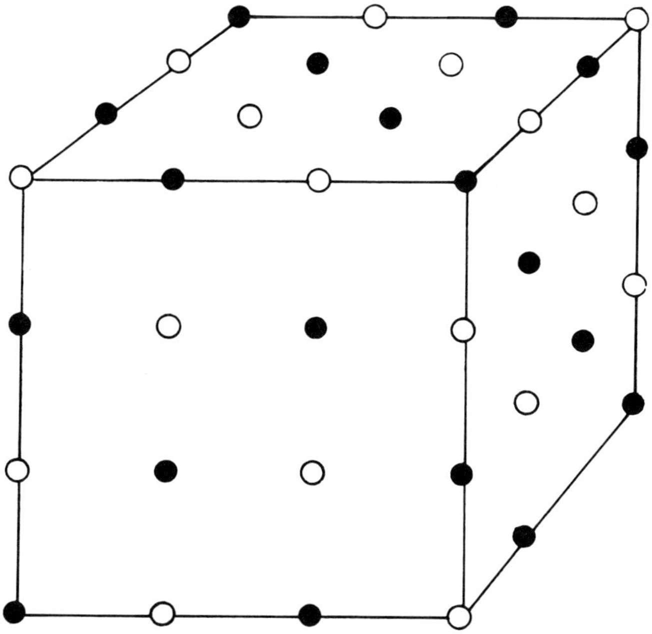

Fig. 1. Space lattice of NaCl.

chloride ions. It must be emphasized that the points do not represent the actual sodium ions and chloride ions; they represent only the positions occupied by the ion centers. In fact, in sodium chloride the ions are of different size and are practically touching each other, as shown in Fig. [2].

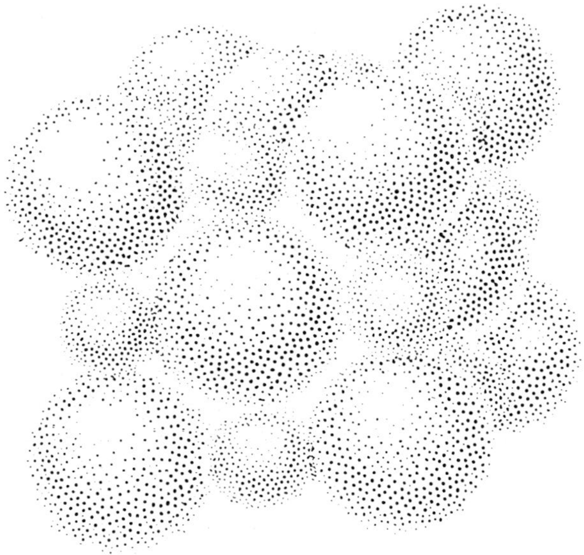

Fig. 2. Space-filling model of NaCl. *Small spheres represent* Na^+ *ions; large spheres,* Cl^- *ions.*

The space lattice has to be thought of as extending in all directions throughout the entire crystal. However, in discussing the space lattice it is sufficient to consider only enough of it to represent the order of arrangement. This small fraction of a space lattice, which sets the pattern for the whole lattice, is called the *unit cell.* It is defined as *the smallest portion of the space lattice which when moved repeatedly a distance equal to its own dimensions along the various directions generates the whole space lattice.* A unit cell of sodium

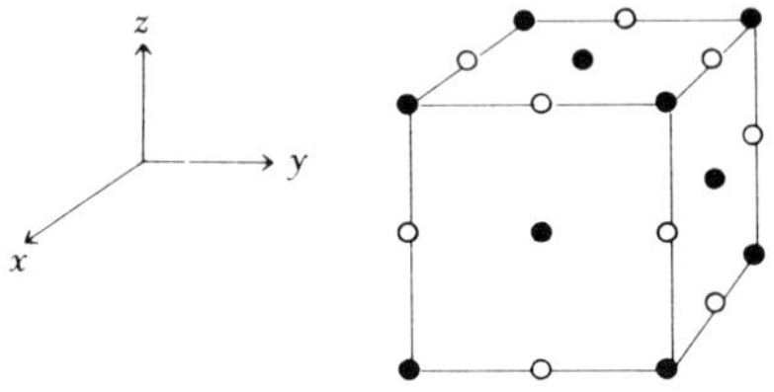

Fig. 3. Unit cell of NaCl.

chloride is the cube shown in Fig. [3]. If this cube is moved through its edge length in the *x* direction, the *y* direction, and the *z* direction many times, the whole space lattice can be reproduced.

Several different kinds of unit-cell symmetry occur in crystalline substances. The simplest, known as *simple cubic*, is shown in Fig. [4]. Each point at the corner of the cube represents a position occupied by an atom or a molecule.

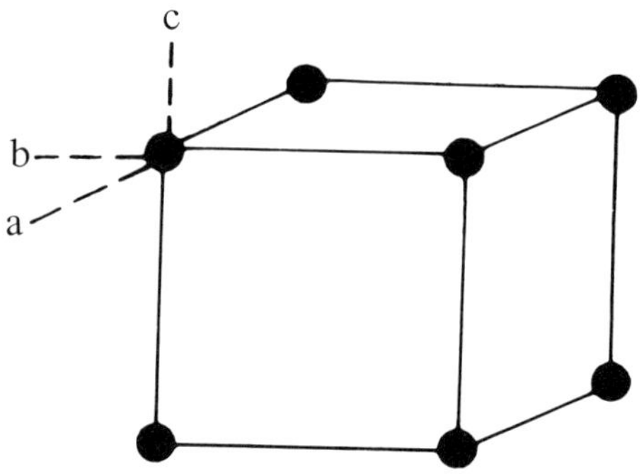

Fig. 4. Simple cubic unit cell.

The dashed lines marked *a, b,* and *c* represent the three characteristic directions of space, or *axes*, along which the structure must be extended to reproduce the entire space lattice. . . . The unit cells just discussed actually concern only points that locate atomic or molecular centers. Atoms are space-filling entities, and structures can be described as resulting from the packing together of representative spheres. The most efficient packing together of equal spheres, called *closest packing*, can be achieved in two ways, each of which utilizes the same fraction (0.7406) of total space. One of these close-packing arrangements is called *hexagonal close-packed.* It can be envisioned as being built up as follows: Place a sphere on a flat surface. Surround it with six equal spheres as close as possible in the same plane. Looking down on the plane, the projection is as shown in Fig. [5a]. Now form over the first layer a second layer of equally bunched spheres staggered as shown in Fig. [5b] so that the second-layer spheres nestle into the depressions formed by the first-layer spheres. A third layer can now be added with each sphere directly above a sphere of the first layer. Succeeding layers follow in alternating fashion *ababab* . . . until the hexagonal close-packed structure has been generated. Figure [6a] shows how the hexagonal close-packed stacking layers are related to each other.

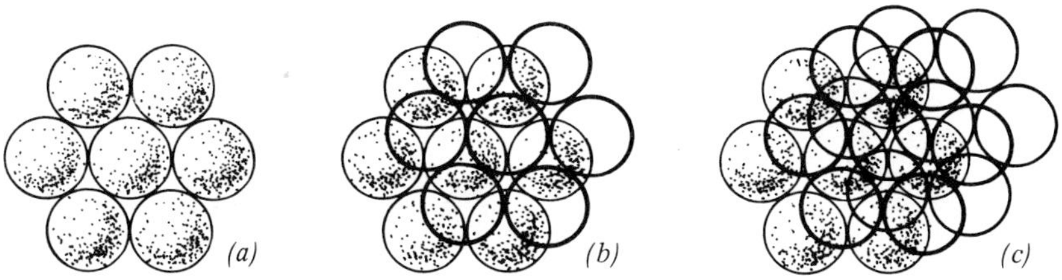

Fig. 5. Close-packing of spheres.

The other close-packed structure, called *cubic close-packed*, results if the buildup of layers *a* and *b* is the same as that described above but with the third layer added as shown in Fig. [5c]. The spheres of the layer *c* (indicated by the black, open circles) are not directly above layer *a* (filled circles) or

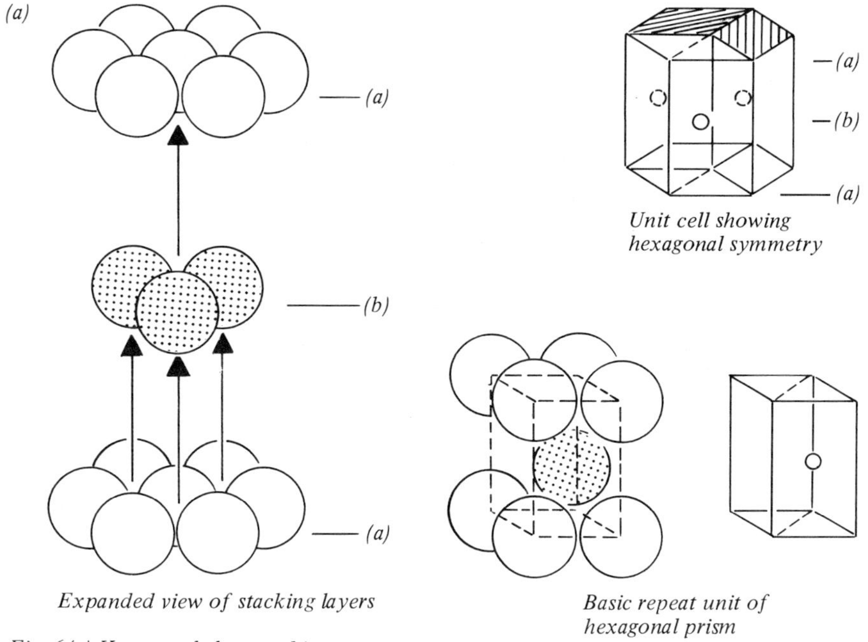

Fig. 6(a) Hexagonal close-packing

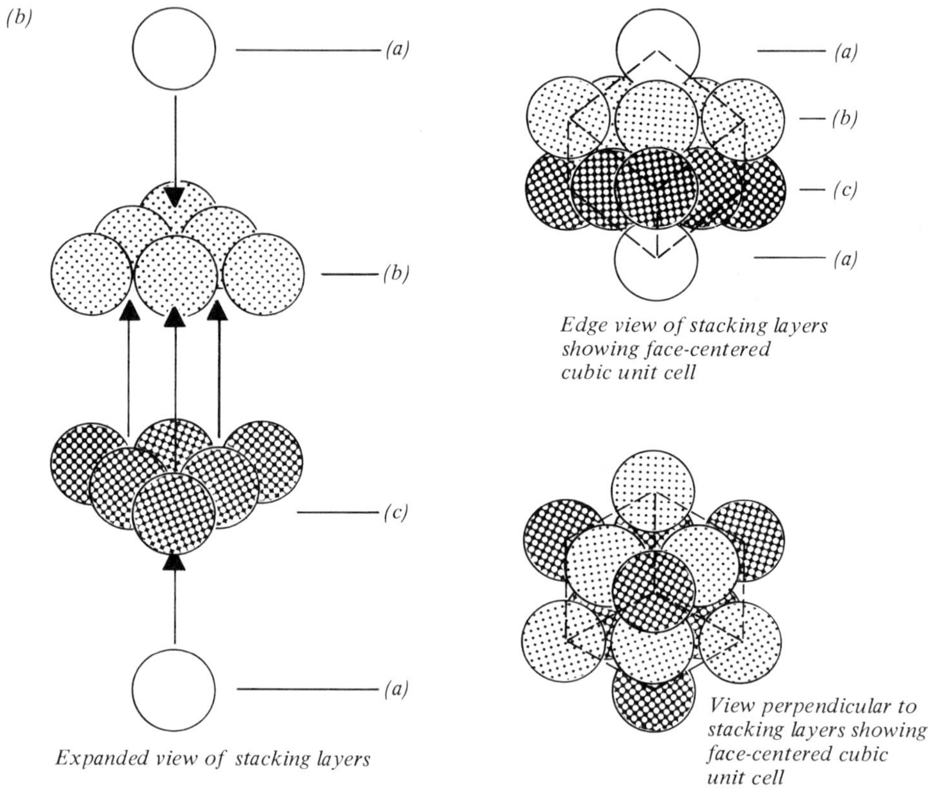

Fig. 6(b) cubic close-packing.

layer *b* (colored circles). For this cubic close-packed structure the sequence of layers is *abcabc*. . . . Figure [6b] shows how the cubic close-packed stacking layers are related to each other. . . .

Among the common materials that crystallize with hexagonal close-packing are many of the metals, such as magnesium, zinc, cadmium, and titanium.

"The calculus." These are not words guaranteed to make the average reader jump up and down with gleeful anticipation, but they can no more easily be stricken from the mathematical underpinnings of nature than the names of Newton and Liebniz. With calculus, Dr. Bronowski asserts, "mathematics becomes a dynamic mode of thought . . . a major mental step in the ascent of man." It has become the mathematics of change, as the following selection illustrates.

The Calculus

Benjamin David (and the Staff Editors of Life Science Library, Mathematics)

Nothing in the world is immune from change. The hardest rock on the driest desert expands or contracts in the shifting sunlight. The steel gauge blocks at the National Bureau of Standards, even though they are stored in temperature-controlled subterranean vaults, are subject to seasonal fluctuations in length thought to be caused by radiation from the surrounding walls. Everything grows or shrinks, warms up or cools down, changes its position, its color, its composition—perhaps even its spots.

Inescapable as the process of change is, and vital as it is to understanding the laws of nature, it is difficult to analyze. Being continuous, it offers no easy point at which the mind can catch it and pin it down. For centuries it baffled mathematicians. Some starts, to be sure, were made toward a mathematics of movements. The Greeks did so when they thought of curves as tracings made by moving points, and when they analyzed curving lines, instant by instant, through the technique of slicing them into infinitesimally fine segments. Descartes did so when he conceived of the items in an equation as functions between *variables*, and most of all when he supplied a way to draw graph-pictures of *fluid* situations and relationships. But by and large the world of mathematics was populated by waxworks—shapes and numbers that stood stock-still.

Then, in 1665 and 1666, England's incomparable Isaac Newton produced a prodigious brain child, now called calculus, which for the first time permitted the mathematical analysis of all movement and change. In calculus Newton combined the fine-slicing technique of the Greeks and the graph system of Descartes to devise a marvelously automatic mental tool for operating on an equation in order to get at infinitesimals. So quickly did calculus prove its effectiveness that in a few years its creators used it to work out the laws of motion and gravitation—the fundamental laws of physics which explain why the solar system acts as it does, or why any moving object reacts as it does to outside forces like gravity, the tension of a spring or the push of a man's hand. By its ability to probe the fleeting mysteries of movement, calculus today has become the principle pipeline between practical science and the reservoir of mathematical thought. Every airplane, every television set, every bridge, every bomb, every spacecraft owes it a tithe of indebtedness.

The different kinds of change which calculus can analyze are as diverse as a queen's wardrobe. If the factors involved in any fluid situation can be put in terms of an equation, then calculus can get at them and uncover the laws they obey. The change under scrutiny may be as dramatic as the gathering speed of a missile lifting from its pad or as quiet as the varying grade of a mountain road. It may be as visible as the pounds added around a once-svelte waistline or as invisible as the ebb and flow of current in a power line. It may be as audible as the crescendo of a Beethoven concerto or as silent as the build-up of flood force behind a dam.

Calculus analyzes all these situations by invoking two new mathematical processes—the first fundamental operations to be added to the canon of mathematics since the laws of addition, subtraction, multiplication, division and finding roots were laid down some 4,000 years ago. These new operations are called *differentiation* and *integration,* and they are the reverse of each other in much the same way that subtraction is the reverse of addition, or division of multiplication. Differentiation is a way of computing the rate at which one variable in a situation changes in relation to another at any point in a process—at a given instant in time, for example, or at a given point in space. The actual method employed in differentiation is to divide a small change in one variable by a small change in another; to let these changes both shrink until they approach zero; then—and this is the key—to find the value which the ratio between them approaches as the changes become indefinitely small. This value is what mathematicians call a "limit," and it is the answer they are seeking, the end result of differentiation—the rate of change at a given instant or point. Integration works back the other way from differentiation; it takes an equation in terms of rate of change and converts it into an equation in terms of the variables that do the changing.

Through differentiation, a mathematician can probe deep into a fluid situation until he finds some unchanging factor that reflects the action of a constant law of nature. In this fashion Newton and later theorists made a discovery which is still not easy for laymen to absorb. This discovery was that the constant factor in many processes of nature is the rate at which a rate of change changes. Deciphering this seeming double talk may appear hopeless. But anyone who drives a car is familiar with the rate of change of a rate of change whether he realizes it or not. The speed of the car—so many miles per hour—is a rate of change of distance with respect to time. In speeding up or slowing down, the car's speed itself changes, and changes at a rate—acceleration or deceleration—which is the rate of change of the rate of change.

In nature, gravity acts to make a falling object move at a rate which increases at a constant rate. For processes involving actual physical movement, Newton defined this rate of a rate as the *acceleration*. And he called the gravity causing it a *force.* He defined force in general as something which causes an object to accelerate. As applied through calculus, this definition—laid down three centuries ago—has enabled scientists to do no less than identify the three fundamental forces of the cosmos: the force of gravitation; the force of magnetism, or electric charge; and the force that binds together the atomic nucleus.

In contrast to the spectacular role that calculus has played in unlocking the secrets of the universe, the nomenclature surrounding differentiation and integration is woefully prosaic. The rate of change of a y or x, found by differentiation, is called a *derivative*—a derivative of y with respect to x, written dy/dx, or of x with respect to y, written dx/dy. A derivative's counterpart, found by integration, is called an *integral*, and is symbolized by $\int$, an elongated letter S, which was short, originally, for "sum" or "summation." When integration is performed on an equation written in terms of derivatives, it converts the equation back into one in which the x and y have doffed their rate-of-change disguises and resumed normal algebraic appearance.

A Definition for Dieters

The handles and hieroglyphs attached to the techniques of calculus may look alien, but the ideas behind them are easily recognized. Being a rate of change, a derivative means, simply, the speed of a process: so many miles per hour or feet per second if it refers to change in position; so many pounds per week if it refers to a triumph in dieting; so many geniuses per childbirth if it refers to

I.Q. statistics; so much cornstalk per candle power of sunshine if it refers to the growth of a corn crop. The integral corresponding to each of these derivatives would be the miles traveled, pounds lost, geniuses gained or length of cornstalk grown.

When used abstractly in an equation, a derivative can most readily be thought of in terms of the curve which represents this equation on a graph. At any point, the curve is rising or falling at a rate of so many y-units per x-unit. This slope up or down is the exact geometric equivalent of the rate of change—the derivative—of y with respect to x. Engineers often express the grade of a hill, the pitch of a roof or the steepness of an airplane's climb in identical terms: as so much altitude gained per unit of horizontal distance traversed. But in these applications the slope is normally conceived of as being measured over some definite span of distance. In calculus, on the other hand, the derivative is thought of as an instantaneous slope at a single point on a curve.

That this elusive concept of instantaneous slope is no figment of the mathematical imagination can be seen in an artillery shell as it arcs toward target. At any single moment the shell is moving in a definite direction. This direction is an instantaneous slope with respect to the ground, a rate of change in the shell's altitude with respect to its horizontal position. In terms of a graph, the speed of the shell, moving up or down, can also be considered as an instantaneous slope on a curve—a rate of change in the shell's altitude with respect to the time that has been elapsing since the shell was fired. A mathematician would normally write such a derivative—the velocity of climb or fall, or the rate of change of vertical distance—as dy/dt, in which the t stands for time.

The counterpart of a derivative, an integral, can also be visualized in terms of a graph. Suppose that y equals some expression of x and that this equation is plotted as a curve. Then the integral of y is the area between the curve and the horizontal line, or axis, running along below it. Why this is so can be seen by imagining that the area under the curve is filled by a picket fence with a scalloped top. . . . As the fence is being built, each new picket adds to the area of the fence. In fact, the height of each added picket is a measure of the rate at which the area of the fence is growing; a six-foot picket, for instance, adds twice as much area as a three-foot picket. The integral of the rate of change, therefore, must be the actual factor in the situation that is changing: namely, the area of the fence itself. The geometric equivalent of each picket is simply the height of a curve—the vertical, or y coordinate of each point on a curve. Integrating y must, therefore, give the total area under the curve.

Many of most practical applications of calculus stem from integration's ability to sum up y-length pickets and determine areas. Through it a mathematician can determine the volume of all manner of irregular shapes, such as airplane fuselages or oil-storage tanks; he can also find the areas of curved surfaces—the amount of sheet metal in a molded car body or the lifting surface on the wings of a jet.

There is one major difficulty in the process of integration—a difficulty so enormous and so recurrent that most of our largest computers today have been built specifically to cope with it. This is the problem of so-called "boundary conditions." When the area of a picket fence is measured, the boundary conditions are established by the two pickets that mark the two ends of the fence. But there are no ends to many of the curves that represent equations. The area beneath this type of curve may be indefinitely large. To give it boundaries, the mathematician erects the equivalent of end posts to mark off the particular part of the area he is interested in. He then integrates the equation represented by the curve between these two verticals. Often, however, in the case of equations which are arrived at experimentally, the proper way to interpret such an equation cannot be found except by integrating it, and the boundary conditions necessary for integrating it cannot be found except by understanding how to interpret it. To avoid this impasse, the scientist, in effect, chooses arbitrary picket posts and lets a computer run off dozens of laborious solutions, which give him insight into the equation and the process of change which it symbolizes.

To work out the rules of calculus, Newton visualized what would happen if one point on a graph-curve slid down into a point nearby. As the slide begins, the average slope of the curve between the two points is the number of y-units separating them vertically, divided by the number of x-units separating them horizontally. As the slide continues, both distances in this fraction diminish toward zero and finally vanish when the two points merge. But this does not mean that the fraction itself vanishes. A ratio of 1:2, for instance, need not suddenly become zero just because its numerator and denominator become indefinitely small. When last heard from, as they disappear arm in arm into the fastnesses of infinity, the numerator may be one zillionth and the denominator two zillionths, but the ratio between them is still 1 to 2.

Finding the value which a fraction approaches as its numerator and denominator both diminish toward nothingness is called "taking a limit." If the numerator equals half the denominator, the limit is one half. If the numerator equals 10 times the denominator, the limit is 10. As two points on a curve slide together, the vertical and horizontal distances between them remain coupled, even as they fade away, by the relationship of y to z expressed in the original equation of the curve. As they merge, therefore, the ratio of their distances approaches a definite limit which can be evaluated in terms of y and x. This limit, ½ or 10 or whatever it may be, is the slope of the curve at the precise spot where the two points merge—the rate of change of vertical y with respect to horizontal x or, put another way, the derivative of y with respect to x.

The subtle train of reasoning which enabled Newton to differentiate equations and find the derivative or limiting value of the ratio—written dy/dx or dy/dt—is the fundamental process of calculus. It can be roughly paraphrased as follows: in a developing situation, the difference between the state of affairs at one moment and the state of affairs at the next moment is an indication of how the situation is shaping up; and if the ratio of the net changes that take place between the two moments is evaluated as a limit—a limit approached when the interval between the two moments is imagined as diminishing toward zero—then that limit shows how fast developments are taking place. The logic of calculus can be applied to moments of time, points on a curve, temperatures in a gas or any state of affairs which can be related by equations; the same rules of differentiation apply to all of them.

A Simple Gift from Galileo

The way these rules work, and the reason for their enormous usefulness, can best be illustrated by applying them to the classically simple equation $y = 16t^2$, devised by the renowned Italian astronomer and physicist, Galileo Galilei. This brief, unpretentious expression is one of the most versatile in all physics because it shows how gravity acts on a freely falling object—an elevator run amuck, a hailstone or a jumper descending to ground. Since almost all movements and changes on earth are heavily influenced by gravity, the equation of free fall indirectly plays a part in innumerable human actions—from taking a step or lobbing a tennis ball to lifting a steel girder or launching an astronaut into orbit.

Timing an object as it falls from a given height is the most straightforward method of gauging the effects of gravity. It was this technique which Galileo used, about 1585, to arrive at his free-fall equation. According to legend, Galileo dropped small cannon balls from the colonnades of the leaning tower of Pisa. According to his own account, he used the less fanciful means of timing cannon balls as they rolled down a ramp. In any event, Galileo ascertained that the equation for free fall was $y = 16t^2$, with y representing the distance fallen in feet and t the elapsed time in seconds after the start of the fall.

By differentiating this equation twice—so as to shave away successive layers of change and inconstancy—Newton uncovered the essential nature of gravitation. Differentiating the equation once, he found that the speed with which a jumper is falling at any moment equals 32 times the number of seconds which he has been falling. Differentiating the equation a second time, he found that the jumper's acceleration—the rate of increase in his speed—is always 32 feet per second, every second.

The fact that in the free-fall equation acceleration equals a constant number, 32, indicates the end of the trail. This 32 need not be differentiated further; it does not change, and its rate of change is zero. It represents a law of nature: that every free-falling object falls to earth with a constant acceleration of 32 feet per second, every second.

Having ascertained this fact by calculus, Newton was able to set his mathematical sights far beyond the earth and to deduce the law of universal gravitation—one of the most important results ever to be achieved by mathematics. It is the law which governs the movements of all celestial bodies—from human beings in orbit to entire systems of stars.

Looking back with awe on what a little deduction could accomplish in the mind of Isaac Newton, later thinkers have ranked him as the greatest physicist and one of the greatest mathematicians the world has ever known. Albert Einstein wrote: "Nature to him was an open book, whose letters he could read without effort." Newton himself said: "I do not know what I may appear to the world; but to myself I seem to have been only like a boy playing on the seashore, and diverting myself in now and then finding a smoother pebble or a prettier shell than ordinary, whilst the great ocean of truth lay all undiscovered before me."

Newton began to use his astounding inventiveness while still a child, to build toys for himself, including a wooden water clock that actually kept time and a flour mill worked by a mouse. His brilliance did not really catch fire, however, until he read Euclid at the late age of 19. The story goes that he rushed impatiently on to Descartes' relatively abstruse *La Géométrie*. Thereafter his progress was meteoric. Five years later, while still a graduate student at Cambridge, he had already worked out the basic operations of calculus—the rules of integration and differentiation, which he called the laws of "fluxions and fluents."

Newton put together his great invention and applied it in a preliminary way to the problems of motion and gravitation in a two-year burst of creativity, while rusticating during the epidemic of plague which swept England in 1665 and 1666. In retrospect it seems as if the whole framework of modern science arose from his mind as miraculously as a jinni from a bottle. But as Newton himself said, he "stood on the shoulders of giants." Many men had wrestled with the same problems; it was his genius to fuse their separate inspirations. The twin processes of differentiation and integration in calculus, for instance, were rooted in two classic questions of Greek antiquity: how to construct a tangent line (a line that just touches a curve at a given point), and how to calculate an area which is bounded on one side by a curve. The problem of the tangent, or "touching" line, was equivalent to the problem of finding the slope of a curve at any point and therefore of finding the derivative of an equation. The area problem was equivalent to the problem of integrating the equation that gives the rate of growth of an area.

A Wine Keg of Infinitesimals

By viewing any curve as a succession of infinitely short segments, or any area as an accumulation of infinitely fine slices, the Greeks—particularly Archimedes—had solved a number of specific problems concerning rates of change. Mathematicians of the 16th and 17th Centuries also used infinitesimal methods, though seldom with rigorous Greek proofs. Kepler, for instance, had employed infinitesimals to give vintners a formula for gauging the volume of wine kegs. In Descartes' time and in the 15 years after his death, his compatriot, Pierre de Fermat, and the Englishman, John Wallis, had begun to cast infinitesimals in the helpful analytic molds of equations. Then, in about 1663, Newton's professor at Cambridge, Isaac Barrow, became the first man to realize that the tangent problem and the area problem are two sides of the same coin—in effect, that integration is the reverse of differentiation.

When Newton first began to unite all these preliminary insights in the single well-knit structure of calculus, he showed Barrow some of his early results. Barrow was so enthusiastic that he generously let it be known about

Cambridge that Newton had done what he himself had failed to do. A few years later, in 1669, when he was retiring, he was instrumental in getting Newton, then 26, appointed as his successor to the Lucasian professorship of mathematics at the university—one of the most desirable chairs of mathematical scholarship in the academic world. Thereafter, honors and inspirations came to Newton in a steady stream. Over the next four decades he formulated the law of gravitation and used it to explain the movements of the planets, moon and tides; analyzed the color spectrum of light; constructed the first modern reflecting telescope; performed innumerable alchemistic experiments; tried to reconcile with Scripture the date of 4004 B.C., which was currently accepted as the time of Adam's creation; served as a member of Parliament; was appointed warden and then master of the British mint; was knighted by Queen Anne in 1705 and was repeatedly elected president of Britain's select scientific club, the Royal Society, from 1703 until his death in 1727.

Strangely enough, Newton revealed his monumental discoveries to only a few of his scientific cronies. Many explanations have been given for his inordinate secretiveness. It has been said that he was always too busy with new ideas to find time to write up old ones, and that he had a passionate distaste for the wrangles and criticism which inevitably raged around scientific pronouncements in those days. Then, too, he was just not much of a talker. While he was in Parliament, his only recorded utterance was a request to open the window. On one occasion the astronomer Edmund Halley came to him, after a discussion with England's most eminent scientists, to ask if he knew what path a planet would take around the sun if the only force influencing it was a force that diminishes according to the square of its distance from the sun. Newton immediately gave the answer: the path would be elliptical. When asked how he knew, he explained casually that he had worked out the problem years before as a graduate student. In other words, he had worked out the fundamental law of the universe and told nobody about it. Encouraged by Halley to re-create his original calculations, he went on to produce his masterwork, the *Principia.*

Newton's *Principia* is generally recognized as the most influential, conclusive and revolutionary scientific work ever to appear in print. In it, he not only explained why the solar system works the way it does but also laid down the laws of dynamics which are still the chief ingredients of practical engineering physics—of missile shoots or thruway construction. Most of these laws Newton had worked out through calculus, but like Archimedes before him, he chose to present his finished work in universally understood mathematics—as a lengthy Greek proof, couched almost entirely in the terms of classical geometry.

Not even the skillful coaxing of Halley could convince Newton to publish his calculus—not, that is, until another mathematician, the German Gottfried Wilhelm von Leibniz, had independently re-created the entire mental machinery. Leibniz invented calculus 10 years after Newton, in 1675, and in 1684 published his account of it, 20 years before Newton was to give the first published explanation of his own version.

Like Newton, Leibniz was as successful and practical as the mathematics he originated. The son of a well-to-do university professor, he learned Greek and Latin by the age of 12, attended university, took a law degree, and went on to become the counsel to kings and princes in an illustrious career that sometimes verged on the shady. He traveled all over Europe tracing dubious lineages to establish the dubious rights of princelings to vacant thrones. He formulated many of our modern principles of international power politics—including the phrase "balance of power." During trips to Paris, he studied algebra and analytic geometry under the great optical physicist Christian Huygens. And while jogging along in coaches on diplomatic missions, he created new mathematics simply for pleasure, including his own version of calculus.

Although Newton accomplished far more with calculus than Leibniz did, Leibniz had a superior notation for it—one he polished so carefully that we still use it. It was Leibniz who first wrote derivatives as dy/dx or dx/dy—forms that suggest the fractional rate-of-change measurements to which they apply. (Newton wrote the derivative of y as $\dot{y}$ and the derivative of x as $\dot{x}$.

The dots in Newton's symbolism led rebellious 19th Century Cambridge students to protest against the "dotage" of English notation and to advocate the "pure d-ism" of continental notation.)

Unfortunately, Newton and Leibniz in their later years became embroiled in a chauvinistic dispute as to who had invented what first. The result was that scholars on the continent, helped by Leibniz' notation, went on to develop calculus further, while English mathematicians, hampered by the less felicitous notation devised by Newton, foundered in a morass of perplexities.

Salute to a Lion's Paw

The supremacy of the continental approach did not emerge, however, while Newton was still alive. At least twice after the rivalry had broken out, Leibniz and his followers posed problems with which they hoped to stump Newton. Each time Newton tossed off the answers in a single evening after coming home from his work at the mint. One of these problems was a particularly devilish one: to find the shape of the curve down which a bead will slide under the influence of gravity so as to move from a higher point to a lower point in the shortest possible time. The problem was important as an early example of "maximizing-minimizing" questions which confront mathematicians today—maximizing industrial productivity or minimizing the amount of fuel required to get to the moon. Newton solved the problem overnight and transmitted his solution anonymously the next morning through the channels of the Royal Society. Upon its receipt, Johann Bernoulli, the disciple of Leibniz who had posed the problem, is reported to have said, *"Tanquam ex ungue leonem,"* which, freely translated, means "I recognize the lion by his paw."

Logicians of the next generation sharply criticized both Newton and Leibniz for having used the equivalents of infinitesimals—for having added up nothings to create the somethings of areas, and for having shaved down rates of change to instantaneous slopes measured in no time at all. The Irish metaphysician Bishop Berkeley, in an essay entitled "The Analyst," examined the logic of Newton's "fluxions" and concluded, "They are neither finite quantities, nor quantities infinitely small, nor yet nothing. May we not call them the ghosts of departed quantities . . .?"

Mathematicians of the 19th Century were to satisfy such critics by invoking new standards of rigor for calculus. But meanwhile it met the test of success—it worked. Scientific problems capitulated to it as the walls of Jericho to Joshua's trumpets. Indeed the chief danger was one of self-satisfaction. Using calculus, scientists explained every process in nature as a sequence of actions and reactions, of causes and effects. Nature, however, cannot be determined in this easy mechanical way. Anyone knows that there are slips between cup and lip—accidents in the forces which produce motion. But the laws of these accidents are also mathematical.

6

The Starry Messenger

Boyle: I seen things, Joxer, that no mortal man should speak about that knows his catechism. Often and often when I was fixed to the wheel with a marlinspike and the wind's blowin' fierce and the waves lashin' and lashin' till you'd think every minute was going to be your last . . . and as it blowed and blowed. I often looked up to the sky and assed meself the question. . . . What is the stars?

Joxer: Ah, thats the question. What is the stars?

Sean O'Casey

Astronomy was built on a number of often conflicting foundations: mathematics and myth, the pseudoscience of astrology, and the rigors of Euclid's geometry. Like Captain Boyle and Joxer in Sean O'Casey's *Juno and the Peacock,* man has long wondered "what is the stars," but until recently he has been able to get only as close to them as a telescope would take him. Out of astronomy spill names like Ptolemy, Kepler, Copernicus, and Galileo. They left myth and magic far behind in their probes of the universe, but in Galileo's case, the faith of his forefathers presented formidable opposition.

In the beginning, there was astrology and the Babylonian's reveled in it, even while they were helping to develop the science of astronomy. The first of the two following readings discusses the astrological contributions of the Babylonians and the second outlines the impact of astrology on the enlightened minds of the Renaissance.

Babylonian Stargazers

Rudolf Thiel

The Babylonian astronomers possessed a charming naïveté and candor, which today enables us to observe the gradual development of star-reading, or astrology, side by side and in pace with astronomy.

It was obvious to everyone that the stars had an influence upon the Earth. By its height in the sky the Sun regulated the seasons and brought drought and rainy months. When the Pleiades reappeared out of the sunshine, about the time of the spring new Moon, there were devastating storms. This effect of the Seven Sisters is described in a very ancient Sumerian hymn: "The winter storms, the wicked demons, engendered in the firmament, raise their heads; mischief they wreak, evil they ponder, day after day; seven their number, heralds of their King Anu; roaring their wrathful hurricane in the sky. Before the radiant Moon they have subsided."

The Moon was itself a weather-maker—and is so considered to this day, despite the lack of meteorological evidence. Certainly the monthly period of women seems to point to some mysterious human relationship with the Moon. The Babylonians believed they had discovered a considerable number of additional ties. A new king, for example, would have a long reign if its first month turned out to be longer than had been calculated, or if two stars set together at the moment the Moon reached its zenith. A ring around the rising Moon foretold the birth of a crown prince. If a new Moon appeared precisely at the beginning of a thirty-day month, it meant good fortune for the entire country.

Similar influences were ascribed to the planets. The Babylonian astronomers must have been astonished and alarmed when they observed that some of the stars were also wanderers and moved in the same paths as the Moon and the Sun—that is, through the zodiac. Two of these, the morning and the evening star, even revealed themselves to be one and the same planet which was overtaken by the Sun, so that it sometimes appeared on the right, sometimes on the left of the Sun. The Babylonians solemnly appointed it their third principal heavenly body, and frequently depicted it with eight rays, alongside the sickle of the Moon and the disc of the Sun. It seemed obvious that a change in the position of the morning star must have an important meaning. A cuneiform tablet dating from around 2000 B.C. declared:

When Ishtar [Venus] is visible in the east on the sixth of Abu, there will be rain and destruction. Until the tenth of Nisan it remains easterly; on the eleventh it vanishes and remains invisible for three months. If, then, it glows in the west again on the eleventh of Du'uzu, there will be fighting in the land, but the fruits of the fields will prosper.

Venus, however, had surprises in store for the Babylonians. With their customary honesty, they recorded their consternation on burnt clay:

When the planet Ishtar enters the Scorpion, floods take place, as is well known. This time, however, there were none, for as soon as Ishtar came up to the breast of the Scorpion, she was snatched away. Although she touched the Scorpion, she did not penetrate it.

This was the Babylonians' first acquaintance with one of the great puzzles of ancient astronomical science: the occasional retrograde motion of the planets. Similar motion was observed in the case of the other planets, and with equal surprise:

Mar-Istar to the King: As to Marduk [Jupiter], I recently reported. My prediction was based on his position in the Anu path [the celestial equator]. I now report: he has been delayed. He has truly run backward; therefore my interpretation was mistaken. But it would not have been mistaken if Marduk had remained on the Anu path. May my Lord King understand this.

In any case, there can be no doubt that the path of scientific astronomy led through the fields of astrology in all ancient civilizations. Even the Chinese observed the retrograde motion of Jupiter only because they considered this bright star a harbinger of good fortune. As soon as they discovered Jupiter's inexplicable irregularity, they decided that it was an omen of famine.

The Babylonians surpassed all other peoples of antiquity in the patience, precision, and impartiality of their observations. Not content with the usual interpretation of the stars based upon empirical rules, they created a different, a dogmatic system of astrology.

Babylonian astrology became a second celestial religion; but it was quite unlike the first, that of ancient China. In china the stars became gods; in Babylon the gods became stars. To the Chinese the mysteries of the cosmos were so sublime that they degraded their traditional popular divinities to demons and created a cult of the stars without priests, myths, or dogmas. The Babylonians, on the other hand, placed their native divinities one after another in the heavens, and transferred the mythic traits of these divinities to the stars. Here was an amazing evolution: for the first and only time a civilized religion rendered the divine beings visible and calculable by identifying them with the seven wandering stars.

The cuneiform script itself expressed that impulse, for its sign of divinity was a star. An age-old Babylonian legend related that the lord of the Earth, Bel, appointed the three gods Shamash, Sin, and Ishtar guardians of the firmament, which they thereafter patrolled as the Sun, the Moon, and Venus. When four more wandering stars were found in the firmament, the Babylonians made bold to repeat the act of Bel. The city-god of Babylon, Marduk, became the planet Jupiter; the god of death, Nergal, became the planet Mars; the god of war, Ninurta, became Saturn; and the god of knowledge, Nabu, became Mercury. Mars was called Star of Judgment upon the Dead. The Tower of Babel, which was simultaneously a sanctuary and observatory, was called tersely the Temple of the Seven Transmitters of Commands from Heaven to Earth.

Thus in Babylon divine worship became equivalent to astronomy, astronomy equivalent to searching out the will of the gods. In observing the movements of the planets and their relationships to one another, the priest was performing the highest rite of his religion. That fact accounted for the enormous power of the priestly class. They did not pray to invisible beings; they associated with them face to face, so to speak. Out of this belief sprang the extraordinarily frank reports of the astronomer-priests to the king, and the king's meek queries as to whether he might end his fast. Moreover, these priests prophesied with a sense of absolute certainty—for *they* did not read the future from deceptive signs; they read it in the eyes of the gods themselves.

The light rays sent forth from the planets were magical glances by which the gods guided activities on Earth. Their influence was predetermined beyond the possibility of doubt by the special traits of the particular god. The system of interpreting the stars no longer depended upon observation and empirical rules amassed in the course of centuries; interpretation flowed simply and directly from the very names of the planets. Each planet's name carried with it the entire body of legend which the old religion had attached to the god.

The transposition of divine characteristics to the wandering planets is the essence of astrology. That is why the position of the planets was held to influence the destiny of individuals. If a child were born while the Moon was rising, his life would be resplendent, long, and happy. If a child were born while Mars was rising, he would be sickly and soon die. If two planets wielded their influence simultaneously, the rising planet operated with greater force. Thus, if Jupiter were rising and Venus setting at birth, a man would have luck in later life, but would abandon his wife. If Venus were rising and Jupiter setting, the man would be ruled by his wife.

Thus the astrological rule was early established that opposition of the planets mutually weakened their influence. Contrarily, in conjunction they strengthened one another, with the higher of the two planets being the stronger. Soon further refinements were added: angles of 60 and 120 degrees, triangles, and hexagons, were considered favorable, squares unfavorable. These various "aspects," multiplied by seven, yielded a complicated doctrinal system which laid the groundwork for a comprehensive craft of prophecy. The aspect of the planets could be consulted for any event in life, for births and the founding of cities, business negotiations, travels, political treaties, battles, harvests, sickness.

The Greeks had the same mystical regard for the heavens as the Babylonians and Mayans did, but they went one step further by imposing a keen sense of rational form and motion on what had gone before. For early Greek astronomers, the only proper motion for the celestial bodies was uniform motion about that most perfect of shapes, the circle. And clearly, for most Greeks, the earth had to be the center of it all. But because of two sticky problems, discussed in the following reading, their early models failed.

The Problem of the Planets

132 *Thomas S. Kuhn*

Unlike the moon and sun, the remaining five planets appear as mere points of light in the heavens. The untrained naked-eye observer can distinguish them from stars with assurance only by a series of observations that discloses their gradual motion around the ecliptic. Usually the planets move eastward through the constellations: this is their so-called "normal motion." On the average, both Mercury and Venus require 1 year for each complete circuit of the zodiac; the length of Mars's cycle averages 687 days; Jupiter's average period is 12 years; and Saturn's is 29 years. But in all cases the time required for any single journey may be quite different from the average period. Even when moving eastward through the stars, a planet does not continue at a uniform rate.

Nor is its motion uniformly eastward. The normal motion of all planets except the sun and moon is occasionally interrupted by brief intervals of westward or "retrograde" motion. . . .Mercury briefly reverses its motion through the stars once every 116 days, and Venus retrogresses every 584 days. Mars, Jupiter, and Saturn show retrograde motion every 780, 399, and 378 days, respectively.

In their gradual eastward motions interrupted by periodic westward retrogressions, the five wandering stars behave quite similarly. But there is an additional characteristic of their motion which divides them into two groups; this is the correlation between their position and the sun's. Mercury and Venus, the two so-called inferior planets, never get very far from the sun. Mercury is always found within 28° of the sun's moving disk, and Venus's maximum "elongation" is 45°. Both planets move in a continuous slow shuttle, back and forth across the moving sun; for a time they move eastward with the sun, then retrogress across its disk, and finally reverse themselves to

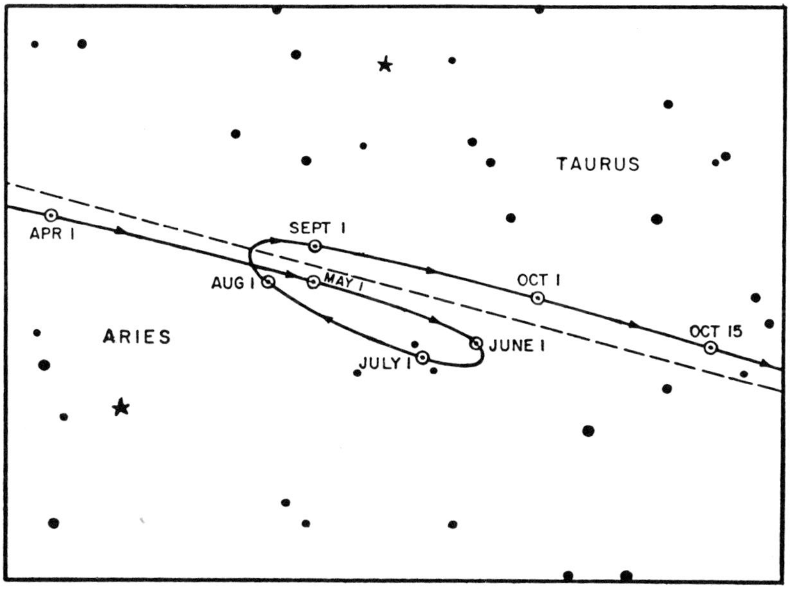

Mars retrogressing in Aries and Taurus. . . . The broken line is the ecliptic and the solid line the path of the planet. Note that Mars does not stay on the ecliptic and that, though its over-all motion is eastward among the stars, there is a period from the middle of June to early August during which it moves to the west. The retrogressions of Mars are always of approximately this form and duration, but they do not always occur on the same date or in the same part of the sky.

overtake the sun once more. When to the east of the sun, either of these inferior planets appears as an "evening star," becoming visible shortly after sunset and then rapidly following the sun below the horizon. After retrogressing westward across the sun's disk, the planet becomes a "morning star," rising shortly before dawn and disappearing in the brilliant light of sunrise. But in between, when close to the sun, neither Mercury nor Venus can be seen at all. Therefore, until their motion was analyzed with respect to the sphere of the stars, neither of the inferior planets was recognized as the same celestial body when it appeared as a morning and as an evening star. For millenniums Venus had one name when it rose in the east shortly before dawn and another when, weeks later, it again became visible just over the western horizon shortly after sunset.

Unlike Mercury and Venus, the superior planets, Mars, Jupiter, and Saturn, are not restricted to the same part of the sky as the sun. Sometimes they are very close to or "in conjunction" with it; at other times they are 180° across the sky or "in opposition" to the sun; between these times they assume all the intermediate positions. But though their positions are unrestricted, their behavior does depend upon their relation to the sun. Superior planets retrogress only when they are in opposition. Also, when in retrograde motion across the sky from the sun, superior planets appear brighter than at any other time. This increased brilliance, which has usually been interpreted (at least since the fourth century B.C.) as indicating a decrease in the planet's distance from the earth, is particularly striking in the case of Mars. Normally a relatively inconspicuous planet, Mars in opposition will frequently outshine every celestial body in the night sky except the moon and Venus. . . .

The problem of the planets is partially specified by the description of the planetary motions sketched in the preceding pages. How are the complex and variable planetary motions to be reduced to a simple and recurrent order? Why do the planets retrogress, and how account for the irregular rate of even their normal motions? These questions indicate the direction of most astronomical research during the two millenniums from the time of Plato to the time of Copernicus. But because it is almost entirely qualitative, the preceding description of the planets does not specify the problem fully. It states a simplified problem and in some respects a misleading one. . . . Qualitatively adequate planetary theories are easily invented: the description above can be reduced to order in several ways. The astronomer's problem, on the other hand, is by no means simple. He must explain not merely the existence of an intermittent westward motion superimposed upon an over-all eastward motion through the stars, but also the precise position that each planet occupies among the stars on different days, months, and years over a long period of time. The real problem of the planets, the one that leads at last to the Copernican Revolution, is the quantitative problem described in lengthy tables which specify in degrees and minutes of arc the varying position of every planet.

It was Ptolemy, working at the Museum of Alexandria, who fully elaborated the Grecian concept that the stars and planets rotated around each other in a kind of "wheels within wheels" array. In Ptolemy's system, the earth was the center of all; he offset the centers of circles from the earth (eccentrics), created uniform motion about points not at the center of their circles (equants), and had small circles riding smaller circles. According to Ptolemy, Mars revolved on an epicycle as the center of the epicycle moved around a large circle (deferent). The combination of motions made the planet reverse direction whenever it came closest to Earth. Jupiter and Saturn he treated similarly and he mounted Venus and Mercury on epicycles whose centers were carried around the earth by the sun. The Ptolemaic system is discussed further here by Gerald Holton and Duane Roller.

Ptolemy's System of Planetary Motion

Gerald Holton and Duane Roller

By adjusting the respective axes, directions of motions, rates and radii of rotations, the number of epicycles, eccentrics, and equants, by fitting his devices to the observed paths in a cut-and-try fashion, Ptolemy assembled the system that was still to prove useful to astronomers and navigators more than 14 centuries later. It was an answer to Plato's original question, and represents a magnificent piece of work carried out by a man of tremendous ability. One striking feature is that the centers of the epicycles of both Mercury and Venus fell on a line connecting the earth and the sun.

But the complexity of the whole scheme was considerable, and subsequent

From Gerald Holton and Duane H.D. Roller, *Foundations of Modern Physical Science*, Addison-Wesley, 1958. Reprinted by permission of the publisher.

observations required modification of some features of the model from time to time (for example, a change of an assigned equant, or the addition of an epicycle having its center on an epicycle, and so on). By the time of Copernicus, in the early part of the 16th century, this geocentric system required

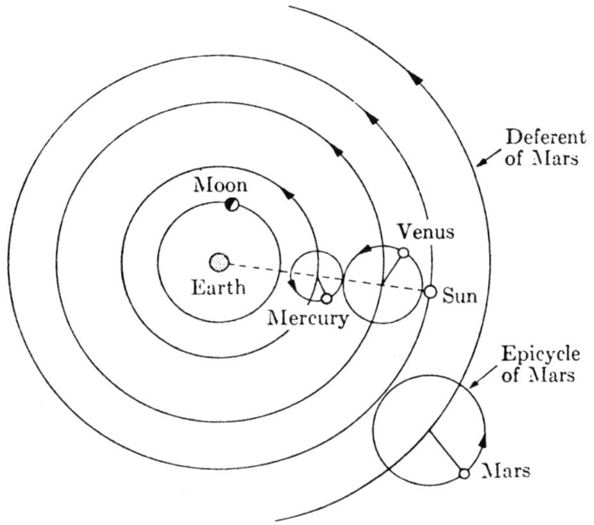

Partial and schematic diagram of the Ptolemaic system of planetary motion.

more than 70 simultaneous motions for the seven celestial bodies. Nevertheless, the system was generally accepted wherever it became known, for five powerful reasons:

(1) The Ptolemaic system gave an accurate enough description of what could be observed with the instruments of the times.
(2) It served well for predicting future positions of the celestial bodies, although only after cumbersome calculations; and when a serious discrepancy between predictions and observations occurred, it was often resolved by tampering a little with the "wheels" of the flexible apparatus.
(3) It provided a natural explanation of why the fixed stars showed no annual parallax.
(4) In most respects it was in accord with Greek philosophic and physical doctrine concerning the nature of the earth and the celestial bodies. Later, when reintroduced to Europe by the Arabs, the Ptolemaic system was given theological significance by medieval churchmen.
(5) It had and still has "common-sense appeal." It is difficult not to feel that we actually can "see" the sun and stars moving around us, and it is both comforting and reasonable to think of ourselves on a stable, immovable earth. Indeed, for many practical purposes, an earth-centered picture is the simplest one, and it is still preferred for calculations in navigation.

Long after Ptolemy's death, his theory of planetary motion was thought to be the last word. It was Copernicus, a Polish astronomer, who in the sixteenth century reversed the field of thought that had dominated man's image of the universe for more than twelve centuries. No longer was the earth the proud center of the universe. Now with the Copernican revolution, the sun acquired that honor and man reached another peak in his intellectual and scientific ascent. The following selection explains the Copernican theory and some of the objections to it.

The Copernican Revolution

Robert T. Lagemann

Nicholas Copernicus . . . was born in Poland. He studied astronomy at the University of Cracow, the classics, astronomy, theology, law, and medicine at the Italian universities of Bologna, Rome, and Padua. In all he spent nine years in Italy, returning occasionally to Poland to visit his uncle, a bishop. In Rome Copernicus took holy orders, and in 1506, with the help and influence of his uncle, he was made a canon of the cathedral at Frauenberg in his own country. There he spent the remainder of his life. He was not a one-sided man. He reported on the coinage system for the government, was physician to his bishop uncle, successfully directed the defense of a besieged castle, and translated Greek works into Latin.

In Italy he had met leading astronomers and had studied both Ptolemy's *Almagest* and Aristotle's thoughts on astronomy. It was clear to him that Ptolemy and Aristotle were sometimes in contradiction, that astronomers were in doubt about the exact length of the year, and that the calendar was in need of reform. He returned to his native land full of dissatisfaction with the state of astronomy. He at once began to formulate his new theory of planetary motion. He worked, he says, for "four times nine years" on the new approach that culminated in the epoch-making treatise *De Revolutionibus* (the Revolutions).

Copernicus was appalled at the inaccuracies of the Ptolemaic scheme and by the diversity of ways in which astronomers were explaining the planetary motions. In particular he objected to the equant, since it did not allow of uniform circular motion of the planets. The system, he wrote, was "not sufficiently pleasing to the mind," an esthetic value judgment that influenced many scientists before and after him.

Although Copernicus clung to the prevailing belief in uniform motion about a center, he brought forward a new point of view. By calculations and arguments unavailable to the early Greeks, he showed that the motion of the celestial objects could be represented by uniform circular motions about the sun rather than the earth. His plan . . . can be summarized as follows:

1. *The sun is assumed to be at rest* (in Copernicus' words: "In the midst of all stands the sun"). *The stars are at rest also. The earth rotates on its axis once each day and revolves about the sun once a year.* He thus

accounts for the apparent diurnal motion of the stars and sun, and opens the way to an explanation of the seasons.

2. *The earth is spherical.* As evidence he cites the change in elevation of certain stars as an observer travels along a N–S meridian, the apparent sinking from sight as a ship leaves the land, the circular arc of the earth's shadow during an eclipse of the moon, and the variation of the time of an eclipse with the position of the observer.
3. *The planets revolve about the sun in planes that are very nearly coincident.* They move in the same direction along their paths; a simpler scheme than that of Ptolemy.
4. *The periods of revolution decrease with the orbits' radii* as one passes from the outer to the inner planets. Saturn accomplishes its revolution in 30 years, he says; then comes Jupiter (12 years); Mars (2 years); the earth accompanied by the moon (1 year); Venus (9 months); and Mercury (80 days).
5. *The planets revolve about the sun* in paths that are combinations of deferents and epicycles.

Copernicus recognized the need to anticipate possible objections to his system and to present his arguments on the same pages as his plan. In his preface he states: ". . . the thought of the scorn which I had to fear on account of the novelty and incongruity of my theory, well-nigh induced me to abandon my project." At the urging of his friends he finally decided to publish his work. The early pages carried a dedication to Pope Paul III that must have made the book more acceptable to the reader of that age. Copernicus was a devout and prominent cleric; he had no thought of using his book to upset the orthodoxy of his day. He felt that his emphasis on simplicity and order was in fact a tribute to the wisdom of God.

To make his system more plausible to the conservative mind, Copernicus used many arguments that appealed to the Scholastics. Most of his defense was qualitative, but he did present enough mathematical discussion to compete with Ptolemy. In this respect he did much better than Aristarchus and others who had put forward a sun-centered system. . . .

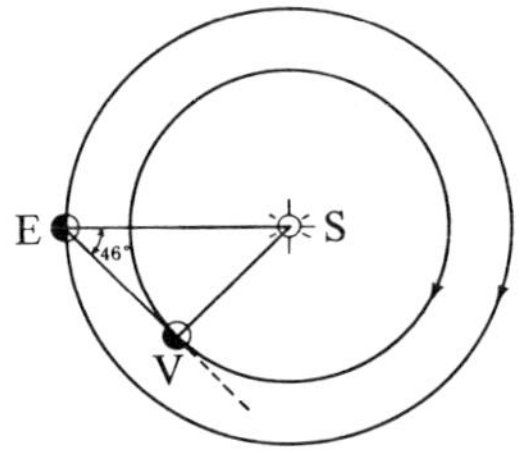

The angle SEV *is about 46° when the planet Venus reaches its maximum angular separation from the sun.*

. . . Suppose one observes the planet Venus, one of the two with orbits between the earth and the sun. The planet will appear to stay near the sun and to oscillate back and forth through an arc whose maximum value can be measured, although many months of observation will be required. Thus Venus . . . , at one stage in its orbit, is not seen below the position *V*.

Well aware of the retrograde motion of Saturn, Jupiter, Mars, Venus, and Mercury, Copernicus compared the relative frequency of the phenomenon for the planets. The cause was to be found in the earth's motion. He did not explain further, perhaps thinking it unnecessary, but no doubt his explanation would have followed the one we use today. . . . The earth requires one year to make a circuit around the sun, whereas Mars consumes 687 sidereal days. The line of sight from the earth to Mars changes direction from day to day and at times leads the observer to note retrograde motion. When the earth and Mars are on the same side of the sun and in positions such as 1, 2, 3, and 4, Mars appears to be moving in the same direction night

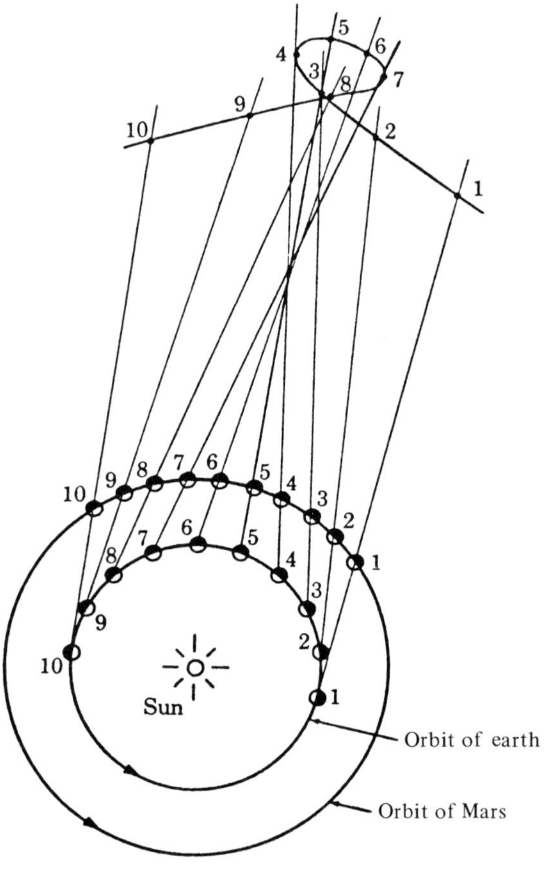

Explanation of the retrograde motion of Mars.

after night. A point is reached, however, illustrated by positions 5, 6, and 7 [in the figure], when it seems to move backward again, in retrograde motion, *i.e.*, moving from west to east. When Mars and the earth are on opposite sides of the sun, the motion of Mars appears to be direct.

Objections to the System

From the arguments advanced by Copernicus himself, it is possible to determine the criticisms he expected, which no doubt were brought forward after his death. To the objection that clouds would be left behind as the earth turned on its axis, he submitted that the water vapor and the atmosphere move with the earth. The question: Why does the earth not fly apart if it rotates? he answered with another: Why does not the larger and more rapidly moving celestial sphere of Ptolemy fly apart? A prominent objection centered on the lack of stellar parallax. As we have said, Copernicus gave the tremendous distances between the stars and the earth as the reason. Today we accept this explanation, but to the men of his day it was absurd. Their religious beliefs placed Hell inside the earth, and, as for Heaven, it could not be as far away as the new theory seemed to demand. The prevalent belief is illustrated by Dante's great poem, "The Divine Comedy." In it Hell is a conical cavity reaching to the center of the earth. Around the sloping sides are placed the places of punishment, and the worst sinners are placed nearest the very center of the earth, where Lucifer dwells. Heaven on the other hand is pictured as distant from the earth, reached by rising through the celestial spheres, each associated with one of the planets, until one comes to the sphere of the fixed stars, and then to the Primum Mobile (the prime source of motion), and, finally, to the tenth sphere, the dwelling of the Deity. This conception is not Ptolemaic but Aristotelian. Surely, it was thought, the stars and Heaven could not be as far away as Copernicus would place them.

An associated argument was that the earth must occupy the central position in God's world. It was inconceivable to most men that God created man only to place him on an earth that moved about some other object, which thereby became more important. For this belief European man could find Biblical support. The leaders of the Protestant Reformation branded the teachings of Copernicus as false. Martin Luther himself said, "The fool will upset the whole science of astronomy, ..." Certain Jewish groups forbade the teaching of the sun-centered system. And the Roman Catholic Church, although it had earlier approved of the work of Copernicus, placed the *Revolutions* on its list of prohibited books, the Index, from which it was not removed until 1822.

The church had little use for Copernicus and his heliocentric theory. Of the poets and dreamers who had expended thousands of images on our heretofore central position in the universe, some were nonplused and others were confused by it all. The poet John Donne wrote, somewhat dejectedly:

And new Philosophy calls all in doubt,
The element of fire is quite put out,
The sun is lost, and th'earth, and no man's wit
Can well direct him where to look for it.

And then came Galileo Galilei, the man who would carry news of the Copernican revolution to a vaster and even more disapproving audience, only to fall victim to the Inquisition. What extended Galileo's view and ultimately, although indirectly, put him within reach of the inquisitional thumbscrew was the telescope. What follows is Galileo's own description of his discovery of the craters of the moon and the apparently erratic behavior of the planet Jupiter.

Galileo and the Starry Messenger

Galileo Galilei (translated by Stillman Drake)

Great indeed are the things which in this brief treatise I propose for observation and consideration by all students of nature. I say great, because of the excellence of the subject itself, the entirely unexpected and novel character of these things, and finally because of the instrument by means of which they have been revealed to our senses.

Surely it is a great thing to increase the numerous host of fixed stars previously visible to the unaided vision, adding countless more which have never before been seen, exposing these plainly to the eye in numbers ten times exceeding the old and familiar stars.

It is a very beautiful thing, and most gratifying to the sight, to behold the body of the moon, distant from us almost sixty earthly radii, as if it were no farther away than two such measures—so that its diameter appears almost thirty times larger, its surface nearly nine hundred times, and its volume twenty-seven thousand times as large as when viewed with the naked eye. In this way one may learn with all the certainty of sense evidence that the moon is not robed in a smooth and polished surface but is in fact rough and uneven, covered everywhere, just like the earth's surface, with hugh prominences, deep valleys, and chasms.

Again, it seems to me a matter of no small importance to have ended the dispute about the Milky Way by making its nature manifest to the very senses as well as to the intellect. . . .

About ten months ago a report reached my ears that a certain Fleming had constructed a spyglass by means of which visible objects, though very distant from the eye of the observer, were distinctly seen as if nearby. Of this truly remarkable effect several experiences were related, to which some persons gave credence while others denied them. A few days later the report was confirmed to me in a letter from a noble Frenchman at Paris, Jacques Badovere, which caused me to apply myself wholeheartedly to inquire into the means by which I might arrive at the invention of a similar instrument. This I did shortly afterwards, my basis being the theory of refraction. First I prepared a tube of lead, at the ends of which I fitted two glass lenses, both plane on one side while on the other side one was spherically convex and the other concave. Then placing my eye near the concave lens I perceived objects satisfactorily large and near, for they appeared three times closer and nine times larger than when seen with the naked eye alone. Next I constructed another one, more accurate, which represented objects as enlarged more than sixty times. Finally, sparing neither labor nor expense, I succeeded in constructing for myself so excellent an instrument that objects seen by means of it appeared nearly one thousand times larger and over thirty times closer than when regarded with our natural vision.

It would be superfluous to enumerate the number and importance of the advantages of such an instrument at sea as well as on land. But forsaking terrestrial observations, I turned to celestial ones, and first I saw the moon from as near at hand as if it were scarcely two terrestrial radii away. After that I observed often with wondering delight both the planets and the fixed stars, and since I saw these latter to be very crowded, I began to seek (and eventually found) a method by which I might measure their distances apart.

Here it is appropriate to convey certain cautions to all who intend to undertake observations of this sort, for in the first place it is necessary to prepare quite a perfect telescope, which will show all objects bright, distinct, and free from any haziness, while magnifying them at least four hundred times and thus showing them twenty times closer. Unless the instrument is of this kind it will be vain to attempt to observe all the things which I have seen in the heavens, and which will presently be set forth. Now in order to determine without much trouble the magnifying power of an instrument, trace on paper the contour of two circles or two squares of which one is four hundred times as large as the other, as it will be when the diameter of one is twenty times that of the other. Then, with both these figures attached to the same wall, observe them simultaneously from a distance, looking at the smaller one through the telescope and at the larger one with the other eye unaided. This may be done without inconvenience while holding both eyes open at the same time; the two figures will appear to be of the same size if the instrument magnifies objects in the desired proportion.

Such an instrument having been prepared, we seek a method of measuring distances apart. This we shall accomplish by the following contrivance.

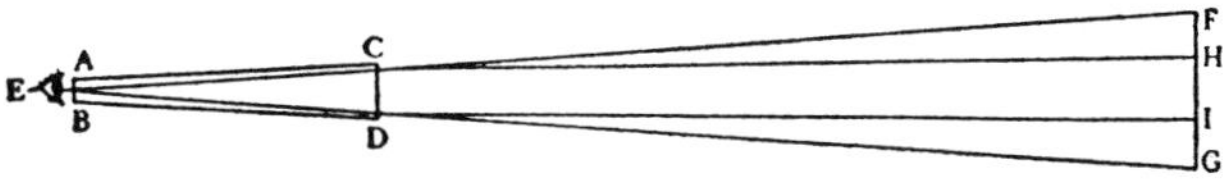

Let ABCD be the tube and E be the eye of the observer. Then if there were no lenses in the tube, the rays would reach the object FG along the straight lines ECF and EDG. But when the lenses have been inserted, the rays go along the refracted lines ECH and EDI; thus they are brought closer together, and those which were previously directed freely to the object FG now include only the portion of it HI. The ratio of the distance EH to the line HI then being found, one may by means of a table of sines determine the size of the angle formed at the eye by the object HI, which we shall find to be but a few minutes of arc. Now, if to the lens CD we fit thin plates, some pierced with larger and some with smaller apertures, putting now one plate and now another over the lens as required, we may form at pleasure different angles subtending more or fewer minutes of arc, and by this means we may easily measure the intervals between stars which are but a few minutes apart, with no greater error than one or two minutes. And for the present let it suffice that we have touched lightly on these matters and scarcely more than mentioned them, as on some other occasion we shall explain the entire theory of this instrument.

Now let us review the observations made during the past two months, once more inviting the attention of all who are eager for true philosophy to the first steps of such important contemplations. Let us speak first of that surface of the moon which faces us. For greater clarity I distinguish two parts of this surface, a lighter and a darker; the lighter part seems to surround and to pervade the whole hemisphere, while the darker part discolors the moon's surface like a kind of cloud, and makes it appear covered with spots. Now those spots which are fairly dark and rather large are plain to everyone and have been seen throughout the ages; these I shall call the "large" or "ancient" spots, distinguishing them from others that are smaller in size but so numerous as to occur all over the lunar surface, and especially the lighter part. The latter spots had never been seen by anyone before me. From observations of these spots repeated many times I have been led to the opinion and conviction that the surface of the moon is not smooth, uniform, and precisely spherical as a great number of philosophers believe it (and the other heavenly bodies) to be, but is uneven, rough, and full of cavities and prominences, being not unlike the face of the earth, relieved by chains of mountains and deep valleys. . . .

142 On the seventh day of January in this present year 1610, at the first hour of night, when I was viewing the heavenly bodies with a telescope, Jupiter presented itself to me; and because I had prepared a very excellent instrument for myself, I perceived (as I had not before, on account of the weakness of my previous instrument) that beside the planet there were three starlets, small indeed, but very bright. Though I believed them to be among the host of fixed stars, they aroused my curiosity somewhat by appearing to lie in an exact straight line parallel to the ecliptic, and by their being more splendid than others of their size. Their arrangement with respect to Jupiter and each other was the following:

East * * O * *West*

that is, there were two stars on the eastern side and one to the west. The most easterly star and the western one appeared larger than the other. I paid no attention to the distances between them and Jupiter, for at the outset I thought them to be fixed stars, as I have said. But returning to the same investigation on January eighth—led by what, I do not know—I found a very different arrangement. The three starlets were now all to the west of Jupiter, closer together, and at equal intervals from one another as shown in the following sketch:

East O * * * *West*

At this time, though I did not yet turn my attention to the way the stars had come together, I began to concern myself with the question how Jupiter could be east of all these stars when on the previous day it had been west of two of them. I commenced to wonder whether Jupiter was not moving eastward at that time, contrary to the computations of the astronomers, and had got in front of them by that motion. Hence it was with great interest that I awaited the next night. But I was disappointed in my hopes, for the sky was then covered with clouds everywhere.

On the tenth of January, however, the stars appeared in this position with respect to Jupiter:

East * * O *West*

that is, there were but two of them, both easterly, the third (as I supposed) being hidden behind Jupiter. As at first, they were in the same straight line with Jupiter and were arranged precisely in the line of the zodiac. Noticing this, and knowing that there was no way in which such alterations could be attributed to Jupiter's motion, yet being certain that these were still the same stars I had observed (in fact no other was to be found along the line of the zodiac for a long way on either side of Jupiter), my perplexity was now transformed into amazement. I was sure that the apparent changes belonged not to Jupiter but to the observed stars, and I resolved to pursue this investigation with greater care and attention.

And thus, on the eleventh of January, I saw the following disposition:

East * * O *West*

There were two stars, both to the east, the central one being three times as far from Jupiter as from the one farther east. The latter star was nearly double the size of the former, whereas on the night before they had appeared approximately equal.

I had now decided beyond all question that there existed in the heavens three stars wandering about Jupiter as do Venus and Mercury about the sun, and this became plainer than daylight from observations on similar occasions which followed. Nor were there just three such stars; four wanderers complete their revolutions about Jupiter, and of their alterations as observed more precisely later on we shall give a description here. Also I measured the distances between them by means of the telescope, using the method explained before. Moreover I recorded the times of the observations, especially when more than one was made during the same night—for the revolutions of these planets are so speedily completed that it is usually possible to take even their hourly variations.

Thus on the twelfth of January at the first hour of night I saw the stars arranged in this way:

East * * O * *West*

The most easterly star was larger than the western one, though both were easily visible and quite bright. Each was about two minutes of arc distant from Jupiter. The third star was invisible at first, but commenced to appear after two hours; it almost touched Jupiter on the east, and was quite small. All were on the same straight line directed along the ecliptic.

On the thirteenth of January four stars were seen by me for the first time, in this situation relative to Jupiter:

Three were westerly and one was to the east; they formed a straight line except that the middle western star departed slightly toward the north. The eastern star was two minutes of arc away from Jupiter, and the intervals of the rest from one another and from Jupiter were about one minute. All the stars appeared to be of the same magnitude, and though small were very bright, much brighter than fixed stars of the same size. . . .

On the twenty-sixth of February, midway in the first hour of night, there were only two stars:

One was to the east, ten minutes from Jupiter; the other to the west, six minutes away. The eastern one was somewhat smaller than the western. But at the fifth hour three stars were seen:

In addition to the two already noticed, a third was discovered to the west near Jupiter; it had at first been hidden behind Jupiter and was now one minute away. The eastern one appeared farther away than before, being eleven minutes from Jupiter.

This night for the first time I wanted to observe the progress of Jupiter and its accompanying planets along the line of the zodiac in relation to some fixed star, and such a star was seen to the east, eleven minutes distant from the easterly starlet and a little removed toward the south, in the following manner:

On the twenty-seventh of February, four minutes after the first hour, the stars appeared in this configuration:

The most easterly was ten minutes from Jupiter; the next, thirty seconds; the next to the west was two minutes thirty seconds from Jupiter, and the most westerly was one minute from that. Those nearest Jupiter appeared very small, while the end ones were plainly visible, especially the westernmost. They marked out an exactly straight line along the course of the ecliptic. The progress of these planets toward the east is seen quite clearly by reference to the fixed star mentioned, since Jupiter and its accompanying planets were closer to it, as may be seen in the figure above. At the fifth hour, the eastern star closer to Jupiter was one minute away.

At the first hour on February twenty-eighth, two stars only were seen; one easterly, distant nine minutes from Jupiter, and one to the west, two minutes away. They were easily visible and on the same straight line. The fixed star, perpendicular to this line, now fell under the eastern planet as in this figure:

At the fifth hour a third star, two minutes east of Jupiter, was seen in this position:

On the first of March, forty minutes after sunset, four stars all to the east were seen, of which the nearest to Jupiter was two minutes away, the next was one minute from this, the third two seconds from that and brighter than any of the others; from this in turn the most easterly was four minutes distant, and it was smaller than the rest. They marked out almost a straight line, but the third one counting from Jupiter was a little to the north. The fixed star formed an equilateral triangle with Jupiter and the most easterly star, as in this figure:

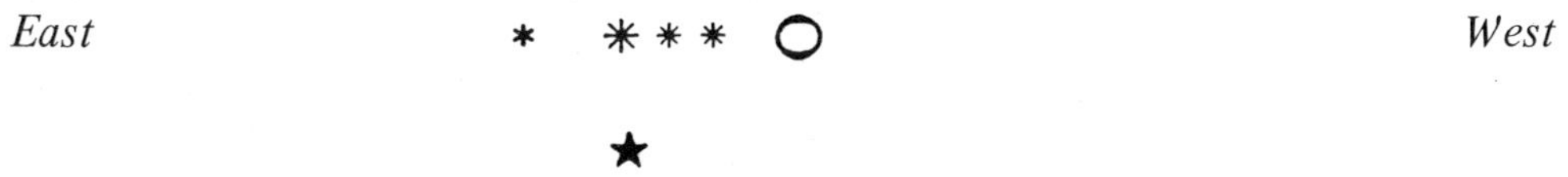

On March second, half an hour after sunset, there were three planets, two to the east and one to the west, in this configuration:

The most easterly was seven minutes from Jupiter and thirty seconds from its neighbor; the western one was two minutes away from Jupiter. The end stars were very bright and were larger than that in the middle, which appeared very small. The most easterly star appeared a little elevated toward the north from the straight line through the other planets and Jupiter. The fixed star previously mentioned was eight minutes from the western planet along the line drawn from it perpendicularly to the straight line through all the planets, as shown above.

I have reported these relations of Jupiter and its companions with the fixed star so that anyone may comprehend that the progress of those planets, both in longitude and latitude, agrees exactly with the movements derived from planetary tables.

7

The Majestic Clockwork

Nature and Nature's laws
Lay hid in Night!
God said, Let Newton be!
And all was Light.

Alexander Pope

Isaac Newton and Albert Einstein – two men who, Dr. Bronowski says, "stride like gods." Newton, an in-turned absolutist, was jealous, vain, and suspicious. Einstein was a pacifist who held an open-minded concern for his fellow human beings. Between them they created the hands of the majestic clockwork, riding the same beam of light in new and revolutionary directions and touching almost every field. Ptolemy saw the movement of the sun and planets as wheels within wheels; Copernicus knocked the earth out of the center of our corner of the universe, and Galileo spread the Polish astronomer's starry message more widely through the world. Then Johannes Kepler, a contemporary of Galileo, "devised the three laws that turned the system of Copernicus from a general description of the sun and the planets into a precise, mathematical formula." In doing so, he had the help of one Tycho Brahe, a Danish astronomer, who was distinguished by the fact that he wore a silver nose. Their partnership is detailed first in the reading from Arthur Koestler's *The Sleepwalkers* and again in the selection by Robert T. Lagemann.

Tycho Meets Kepler

Arthur Koestler

The town and castle of Benatek were twenty-two miles, a six-hours' journey, to the north-east of Prague. They overlooked the river Iser, which often flooded the surrounding orchards, hence the name "Bohemian Venice". Tycho had selected Benatek among the three castles which the Emperor had offered for his choice—perhaps because the watery surroundings reminded him of Hveen. He had taken possession of the castle in August 1599—six months before Kepler's arrival—and had started at once tearing down walls and erecting new ones, intending to build another Uraniburg, and announcing his intention in high-flung poems which were inscribed over the entrance of the future observatory. There was also to be a separate gate for the Emperor, who had reserved an adjoining building for his visits.

But everything seemed to go wrong from the beginning. The salary of three thousand florins which the Emperor had granted Tycho beat all previous records; "there was nobody at Court, not even among counts and barons of long service, who enjoyed so large an income". Both the mind of Rudolph II, and his finances, were in a highly disordered state; and his court officials effectively sabotaged the carrying out of his extravagant royal promises. Tycho had to fight for his salary and to be content if he could squeeze half of it out of the Exchequer; when Kepler succeeded him, he would get only a dribble of what was due to him.

By the time Kepler arrived at Benatek, Tycho had already quarrelled with the Director of the Crown Estates who held the purse-strings, complained to the Emperor, threatened to leave Bohemia, and to tell the world the reasons. Also, several of Tycho's assistants, who had promised to join him at the new Uraniburg, had failed to turn up; and the largest instruments were still delayed on the long trek from Hveen. Towards the end of the year the plague had broken out, obliging Tycho to sit it out with Rudolph at the imperial residence in Girsitz, and supply him with a secret elixir against epidemics. To add to Tycho's worries, Ursus, who had disappeared from Prague on Tycho's arrival, now returned again, trying to create trouble; and Tycho's second daughter, Elisabeth, was having an illicit love affair with one of his assistants, the Junker Tengnagel. Young Kepler, in the provincial backwoods of Gratz, had dreamed of Benatek as a serene temple of Urania; he arrived at a mad-house. The castle was teeming with workmen, surveyors, visitors, and the formidable de Brahe clan including the sinister dwarf Jepp—huddled under the table during the interminable, tumultuous meals, and finding an easy butt for his sarcasms in that timid scarecrow of a provincial mathematicus.

Kepler had arrived in Prague in the middle of January. He had at once written to Benatek, and a few days later received an answer from Tycho, regretting that he could not welcome Kepler in person because of a forthcoming opposition of Mars and Jupiter, to be followed by a lunar eclipse; and inviting him to Benatek "not so much as a guest, than as a very welcome friend and colleague in the contemplation of the skies". The bearers of the

letter were Tycho's eldest son and the Junker Tengnagel, both of whom were jealous of Kepler from the start, and remained hostile to the end. It was in their company that Kepler completed the last lap of his journey to Tycho—but only after a further delay of nine days. Tengnagel and Tycho Junior were probably having a good time in Prague, and were in no hurry to get back.

At last, then, on 4 February, 1600, Tycho de Brahe and Johannes Keplerus, co-founders of a new universe, met face to face, silver nose to scabby cheek. Tycho was fifty-three, Kepler twenty-nine. Tycho was an aristocrat, Kepler a plebeian; Tycho a Croesus, Kepler a church-mouse; Tycho a Great Dane, Kepler a mangy mongrel. They were opposites in every respect but one: the irritable, choleric disposition which they shared. The result was constant friction, flaring into heated quarrels, followed by half-hearted reconciliations.

But all this was on the surface. In appearance, it was a meeting of two crafty scholars, each determined to make use of the other for his own purposes. But under the surface, they both knew, with the certainty of sleepwalkers, that they were born to complete each other; that it was the gravity of fate which had drawn them together. Their relationship was to alternate all the time between these two levels: *qua* sleepwalkers, they strolled arm in arm through uncharted spaces; in their waking contacts they brought out the worst in the other's character, as if by mutual induction.

Kepler's arrival led to a reorganization of work at Benatek. Previously, Tycho's younger son, Joergen, had been in charge of the laboratory, the senior assistant, Longomontanus, was assigned the study of the orbit of Mars, and Tycho had intended to put Kepler in charge of the next planet to be taken up for systematic observation. But his eagerness, and the fact that Longomontanus got stuck with Mars, led to a re-distribution of the planetary realm among the Tychonites: Kepler was given Mars, the notoriously most difficult planet, while Longomontanus was switched to the moon. This decision proved of momentous importance. Kepler, proud to be entrusted with

Mars, boasted that he would solve the problem of its orbit in eight days, and even made a bet with this deadline. The eight days grew into nearly eight years; but out of the struggle of these years with the recalcitrant planet emerged Kepler's *New Astronomy or Physics of the Skies*.

He knew, of course, nothing of what lay ahead of him. He had come to Tycho to wrench from him the exact figures of the eccentricities and mean distances, in order to improve his model of the universe built around the five solids and the musical harmonies. But though he never discarded his *idée fixe*, it was now relegated into the background. The new problems which arose out of Tycho's data "took such a hold of me that I nearly went out of my mind". Himself no more than an amateur observer with the coarsest of instruments, an armchair astronomer with the intuition of genius but still lacking in intellectual discipline, he was overwhelmed by the wealth and precision of Tycho's observations, and only now began to realize what astronomy really meant. The hard facts embodied in Tycho's data, the scrupulousness of Tycho's method, acted like a grindstone on Kepler's fantasy-prone intellect. But although Tycho did the grinding, and the process seemed to be more painful for Kepler than for him, in the end it was the grindstone which was worn down, while the blade emerged sharp and shining from it.

Soon after his arrival in Benatek, Kepler wrote:

Tycho possesses the best observations, and thus so-to-speak the material for the building of the new edifice; he also has collaborators and everything he could wish for. He only lacks the architect who would put all this to use according to his own design. For although he has a happy disposition and real architectural skill, he is nevertheless obstructed in his progress by the multitude of the phenomena and by the fact that the truth is deeply hidden in them. Now old age is creeping upon him, enfeebling his spirit and his forces.

There could be no doubt regarding the identity of the architect in Kepler's mind. Nor was it difficult for Tycho to guess Kepler's true opinion of him. He had amassed a treasure of data as nobody before him; but he was old, and lacking the boldness of imagination to build, out of this wealth of raw material, the new model of the universe. Its laws were there, in his columns of figures; but "too deeply hidden" in them for him to decipher. He must also have felt that only Kepler was capable of succeeding in this task—and that nothing could prevent him from succeeding; that it would be this grotesque upstart, and not Tycho himself, nor the hoped-for Tychonides of the Uraniburg mural, who would reap the fruit of his lifelong labours. Half resigned to, half appalled by his own fate, he wanted at least to make it as difficult for Kepler as possible. He had always been most reluctant to disclose his treasured observations; if Kepler had thought he could simply grab them, he was woefully mistaken—as the indignant complaints in his letters show:

Tycho gave me no opportunity to share in his experiences. He would only, in the course of a meal, and in between conversing about other matters mention, as if in passing, today the figure for the apogee of one planet, tomorrow the nodes of another.

One might add: as if he were handing bones to Jepp under the table. Nor would he allow Kepler to copy out his figures. In exasperation, Kepler even asked Tycho's Italian rival, Magini, to offer his own data in exchange for some of Tycho's. Only gradually, step by step, did Tycho yield; and when he put Kepler in charge of Mars, he was forced to disgorge his Mars data.

Kepler had spent barely a month at Benatek when Tycho, in a letter, first hinted at difficulties that had arisen between them; another month later, on 5 April, the tension blew up in an explosion which might have shattered the future of cosmology.

The immediate cause of the row was a document which Kepler had drafted, and in which the conditions of his future collaboration with Tycho were laid down in unpleasant detail. If he and his family were to live permanently at Benatek, Tycho must provide them with a detached apartment because the noise and disorder of the household were having a terrible effect on Kepler's gall, and provoked him to violent outbursts of temper. Next, Tycho must obtain a salary for Kepler from the Emperor, and in the meantime pay him fifty florins a quarter. He must also provide the Keplers with specified quantities of firewood, meat, fish, beer, bread and wine. As for their collaboration, Tycho must leave Kepler his freedom to choose the time and subject of his work, and only ask him to undertake such researches which were directly connected with it; and since Kepler was "not in need of a spur but rather of a brake to prevent the threat of galloping consumption due to overwork", he must be allowed to rest in the day-time if he had worked deep into the night. And so on, for several pages.

This document was not meant for Tycho's perusal. Kepler handed it to a guest, a certain Jessenius, professor of medicine in Wittenberg, who was to serve as an intermediary in the negotiations between Tycho and himself. But whether by chance or intrigue, Tycho got hold of the document, which he could hardly regard as flattering to himself. Nevertheless he took it with that good-humoured magnanimity which lived side by side in the Danish *grand seigneur* with jealousy and bullying. He remained a benevolent despot so long as nobody challenged his rule; and socially, Kepler was so much his inferior that his carping and bickering demands did not affect Tycho as a challenge. One of the reasons for Kepler's bitterness was, incidentally, that he had been assigned an inferior position at the dinner table.

But above all, Tycho needed Kepler, who alone could put his life-work into proper shape. Hence he sat down to negotiate with Kepler in the presence of Jessenius, patiently rubbing ointment on his nose, a paragon of paternal moderation. This attitude grated even more on Kepler's inferiority complex, and he attacked Tycho, in the latter's words "with the vehemence of a mad dog, to which creature he, Kepler himself, so much likes to compare himself in irritability."

Immediately after the stormy session Tycho, who always had an eye on posterity, wrote down the minutes of it, and requested Jessenius to endorse them. However, when his temper had cooled down, he entreated Kepler to stay on, at least for another few days, until an answer arrived from the Emperor, whom Tycho had approached concerning Kepler's employment. But Kepler refused to listen, and on the next day departed in the company of Jessenius to Prague, where he took quarters with Baron Hoffman. Just before his departure, Kepler had another choleric outburst; at the moment of farewell, he was overcome with remorse and apologized; while Brahe whispered into Jessenius' ear that he should try to bring the *enfant terrible* back to reason. But as soon as they arrived in Prague, Kepler wrote another abusive letter to Tycho.

He must have been in a dreadful state of hysteria. He was suffering from one of his recurrent obscure fevers; his family was in faraway Gratz; the persecution of the Protestants in Styria, and the debacle at Benatek had made a shambles of his future; and the data on Mars remained inaccessibly in Tycho's hands. Within a week, the pendulum swung to the other extreme: Kepler wrote a letter of apologies to Tycho which sounds like the ravings of a masochist against his own guilty ego:

The criminal hand which, the other day, was quicker than the wind in inflicting injury, hardly knows how to set about it to make amends. What shall I mention first? My lack of self-control, which I can only remember with the greatest pain, or your benefactions, noblest Tycho, which can neither be enumerated nor valued according to merit? For two months you have most generously provided for my needs . . . you have extended to me every friendliness, you have allowed me to share in your most cherished possession. . . . Taken all in all, neither to your children, nor to your wife, nor to yourself did you devote yourself more than to me. . . . Therefore I think with the deepest dismay that God and the Holy Ghost delivered me to such an extent to my impetuous attacks and to my sick mind that instead of displaying moderation, I indulged during three weeks with closed eyes in sullen stubbornness against you and your family; that instead of thanking you, I displayed blind rage; that instead of showing you respect I displayed the greatest insolence against your person which by noble descent, prominent learning and great fame deserves all respect; that instead of sending you a friendly greeting I let myself be carried away by suspicion and insinuation when I was itching with bitterness . . . I never considered how cruelly I must have hurt you by this despicable behaviour. . . . I come to you as a postulant to ask, in the name of Divine pity, for your forgiveness of my terrible offences. What I have said or written against your person, your fame, your honour and your scientific rank. . . . I retract in all parts, and declare it voluntarily and freely as invalid, false and unsound. . . . I also promise sincerely that henceforth at whatever place I shall be I shall not only refrain from such foolish acts, words, deeds and writings, but I shall also never and in no way unjustly and deliberately offend you. . . . But since the ways of men are slippery, I ask you that whenever you notice in me any tendency towards such unwise manner of behaviour, to remind me of myself; you will find me willing. I also promise . . . to oblige you by all kinds of services and . . . thus to prove by my acts that my attitude towards your person is different, and al-

ways was different, from what one may conclude from the reckless condition of my heart and body during these last three weeks. I pray that God may help me to fulfil this promise.

I have quoted this letter at some length, because it reveals the tragic core of Kepler's personality. These turns of phrase do not seem to come from a scholar of repute, but from a tortured adolescent, begging to be forgiven by a father whom he hates and loves. . . . At the base of his iridescent, complex character, Kepler always remained a waif and stray.

But Tycho was no less dependent on Kepler than Kepler on Tycho. In their worldly contacts, Tycho was the old man of the tribe, Kepler the nagging, ill-mannered adolescent. But on that other level, the rules were reversed: Kepler was the magician from whom, Tycho hoped, would come the solution of his problems, the answer to his frustrations, the salvation from ultimate defeat; and however foolishly they both behaved, *qua* sleepwalkers, they both knew all this.

Therefore, three weeks after the row, Tycho turned up in Prague and drove Kepler back to Benatek in his coach—one can almost see Tyge's great fat arm in the leg-of-mutton sleeve, crushing in an affectionate embrace Kepler's skimpy bones.

On the 13th of October, 1601, Tycho was a guest at supper at Baron Rosenberg's table in Prague. Among the other guests was an Imperial Councillor, so it must have been an illustrious company; but since Tycho had been in the habit of entertaining royalty, and was accustomed to vast amounts of drink, it is difficult to understand why he was unable to cope with the predicament in which he found himself. Kepler has carefully recorded what happened in the Diary of Observations—a kind of logbook where all important events of the Brahe household were entered:

On October 13, Tycho Brahe, in the company of Master Minkowitz, had dinner at the illustrious Rosenberg's table, and held back his water beyond the demands of courtesy. When he drank more, he felt the tension in his bladder increase, but he put politeness before his health. When he got home, he was scarcely able to urinate.

At the beginning of his illness, the moon was in opposition to Saturn . . . [follows the horoscope of the day].

After five sleepless nights, he could still only pass his water with the greatest pain, and even so the passage was impeded. The insomnia continued, with internal fever gradually leading to delirium; and the food he ate, and from which he could not be kept, exacerbating the evil. On October 24, his delirium ceased for several hours; nature conquered and he expired peacefully among the consolations, prayers and tears of his people.

So from this date the series of celestial observations was interrupted, and his own observations of thirty-eight years have come to an end.

On his last night in his gentle delirium, he repeated over and again these words, like someone composing a poem:

Let me not seem to have lived in vain.

The Astronomical Partnership of Tycho and Kepler

Robert T. Lagemann

The Danish astronomer Tycho Brahe (1546–1601) developed instruments and techniques for accurate astronomical measurement and accumulated extensive data on planetary motions.

Born three years after the death of Copernicus, Tycho became interested in astronomy when at the age of 14 he saw an eclipse of the sun. For three years he studied law with the support of his wealthy uncle, at the same time secretly buying and studying books on astronomy. When Tycho was 19 his uncle died and made Tycho his heir. Free now to pursue his interest in the stars, Tycho went to Germany where he built a gigantic quadrant for the careful observation of the positions of celestial bodies. It was carefully constructed and firmly mounted on a solid foundation. Twenty men were required to carry it to its place and erect it. The observations he made with this instrument and his knowledge and skill as an observer brought him an invitation from Frederick II, king of Denmark, to establish an observatory in that country. Given an annual pension and funds for construction of instruments and buildings, Tycho, at the age of 30, built the most elaborate and most useful observatory achieved up to that time. Named Uraniborg, the observatory was carefully described by Tycho in one of his works and recent archeological excavations confirm his account. It was there that he observed for 21 years, gathering more accurate data than anyone before him, data that were to help spell the doom of the Ptolemaic system.

His magnificent observatory contained living quarters, storerooms, a printing shop, and a machine shop in which were built the instruments he invented. The observatory had circular porches upon which large observing instruments were mounted under sliding conical roofs. These instruments were noted both for their large size and for the skill with which they had been constructed. The chief observing instrument used by Tycho was the quadrant . . . a quarter-circle of wood or metal. Usually one of its arms was precisely horizontal and the other exactly vertical. When a sighting arm, or alidade, was pointed at a planet or star, the altitude of the object above the horizon was read off the 90° arc. The entire instrument was usually pivoted so that it could be rotated 360°, permitting the determination of yet another direction used to establish the position of the object. On one of the quadrants he was able to note the altitudes of celestial objects as they crossed the N–S meridian. His chief contribution to the construction of astronomical instruments was probably the special sight, a great improvement over the simple peephole formerly used. The sight consisted of two slits placed near the eye, and a front cylinder, exactly as wide as the separation of the slits, placed some distance away. . . . The observer sighted alternately through one slit and then the other as the alidade arm was moved, until he saw the star just touching each side of the cylinder. At that point a reading of the angle was taken. Whereas Ptolemy had measured the position of a star to within 30′ of arc, Tycho was able to do so to about 5′ of arc.

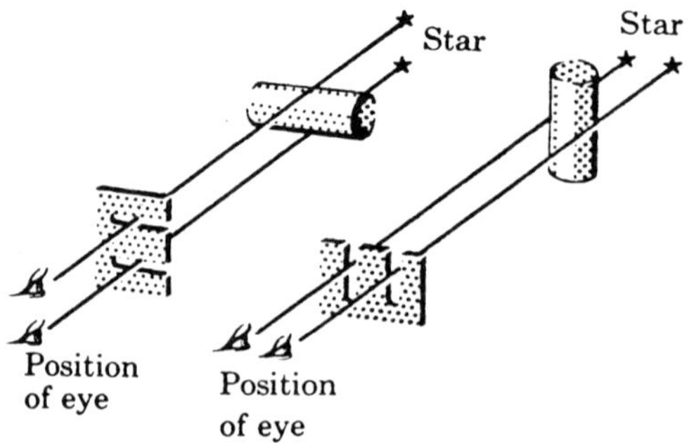

Tycho's sighting method used a cylinder as a front sight and two slits, with a separation equal to the diameter of the cylinder, as a rear sight. The arrangement at the left was for measuring the altitude of a celestial object; that to the right for measuring azimuth.

With quadrants Tycho plotted the motions of the planets visible to the unaided eye: Mercury, Venus, Mars, Jupiter, and Saturn. He established exact positions for 777 fixed stars, more than three-fourths of those visible with the naked eye at the location of his observatory. Both in Germany and in Denmark Tycho was an untiring observer for 33 years. He was convinced that the true scheme of the celestial universe would not be known until the exact positions of the stars and planets had been discovered.

At the death of his Danish patron, Tycho's support was withdrawn and he moved to Prague as imperial mathematician of the German Emperor Rudolph. Here Tycho took on as an assistant Johannes Kepler, who inherited his position and his extensive observational data when Tycho died about two years afterward.

Johannes Kepler

The young German formed a remarkable contrast to his mentor. Tycho was devoted to careful observations and measurements; Kepler's main concern was mathematics and speculation. Tycho for most of his life was wealthy; Kepler was frequently beset by financial worries. Their common, absorbing interest was the motion of Mars. It could not be accurately predicted by using the new data of Tycho, even though one used the Copernican system as Kepler did. Was there, asked Kepler, some path that described the motion of Mars, one of the most misbehaving of the planets, and that did not involve deferents and epicycles? His answer to that question resulted in the three important laws we are about to consider.

Kepler's First Law

Having Tycho's data at hand, Kepler attempted to choose such an orbit and such a law of speed, for both the earth and Mars, that the line connecting them would always point to the apparent position of Mars as seen from the earth. He tried all sorts of circular orbits for the earth and Mars, placing them in various positions with respect to the sun. Once, after four years of labor, using the concentric circles and the limited number of epicycles allowed by Copernicus, he came within 8 minutes of predicting the position of

Mars. But he knew that the careful work of Tycho would not have permitted that much experimental error, and he kept on.

All kinds of combinations of orbit form and relative positions of the earth and Mars had to be tried, and the relative position of each of the two planets had to be worked out for each combination and compared with Tycho's observations. It was a case of cut and try, or, as the philosopher of science would say, a case of adopting one hypothesis (conjecture) and testing it, and then another, and so on. He tried ovals. He tried circles. Finally, he discovered that he could fit an ellipse to the observations by placing the sun at one of the foci. . . . The other planets were also found to move in elliptical orbits with one common focus. The first law of Kepler may be stated: *Each of the planets moves about the sun in an orbit that forms an ellipse, with the sun at one of the foci.*

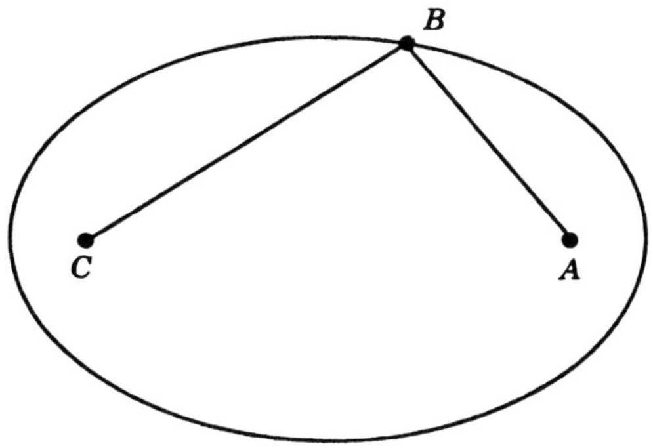

The ellipse—that curve for which the distance (AB + BC) *is a constant as point* B *moves along the ellipse. Points* A *and* C *are called the* foci *of the ellipse.*

The first law, an amendment to the Copernican theory, represented simply and correctly a motion that previously had to be described with epicycles and eccentrics. Later, when Uranus, Neptune, and Pluto were discovered, it was found that they too move in ellipses. Kepler did not explain why the planets move in elliptical orbits; the explanation had to await the work of Newton, but he did tell us how they move. He applied the mathematics of ellipses, as developed by the Greeks, to his own special uses.

Kepler's Second Law

Kepler's first law taken alone leaves something to be desired. It does not predict the later position of Mars or another planet along its ellipse from its present position. It also does not connect the speed of the planet at one point with its speed at another. In thinking about this problem, Kepler was reminded of a device he had attempted to use earlier, the division of the orbits of uniform circular motion into triangles such that the area of each triangle was equal to those of the others. Indeed, once he had discovered the elliptical orbit he could apply to it the scheme he had used for the eccentric circle. It succeeded, and he formulated his second law, which he had really discovered before the first: *During time intervals of the same length, the line connecting a planet and the sun sweeps out equal areas.*

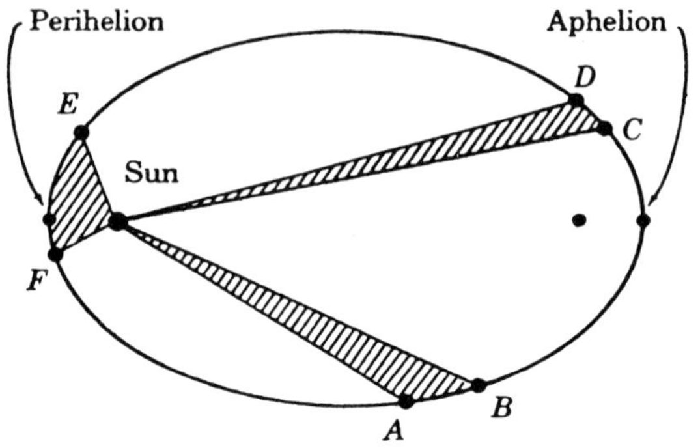

Elliptical path of a planet about the sun. Distances AB, CD, *and* EF *are traversed in equal time intervals. The areas swept by the radius vector as the planet moves from* A *to* B, C *to* D, *etc., are then equal.*

The law is illustrated . . . [by a drawing in which] the ellipse represents the orbit of the planet. The sun is at one of the foci. Three areas under consideration are shaded. The first area is swept out by the radius vector (the line connecting the sun and the planet) as the planet moves from *A* to *B*. An equal area is swept out as the planet moves from *C* to *D*, and similarly from *E* to *F*. The planet takes the same time to move from *A* to *B* as from *C* to *D* and *E* to *F*. It may be seen that a planet moves fastest when it is at perihelion [nearest to sun] and slowest when at aphelion [most distant from the sun].

This law (as well as the other two) applies to any planet as it moves about the sun, to any moon of a planet as it circles the parent planet, and to man-made satellites as they revolve about the sun, a planet, or a moon. It is sometimes called the law of equal areas.

Kepler's Third Law

Kepler worked on the orbit of Mars with little interruption from 1600, when he was appointed to assist Tycho, to 1609, when he published his first and second laws in his book *New Astronomy*. It was not until 1618 that he expressed his third law. It was the capstone of the three. The first two described each planet's motion, but did not connect the motion of one planet with that of another. Although each planet moved in space oblivious of the others, all had the sun as a common focus, and Kepler was obsessed with the belief that some relationship existed. And so it was with great elation that Kepler found his long-sought mathematical connection. It was this: *The square of the time (T) for a planet to complete one revolution about the sun is proportional to the cube of its average distance ($\overline{R}$) from the sun.*

Appraisal of Kepler's Accomplishments

Kepler's work shows certain characteristics comparable to those of modern science:

1. He depended greatly upon accurate, observed facts, and allowed the facts to be the final arbiter of the validity of the various systems he posed for consideration.
2. He represented the paths of the planets by a mathematical equation and used another to express his third law.
3. He described "how" the planets moved and gave little consideration to "why" they moved, suggesting (incorrectly) that some sort of magnetic force emanating from the sun was the cause.
4. He had a deeply entrenched belief in the simplicity of nature, in its order and harmony.
5. He corresponded with other scientific workers and exchanged information freely and unselfishly.
6. He adopted many temporary hypotheses as to how a system operated. Often they were wild speculations. But he was not reluctant to give up a false hypothesis, once convinced of its worthlessness, even though he might have spent years developing and testing it.
7. His progress toward his goal was not always "logical" or straightforward.

Isaac Newton, noted for the rigor of his mathematical probes of the worlds of light and gravity, was not an ordinary or a dull man. Dr. Bronowski speaks of him as "a most extraordinary wild character," an alchemist who wrote prolific treatises on the Book of Revelation when he was not conjuring up new theories of mathematics. Newton was intrigued with the Pythagorean notion of the "music of the spheres" (see Chapter 5) and philosophized that there was a harmony of sorts between natural and divine knowledge. This notion is explored further in the following selection.

Newton and the Pipes of Pan

J.E. McGuire and P.M. Rattansi

The personal testimony of David Gregory confirms the importance of this set of ideas for Newton's philosophy. His memorandum of 21 December 1705 tells us that Newton would answer the question, 'What cause did the ancients assign to gravity?' (in the projected Latin edition of the *Opticks*) by saying that, 'they reckoned God the cause of it, nothing else, that is no body being the cause; since every body is heavy'. Thus we have in the intended Query, an expression of the Newtonian distinction between passive and active principles in an orderly universe, and the complete dependence of matter, for its existence and motion, on the will of God; and all of this expressed by the ancients through the idea of 'Harmony'.

The draft Scholium to Proposition IX develops the same theme in greater detail. It starts with Newton's customary abjuring of causal explanations, and concludes with an eloquent passage in which the ancient dieties *[sic]* are assimilated into the one true God.

So far I have expounded the properties of gravity. Its cause I by no means recount. Yet I shall say what the ancients thought about this subject. Thales

From J.E. McGuire and P.M. Rattansi, "Newton and the Pipes of Pan," *Notes and Records of the Royal Society of London,* Vol. 21, 1966. Reprinted by permission of the author and The Royal Society.

regarded all bodies as animate, deducing that from magnetic and electrical attractions. And by the same argument he ought to have referred the attraction of gravity to the soul of matter. Hence he taught that all things are full of Gods, understanding by Gods animate bodies. He held the sun and the Planets for Gods. And in the same sense Pythagoras, on account of its immense force of attraction, said that the sun was the prison of Zeus, that is, a body possessed of the greatest circuits. And to the mystical philosophers Pan was the supreme divinity inspiring this world with harmonic ratio like a musical instrument and handling it with modulation, according to that saying of Orpheus 'striking the harmony of the world in playful song'. Thence they named harmony God and soul of the world composed of harmonic numbers. But they said that the Planets move in their circuits by force of their own souls, that is, by force of the gravity which takes its origin from the action of the soul. From this, it seems, arose the opinion of the Peripatetics concerning Intelligences moving solid globes. But the souls of the sun and of all the Planets the more ancient philosophers held for one and the same divinity exercising its powers in all bodies whatsoever, according to that of Orpheus in the Bowl.

Cylennius himself is the interpreter of divinity to all:
The nymphs are water. Ceres corn, Vulcan is fire.
Neptune is the sea striking the foaming shores.
Mars is war, kindly Venus is peace, the Bull-born
Horned Bacchus frequenting gladsome feasts
Is to mortals and to gods relief of mind from care.
Golden Themis is guardian of Justice and right
Next Apollo is the Sun, hurling his darts
From afar, circling round, the Divines and the Soothsayers
The Epidaurian God is the expeller of diseases: these things
All are one thing, though there be many names.

For the material of this passage, Newton drew heavily on Macrobius, Cicero, Virgil, Porphyry, and the Orphic hymns. In it, he completes the view of nature which was developed in the earlier Scholia. In those, the universe was seen as comprising innumerable worlds, composed of immutable atoms, held together by gravity, moving in an absolute void. Now the immaterial, 'immechanical' cause of it, is seen to be God himself. Newton states this conception clearly in another manuscript intended for the same unimplemented edition of the 1690's:

. . . *those ancients who more rightly held unimpaired the mystical philosophy as Thales and the Stoics, taught that a certain infinite spirit pervades all space* into infinity, *and contains and vivifies the entire world. And this spirit was their supreme divinity, according to the Poet cited by the Apostle. In him we live and move and have our being.*

Thus the more ancient philosophers, such as Orpheus, who were closer to the true philosophy, held that gravity was a direct result of the exercise of divine power. Later philosophers such as the Ionics, the Italics and Plato reveal themselves to have partial knowledge of this, if their utterances are properly interpreted. For instance, Newton says of Plato:

Hence after Plato has, by succession from Pythagorean doctrine and by the divine profundity of his own genius, shown that apart from these ratios (i.e. musical ratios) there can be no possibility of conjunction: in his Timaeus, he constitutes the soul of the world by means of the composition of those ratios, by the ineffable providence of God the craftsman. Consequently the soul of the world, which propels into movement this body of the universe visible to us, being constructed of ratios which created from themselves a musical concord, must of necessity produce musical sounds from the movement which it provides by its proper impulse having found the origin of them in the craftsmanship of its own composition.

There is little doubt that Newton saw in analogy to musical harmony, the principles of law and order in the natural world. Such harmony was the profoundest expression of cosmos. But for Newton, nature operating according to these divine ratios, could scarcely be dependent on the guidance of an intermediate world soul. Rather . . . the exquisite structure of things immediately bespoke the providential governance of a Divine power actually present in the world.

Isaac Newton once said, "If I have seen farther, it is by standing on the shoulders of giants." It was a humble admission for one who was not considered a humble man. It might also be said that in the ascent of man, giants give birth to giants; certainly Albert Einstein was a true descendant of Newton in terms of his vision and accomplishments. Yet there was a rigidity in Newton's ideas that almost challenged others to attack them. So it happened that the poet Johann Wolfgang von Goethe (1749–1832), who was born twenty-two years after Newton's death, rejected almost everything that Newton stood for. Goethe was part of the spirit of Romanticism that permeated German art and letters at the time, leaving little room for the exactness of Newton's theories of light. The following reading from Schroeer's *Physics and its Fifth Dimension: Society* illustrates the dimensions of the clash between Goethe and the spirit and teachings of Newton.

Goethe's Theory of Light
Dietrich Schroeer

The German poet Goethe is an excellent example of the attitudes of the Romantics and of their relationship with the sciences against which they rebelled. Johann Wolfgang von Goethe is important because he is the German equivalent of Shakespeare, with the added fact that his thoughts have been much more influential not only in literary developments but in all spheres of German thought. And he is interesting here not only because he characterizes some aspects of Romanticism, but also because he made an attempt to

From Dietrich Schroeer, *Physics and Its Fifth Dimension: Society*, Addison-Wesley, 1972. Reprinted by permission of the publisher.

unite in one person the two cultures. It was unfortunately an attempt which failed and which may ultimately have split the two cultures even further apart.

Goethe was born in Frankfurt/Main in 1749. By the age of 18 he had written a lot of "Rococo" or "Enlightened" poetry, with witty phrases and stylized "sheperd and maiden" situations. Then he came in contact with the medieval literature of Paracelsius and other magical and mystical animistic writings, became acquainted with several pietists, and finally met Herder. These experiences were immediately reflected in his writings. In 1774 he achieved instantaneous world-wide fame when he published *The Sorrows of Young Werther.* This very autobiographical novel deals with a young man who desperately wants to become one with nature, to dissolve in nature. The style of the writing makes everything totally subjective and nonrational; the novel consists of letters written in the very personal "I" form. The letters are never answered in the book; the external world never replies, so the reader sees the whole universe mirrored in one soul. In fact, the whole novel describes an attempt to take all the world into one individual, to dissolve subject and object. And there is something very erotic in this; it is a love affair with organismic nature. It is in a sense a protest against the Establishment by a young soul which refuses to be restricted; the failure to erase the boundary between subject and object leads to death. There is a trancelike love affair, which starts in spring. Nature's seasons then parallel the development of the relationship. Werther finally and inevitably loses out to a rival who is a stable and productive individual. And at 11 p.m. on December 23, when nature is closest to death, Werther shoots himself.

The Romanticism of this book struck a very responsive chord in the youth of Europe. It became the most widely read book of the 18th century; in 20 years there appeared 20 English and 20 French translations of it, and it was, in fact, the first European book ever translated into Chinese. Furthermore, as a result of this book there spread throughout Europe a deadly illness—the so-called *Wertherfever*; young men went into the woods dressed in blue pants and a yellow vest, and with a copy of a certain play nearby, shot themselves just as Werther had committed suicide. It became so bad that Goethe, in the foreword to a later edition, asked the reader not to take Werther's example too seriously: "Be a man and don't follow me."

In Goethe's work and thought and in the influence he had there was a very clear contrast between an Aristotelian-humanistic view of the world as something organic as opposed to the world-machine view of the Newtonian tradition. The desire of Werther to be an integral part of nature was a reflection of a belief in something greater than any individual and an opposition to dry scientific knowledge. This antimechanistic attitude permeates the most famous and most puzzling of Goethe's works, namely, his play *Faust*.

Faust is an ancient figure dating from the Middle Ages. Christopher Marlowe, almost 200 years before Goethe, also wrote a play, *Doctor Faustus*, on the same theme; this is the play in which there occurred the famous lines concerning Helen of Troy: "Was this the face that launched a thousand ships, and burnt the topless towers of Ilium?" Marlowe in the very beginning of that play expresses the Aristotelian origins of the Faustian view of science when he has him say:

Settle the studies, Faustus and begin
To sound the depth of that thou wilt profess.
Having commenced, be a divine in show,
Yet level at the end of every art
And live and die in Aristotle's works:
Sweet analytics, 'tis thou hast ravished me!

This attitude toward nature as something more than cold rationalism, as something Aristotelian, carries over to Goethe's version. The Goethe play presents Faust as a seeker after the meaning of life. As the play begins, Faust has been searching for years for this meaning in study and in the sciences, and he finally realizes that it may all be meaningless, that after all this time he is no wiser than he was before. In a final effort to find this meaning, he signs a contract with Mephistopheles, a contract according to which he is to be shown all the attractions the world has to offer. The collection of Faust's soul is to take place if he ever says: "this moment I like; I wish it would last forever." Most of the play then deals with the attractions which Mephistopheles invokes in order to tempt Faust. Throughout the play there are reflections (both subtle and blunt) of Goethe on science as a soul-deadening activity:

Unless you feel it, you will never achieve it.
If it doesn't flow from your soul
With natural easy power,
Your listeners will not believe it. . . .
How can anyone cling to such trash
Keep any hope in his head?
With greedy hands he digs for treasure,
And is happy when he finds earthworms instead. . . .
Gray and ashen, my friend, is every science,
And only the golden tree of life is green.

Goethe made it clear in many ways that he considered the sciences to be a cultural wasteland. As then practiced, the Newtonian sciences seemed to him to run completely contrary to man's relationship to nature. The whole of *Faust* is in a sense an exploration of cultural alternatives to musty science.

Goethe as a Scientist

As a man who fancied himself a universal man, as a man who was a statesman and administrator as well as a poet, Goethe also attempted to make a contribution to the sciences. He expected, of course, that this contribution would substantiate the scientific correctness of his view of the world as an organic and purposive whole. He did actually gain some success with this view in his biological studies, studies which were then still almost purely descriptive. He found, for example, on intuitive grounds, similarities between certain bones of animals and of man; in fact, he predicted a previously unknown bone in man (the *os intermaximare*). In the field of biology, his attentiveness, good memory, and fertile imagination could couple successfully with his organically teleological approach.

But then in 1786 Goethe traveled to Italy and was enormously impressed, like so many before him, by the colors of the many paintings and of the countryside. He then felt the need to develop a theory of color. Newton himself had already proposed a theory of color, but to Goethe that theory was clearly unsatisfactory. After all, Newton's theory was based on many complicated experiments which were designed to take human subjectivity out of the data. One of Newton's conclusions was that white light contained all the colors in it and that a prism could be used to separate these color components. This idea disturbed Goethe greatly. He was committed to an Aristotelian-like preference of opposites, in this case of light versus dark rather than of a whole range of colors. The breakthrough came for him when he inserted the "I" of the observer back into the experiment by looking at a white object through a prism. Instead of seeing this object in all the colors, as he interpreted Newton's theory to predict, he saw color only on the edges. "I immediately exclaimed that Newton's theory is false," he wrote. Goethe's interpretation was that the eye creates the colors out of the various admixtures of the two opposites of light and dark. This makes man the center of the observation process. Goethe felt that "Man himself, if he uses his common sense, is the best and the most exact physical apparatus possible." The problem with the Newtonian physics was that it wanted to recognize nature only when "restricted in artificial apparatus."

Goethe spent much time on his optical studies. This resulted in the book, *The Theory of Colors*, of which he was so proud that at the end of a life rich in superlative literary accomplishments, he said:

As for what I have done as a poet, I take no pride in it whatever. . . . Excellent poets have lived at the same time with me, poets more excellent lived before me, and others will come after me. But that in my century I am the only person who knows the truth in the difficult science of colors—of that, I say, I am not a little proud, and here I have a consciousness of superiority to many. (Conversations with Eckerman)

This is a little like Newton saying "my laws of motion are nothing; but my chronology of the Bible, that will make me immortal."

The Theory of Colors included a historical part about previous color theories, with the emphasis on Greek and medieval sources and no clear analysis or understanding of Newton's contribution. Goethe's comment is self-explanatory. "I would rather be wrong with Plato than be right with his enemies." The third part described Goethe's experiments and his analyses of them; many interesting visual effects are described there. But from our distant perspective, the most interesting part is the middle section, the polemical part; it is entertaining just because of its language. In it Goethe charged

Newton with obdurate resistance to the light of evidence, labeled him a band-leader of Cossacks, called him a mere twaddle, and described Newton's version of the scientific method as follows:

First he finds his theory plausible, then he convinces himself with excessive haste, even before he realizes how contrived a sleight-of-hand he will need to apply his hypothetical insight to the real order of things. But once he has openly expressed it, he does not fail to make use of all his mind's devices to see his thesis through. In this he asserts with unbelievable sangfroid before the whole world, the whole absurdity as the obvious truth. (The Theory of Colors: Historical Part)

Physicists were spoken of as Newton's herd. And Goethe decried their eagerness to

. . . cement, patch-up, and glue together, as witchdoctors do, the Newtonian doctrine, so that it could, as an embalmed corpse, preside in the style of ancient Egyptians, at the drinking bouts of physicists.

This may have been an entertaining commentary, but it was not a very responsible attack on a substantial theory by a most reputable scientist.

Goethe's theory of colors was, of course, rejected by all physicists—in its scientific sense. His frantic fear that mathematics destroyed the beauty and immediacy of personal observation was an echo of the attacks by Aristotle on the Pythagoreans. Here, however, the challenge was against the complete and successful Newtonian system. His demand that light be left alone as an indivisible phenomenon, and his feeling that the various aspects of optics are subjective, were cries that went unheeded by the mainstream of physics. Later generations of physicists have, nevertheless, tried to be tolerant and to see some good in Goethe's struggle against the coldness inherent in the unfeeling mechanics. For example, Max Born, Nobel prize winner in physics, asked that we "learn from them [Goethe and his followers] not to forget the meaning of the whole amidst the fascination of details." Goethe is simply too imposing a figure to be ignored, even when totally wrong.

For billions of dollars, man went to visit the hills and valleys of the moon that Galileo viewed through his telescope. Whether or not the information that the astronauts brought back with them, along with their satchels full of rocks, was worth the cost is still debatable, but there was once a time when great strides in understanding nature could be bought for much less. The equipment used by Albert Michelson and Edward W. Morley to do the experiment that inspired Einstein's theory of relativity cost only ten thousand dollars. The Michelson–Morley experiment is described here by Bernard Jaffe.

Michelson and the Speed of Light

Bernard Jaffe

Lord Kelvin and Lord Rayleigh had asked Michelson to test the effect of motion on the speed of light. Michelson decided to use water as the moving medium and mentioned his plan to Morley. Morley invited him to use his laboratory. It was ideal for the project, a large basement room. Morley was not a trained physicist, but he was bright, quick, resourceful, and enthusiastic. While a student at Williams College in 1860, he had done some astronomical work. When Michelson outlined the problem to him and the apparatus he had in mind, Morley was all for starting at once. They were only getting under way when, on a morning in September 1885, Michelson appeared at the laboratory looking haggard and distraught. He was nervously exhausted, he said, and needed a long rest. He had to get away from Cleveland for at least a year. Could he prevail on Morley to finish the apparatus, go on with the experiment, and publish the results? He turned over some of the money he had received for the project and added a hundred dollars of his own.

Morley next heard from Michelson in a letter from a New York City hotel. They corresponded about the experiment. Four months later Michelson showed up unexpectedly and suggested a partnership. His health was restored sufficiently to complete the experiment. In 1886 they submitted a joint paper to the *American Journal of Science*, "Influence of Motion of the Medium on the Velocity of Light." The team of Michelson and Morley had found that the "luminiferous ether" was quite unaffected by the motion of the water. They confirmed Fizeau's findings of 1851. Both Western Reserve and Stevens Institute of Technology conferred the Ph.D. degree on Michelson, his first degree. . . .

Now Michelson had the improved apparatus and the broader experience he had wanted for another go at the ether drift experiment, so long delayed. Morley would be in on it, too. On April 17, 1887, Morley conveyed their optimism in a letter to his father, "Michelson and I have begun a new experiment, to see if light travels with the same velocity in all directions. I have no doubt we shall get decisive results." Morley's brief description was not, of course, the whole story. Michelson and Morley were about to try to pin down the elusive ether, once and for all. From a positive result science would obtain not only a measurement of the earth's motion in its orbit relative to the ether but also of its rotation on its axis and perhaps even a method to determine the velocity of the whole solar system through space. It would be the first attempt by an American scientist to demonstrate absolute motion of the earth in space, insofar as space was identified with the ether, in an experiment with a local optical phenomenon.

The M–M Apparatus

The device they constructed was massive. It consisted of a stone about 5 feet square and 14 inches thick, on which were mounted four mirrors made of speculum metal (an alloy of copper, tin, and arsenic) and other equipment, including an Argand lamp. To make sure that the axis of rotation of the stone was rigorously horizontal and to prevent any errors due to vibrations, stresses, and strains, they had the stone floated in liquid mercury, which Morley had collected and purified. The mercury was poured into an annular, or ring-shaped, cast-iron trough about 0.6 inch thick; on the mercury floated a doughnut-shaped piece of wood, and the stone rested on the wood. A pivot

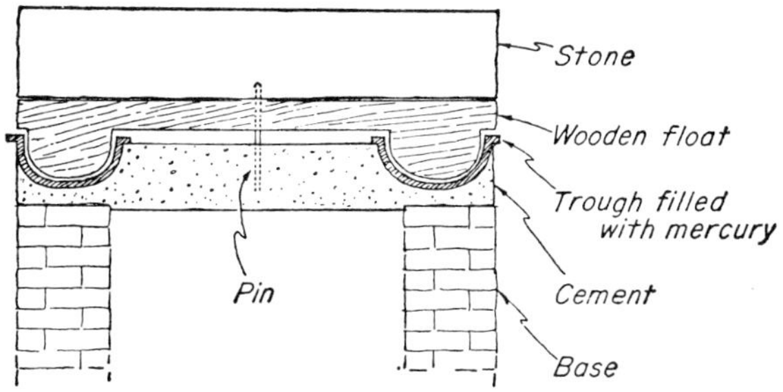

Michelson-Morley apparatus, seen in cross section here, was a sensitive combination of ponderous elements. The brick pier had bedrock for a footing. The large and very heavy stone rested on a wooden float that rotated on liquid mercury. The trough holding the mercury was shaped like a doughnut. Because the stone and its wooden raft were floating, they remained level.

in the center made the float concentric with the trough, which was of such dimensions that it left a clearance of less than half an inch around the float. The cast-iron trough rested on a bed of cement on a low brick pier built in the form of a hollow octagon. They dug down to bedrock to set the supporting column of the interferometer; it could not be safely anchored in soil. Around the trough were sixteen equidistant markings. A wooden cover protected the whole of the optical portion—one mirror on each corner of the stone—from air currents and any rapid changes of temperature.

Resistance to the movement of the heavy apparatus was kept to such a minimum that only the slightest force on its circumference was required to rotate it slowly, smoothly, and continuously. The rate of rotation was held at about six minutes for one complete turn. The observer walked around the trough with the moving stone, and at intervals would stop and look through a small telescope to watch for any displacement of the interference fringe system. Such a shift would indicate a change in the velocity of light in that direction. . . .

It took several months to get this unique piece of apparatus so delicately adjusted that it could record so tiny a measurement as the slight displacement of the fringe pattern in a single wavelength of light. Michelson and Morley took turns peering through the eyepiece of the telescope of the optical setup as they slowly pushed the stone around its course. They reasoned that in the course of a year there would be two days when the maximum effect, if indeed there was any effect, could be observed. On one day the earth would be traveling in exactly the opposite direction from its path on the other day. Hence, on one of these days the earth would be moving against an ether wind, and exactly six months later it would be rushing along with the ether wind. They made thousands of observations both at noon and at 6 P.M., day after day, for sixteen different directions, while the earth was constantly changing its position. They strained their eyes looking for interference bands and for any shift in the pattern of interference bands.

The final observations were made in July 1887. When all the data had been recorded and analyzed, and all the calculations had been made, checked, and rechecked, they had to face an obstinate fact that had destroyed a beautiful theory. No shift had been recorded of the order of magnitude demanded by the hypothesis of a stationary ether. It looked like a verdict of death to the notion of a calm ether sea. This was contrary again to all expectations. Michelson's thinking had taken kindly to the idea of an ether that could be detected. How otherwise could electromagnetic waves such as light waves travel? Once again a beautifully planned and executed experiment bewildered Michelson.

"The Greatest of All Negative Results"

Michelson and Morley sent their report to the *American Journal of Science* for publication. It was entitled "On the Relative Motion of the Earth and the Luminiferous Ether." It also was printed in the *Philosophical Magazine* in England that year. Scientists all over the world could read the conclusion. There was no observable difference in the velocity of light in whatever direction the observer was moving. What it meant, incredible as it might appear, was that no matter how fast you ran after light you could not catch up with it. It still leaped away from you at the speed of about 186,000 miles per second. It was contrary to experience in another way. A plane flying at 400 miles per hour with a tail wind of 50 miles per hour would actually be traveling at 450 miles per hour in relation to a fixed point. A plane flying at 400 miles and bucking winds of 50 miles per hour would be slowed down to 350 miles per hour. Now, since the earth moves around the sun at about 18 miles per second, the speed of a beam of light traveling *with* the earth's orbital motion should be greater than that of a beam traveling in the *opposite* direction. Yet Michelson's experiment denied this assumption.

The British physicist and writer John D. Bernal has called the Michelson–Morley findings "the greatest of all negative results in the history of science." But the outcome did not frustrate Michelson. Even though it ruled out the existence of a *stationary* ether, the results of his experiments, wrote Michelson, "could still be accounted for by the assumption that the earth drags the ether along at nearly its full speed, so that the relative velocity between the ether and the earth at the surface is zero or very small."

Ten years after the appearance of this historic paper Michelson tested his alternative hypothesis by sending two light beams around a vertical rectangle measuring 50 by 200 feet. The results indicated that the ether was not carried along by the moving earth to any significant degree. Sir Oliver Lodge tackled the same problem in another way. He used two heavy steel disks mounted close together on a common axis and moving in opposite directions. A beam of monochromatic light was divided, one division traveling around a closed loop between the rotating disks and the other in the opposite direction around the loop. Then the two parts of the original light beam were reunited to produce interference fringes. Lodge could observe no significant change. The ether was at rest; it was stagnant.

Michelson was still not convinced of his "failure." "Since the result of the original experiment was negative, the problem is still demanding a solution," he insisted publicly. And he consoled himself with the rather unexpected observation that "the experiment is to me historically rich because it was for the solution of this problem that the interferometer was devised. I think it will be admitted," he added, "that the problem, by leading to the invention of the interferometer, more than compensated for the fact that this particular experiment gave a negative result." A strange rationalization from an "objective" scientist!

Many years later, while addressing a distinguished audience of scientists at Mount Wilson Observatory, Michelson reconsidered his relative evaluation of the ether experiment and the invention of the interferometer. He acknowledged that his assumption of the superior worth of the instrument contradicted "some other important theoretical considerations" which had convulsed the world of science. Unwittingly, Michelson, as it turned out, had supplied the raw material for one of the great structures of science—a synthesis which was to be completed overseas. This was one of the very few instances when a basic discovery was made in America for European exploitation. Almost always it was the other way around.

"Physics," says Dr. Bronowski, "is not events but observations," observations of the processes of nature. "Relativity is the understanding of the world not as events but as relations." In his Theory of Relativity, Einstein gave us a fresh appraisal of space, time, and our universe. This theory, which revolutionized philosophy as well as physics, is discussed in the following selection.

The Theory of Relativity

Francis W. McCarthy, Joan H. Brenner, and Maurice Temple

In the early 1900's, it was discovered that some of our "basic" "common-sense" ideas about space and time were not quite right. It was found that when bodies move at speeds approaching the speed of light our ordinary concepts of distance, time and simultaneity break down. The story of this discovery and the experiments that led to it is a fascinating study of a scientific "establishment" having to change its rules when faced with facts that it could not convincingly explain. The story is, however, too involved to relate in detail here. We will instead indicate briefly the kind of thinking that went into the formulation of the new theories of space and time, namely, the theory of relativity.

The experiments that led to the discovery of inconsistancies [*sic*] in our common-sense notions of space and time were concerned primarily with the relationship between observations made by observers moving with respect to one another. Accordingly, we will first consider frames of reference and in particular so-called inertial frames. For our purposes, a frame of reference is a coordinate system (x, y, and z-axis) and a clock, which allows us to measure distances and time.

Inertial Frames

An *inertial frame* is a frame of reference for motion in which Newton's laws hold. In particular, it is a system in which Newton's first law—the law of inertia—holds. As a reminder, Newton's first law is: A body will continue in a state of rest or in uniform motion in a straight line unless acted upon by an externally applied force. For all practical purposes, the earth can be considered an inertial frame, i.e., Newton's laws hold on the surface of the earth. Any frame of reference moving with constant velocity with respect to an inertial frame will also be an inertial frame. For example, a person in a railroad car traveling with constant speed over the surface of the earth would be in an inertial frame. Any frame of reference that is accelerating with respect to an inertial frame is called a noninertial frame. For example, a person on a merry-go-round on the surface of the earth would be in a noninertial frame.

We believe that the laws of mechanics hold in the same form in any inertial frame of reference. This means that the observed laws of mechanics cannot be used to determine the inertial frame one is in. Thus, for example, Newton's second law $F = ma$ is true in all inertial systems, regardless of the point of observation. When a force is applied to an object, all observers, regardless of which frame of reference they occupy, would say that the object was accelerating. If a collision occurs in which momentum and kinetic energy are conserved in one inertial frame of reference, then they will appear to be conserved in all inertial frames.

As you look from one inertial frame to another, you will only be able to tell that the second inertial frame is moving with respect to your frame with some constant velocity. You will not be able to find the absolute velocity in any sense for either frame. The usual way of putting this is to say that there is no preferred frame of reference. There should be no experiment for which the outcome depends on the inertial frame in which the observations are made. From the point of view of physics, it is just as valid to say that the rest of the world is moving while the man on the railroad car is standing still as it is to say the man on the railroad car is moving. There should be no experiment that singles out one inertial frame from another.

Now consider the situation in Fig. 1. An observer is measuring the force between a charge Q and a positively charged wire, both of which are at rest in his frame. The force between these two charged objects will be purely electrostatic. Next, consider what observation would be made if the ob-

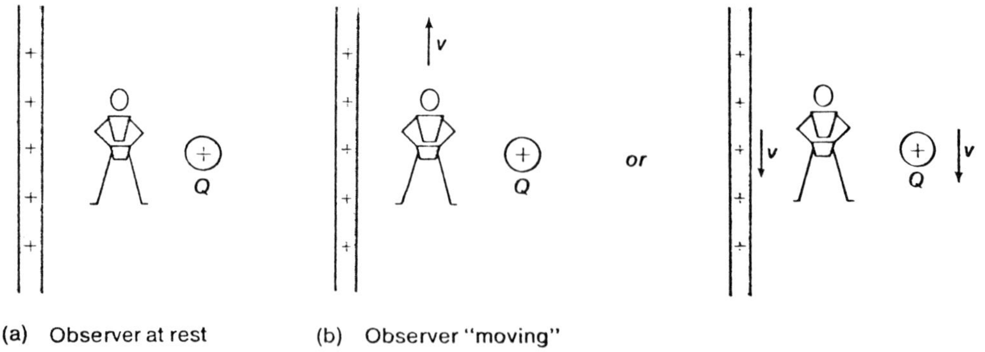

Fig. 1. Charged wire and charge Q as seen by an observer at rest and a "moving" observer.

server were moving with a velocity v in a direction parallel to the charged wire. The observer would still know that the wire was charged and would expect the charge Q to be acted on by the same electrostatic force as would be predicted by the observer at rest. Since he is moving with respect to the wire, however, he would see the wire not only as a charged object but as a moving line of charge (i.e., a current) as well. This observed current should produce a magnetic field about the wire; and since this observer also sees the charge Q as moving, he would predict that a magnetic force was exerted on the charge Q as well as the electrostatic force. To summarize, the observer at rest with respect to the wire and the charge Q sees Q acted on by only an electrostatic force. The moving observer sees a magnetic force on Q as well. Thus, in this situation the relative motion between the charged wire and the observer will alter the observed forces. The laws of electricity and magnetism appear to depend in some way on the inertial frame in which an observer finds himself. In particular, the frame in which the charge and the wire are at rest is singled out or preferred in the sense that it is the only frame in which the forces acting appear purely electrostatic.

One of the fundamental postulates in physics, however, is that the laws of physics shall be the *same* in all inertial frames. This postulate can only be made to hold, in the light of the above discussion of electricity and magnetism, by modifying in some way the laws of classical physics. We could, for example, abandon the postulate for electricity and magnetism and assume that there is some *one* frame in which the laws of electricity hold, or perhaps we should change the laws of electricity and magnetism. As it turns out, the basic postulate was retained and the definition of mass, momentum, and energy, and the concepts of space and time were modified by Einstein, so that both the laws of mechanics *and* the laws of electricity and magnetism obey this fundamental postulate. The theory that incorporates these ideas is known as The Special Theory of Relativity.

Postulates of Special Theory of Relativity

The two major postulates of the special theory of relativity are

1. The laws of physics can be written in the same form for all inertial frames of reference.
2. The velocity with which light travels in a vacuum is a constant and is the same regardless of any relative motion between a source of light and an observer.

The first of these postulates we have already discussed. The second is necessary to preserve the form of the laws of electricity and magnetism as we go from one inertial frame to another.

The second postulate contains a very startling statement. Whether you are moving toward a source of light or are standing still with respect to it, or moving away from it, if you measure the velocity with which the light is coming toward you, you will get exactly the same number. This is contrary to any situation with which you are familiar.

In all situations with which you are familiar, the velocity of an object as observed by one observer differs from the velocity of that object as seen by another observer by an amount equal to the relative velocity of the two observers. In other words, the velocity of an object as measured in one frame equals the velocity of the object as measured in another frame plus the velocity of the second frame relative to the first. In equation form, this can be written as

$$v_1 = v_2 + v$$

where:

v_1 is the velocity of the object as measured in the first system
v_2 is the velocity of the object as measured in the second system
v is the velocity of the second system with respect to the first system.

170 For example, if a man on a railroad train rolled a ball forward along the floor of the car at 10 feet per second while the train was traveling at 80 feet per second, an observer on the ground would say the speed of the ball was 90 feet per second with respect to the ground. (This differs from 10 feet per second by 80 feet per second, the relative velocity of the two observers.) If, however, the man on the railroad car flashed a flashgun and measured the speed with which the light left the flashgun, he would come up with the "speed of light". An observer on the ground would come up with *exactly the same number* as the man in the railroad car if he measured the speed with which light left the flashgun, in spite of their relative velocities.

One word of caution is in order–it is not only electromagnetic waves (of which light is an example) that obey the second postulate of the theory of relativity. Any object moving at the speed of light would appear to have the same velocity in any inertial frame.

Consequences of Relativity

These postulates lead to many startling consequences. Within the framework of relativity, if you wish to add together the velocity of 100 miles an hour and 600 miles an hour, the answer will not be 700 miles an hour but will be 699.99999 . . . . miles per hour.

Addition of Velocities

The formula for addition of velocities in relativity is

$$v_1 = \frac{v_2 + v}{1 + \frac{v_2 v}{c^2}}$$

where:

v_1 is the velocity of an object in the first system
v_2 is the velocity of the object as measured by the observer in the second system
v is the velocity of the second system with respect to the first system
c is the speed of light.

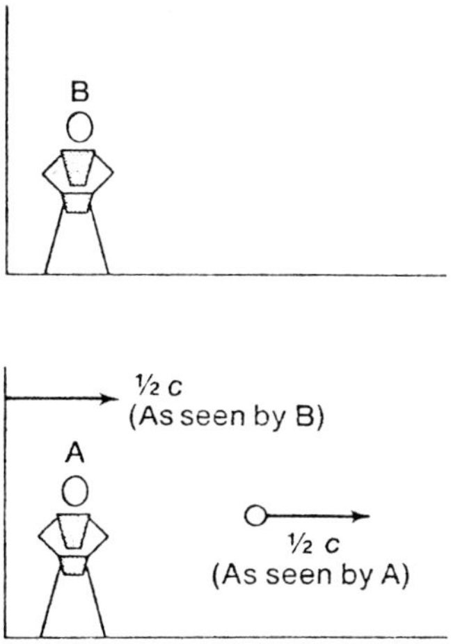

Fig. 2

This new way of adding velocities gives almost the same results as classical physics at low speeds. At high speeds, however, it gives quite different results. For example, if a ball were traveling at half the speed of light according to observer A, and if observer A were moving at half the speed of light with respect to observer B's system, . . . the speed of the ball according to observer B will be

$$v_1 = \frac{\tfrac{1}{2}c + \tfrac{1}{2}c}{1 + \frac{(\tfrac{1}{2}c)(\tfrac{1}{2}c)}{c^2}}$$

$$= \frac{c}{1 + \tfrac{1}{4}} = \frac{4}{5}c \ .$$

Classically, you would expect B to say the ball is moving at one c—the speed of light.

Length Contraction

Another strange result is that lengths and time intervals are not absolute quantities but depend on relative motion. This statement, if followed through to its implications at first seems rather nonsensical. We are saying here that the size or length of something is not an absolute but would depend on the relative velocity between the object and the measurer. Similarly, the length of time between two events happening is not an absolute but, once again, depends upon the relative motion between the system in which the events occurred and the person measuring these events.

In relativity theory, the length of an object is always longest to an observer when the object is at rest with respect to the observer. (This does not imply in any way that both the object and the observer are at rest, but it says that they are both moving with the same speed. Their relative velocity is zero.) However, if the object is moving at speed v, with respect to the observer, its dimension *in the direction of travel* will appear to be contracted by a factor $(1 - v^2/c^2)^{1/2}$. This phenomena is known as *length contraction*.

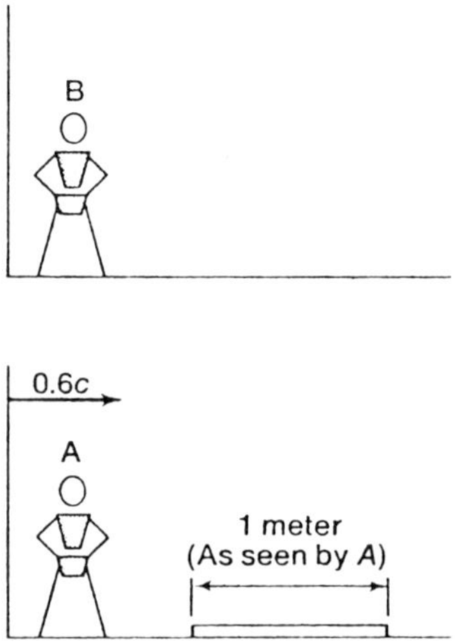

Fig. 3

As an example, suppose a man had a meter stick one meter long in his system (see Fig. 3). How long would this stick appear to an observer traveling at a speed of six-tenths the speed of light with respect to the first system. Length measured by the "moving" observer is $(1 - v^2/c^2)^{1/2}$ times the length measured by observer at rest with respect to the stick where:

v = velocity between systems

c = speed of light.

So

$$\begin{aligned} L &= L_0 (1 - (v/c)^2)^{1/2} \\ &= (1\text{ m}) \times [1 - (.6c/c)^2]^{1/2} \\ &= (1\text{ m}) (.64)^{1/2} \\ &= .8 \text{ meters.} \end{aligned}$$

Thus, the "moving" observer "measures" the length of the stick to be only eight-tenths of a meter.

Now suppose the "moving" observer had a stick in his system that was one meter long. How long would the observer "at rest" say the stick was? Once again, the relative velocity between the two systems is six-tenths c, and the observer who is "at rest" will say that the stick in the "moving" system is eight-tenths of a meter long.

Thus, each thinks that his own meter stick is the longest and that the other man has a short meter stick.

This effect is very small except for objects traveling at speeds near the speed of light. For example, a car five meters long, traveling at 30 meters per second (about 60 miles per hour) would appear to shrink 2.5×10^{-14} m, which is too small to be measured by any known means. This is, of course, why the effects of special relativity were not recognized until fairly recently.

Time Dilation

Somewhat similar to the concept of length contraction is time dilation. The time between two events depends upon the relative velocity between the frame in which the two events occur and the observer's frame. For example, suppose a light bulb were connected to a battery through a timer which turns the bulb on and off. Suppose also that an observer at rest with respect to the light bulb found that it turned on for ten seconds and then turned off. If the light bulb, battery and timer were placed on a platform moving past another observer with speed v, the theory of relativity predicts that this second observer would say that the light stayed on longer than ten seconds. This effect is known as *time dilation*. The apparent time interval between two events occurring at the same place in one frame of reference depends on the relative velocity between the observer and the rest frame of the events. The time intervals measured by an observer in the rest frame and a "moving" observer are related by the equation:

$$T_1 = T_2 \frac{1}{(1 - v^2/c^2)^{1/2}}$$

where:

T_2 is the time between the events as measured in the rest frame of the two events

T_1 is the time between the events as measured by an observer moving with speed v.

Another interesting prediction of relativity is that events appearing to happen simultaneously in one system will not necessarily appear to be simultaneous in any other system. This will happen whenever the two events are separated by some distance in the frame in which they happen simultaneously. Indeed, sometimes the apparent order of events will seem reversed when considered by different observers.

Relativistic Mechanics

In addition to changes in our concepts of length and time, changes have to be made in the quantities met in mechanics, such as mass, momentum, and energy. It appears that the mass of an object is not an invariant but increases with velocity by a factor $(1 - v^2/c^2)^{1/2}$. Thus the formula for the "increased mass" or "relativistic mass" is:

$$m = \frac{m_0}{(1 - v^2/c^2)^{1/2}}$$

where:

m is the observed mass of a moving object
m_0 is the mass of the object when at rest (termed the rest mass, or proper mass)
v is the velocity of the moving mass

c is the speed of light

As you can see in the above formula, if the speed of the mass gets very close to the speed of light, the denominator will become very close to zero, which means the effective mass of the moving object will be extremely large. This means that, as the speed of an object increases, more force is necessary to make it accelerate at the same rate. More simply put, the faster something goes the harder it is to make it go still faster. This reasoning often leads to the statement that the speed of light is the ultimate speed.

As with all the other effects discussed, this apparent increase in mass becomes important only at speeds close to the speed of light, say speeds greater than one-hundred million miles per hour (100,000,000 mph). Of course, such speeds are not met in ordinary situations but may be produced for atomic sized particles in accelerators or "atom smashers". Studies of particles in such accelerators confirm the theory of relativity in every respect.

The energy of the object, i.e., the total energy, as given by the special theory of relativity, is $e = mc^2$, where m is the relativistic mass just defined and c is the speed of light. The kinetic energy of a particle is its total energy minus its so-called rest energy, m_0c^2.

$$\begin{aligned} \text{KE} &= mc^2 - m_0c^2 \\ &= m_0c^2\left(\frac{1}{(1 - v^2/c^2)^{1/2}} - 1\right) \end{aligned}$$

It can be shown that, when v is much smaller than c, this expression reduces to $\frac{1}{2} mv^2$ (the classical expression for kinetic energy).

Another result of the theory of relativity is that mass may be converted to energy and vice-versa. If a mass m_0 is converted to energy, the amount of energy which results is $E = m_0c^2$. Notice that this is the same as what we have called the rest energy. Hence it is often said that mass and energy are equivalent.

With these new definitions in mechanics and the new concepts of space and time, a more general mechanics than Newtonian mechanics can be formulated. This is called *relativistic mechanics*. This theory presents a unified system of mechanics for all possible velocities (those velocities less than the speed of light). Classical physics, incorporating Newton's laws and many other concepts, gives the same results as the theory of relativity for low speed motion.

8

The Drive for Power

Now that is the wisdom of a man,
in every instance of his labor,
to hitch his wagon to a star,
and see his chore done
by the gods themselves.
That is the way we are strong,
by borrowing the might
of the elements.

Ralph Waldo Emerson

The Industrial Revolution was a peculiarly English revolution, a peaceful contrast to the blood and thunder of the French and American Revolutions. But, like all revolutions, much of it was inspired by human misery, fired by the belief of a small number of men and women that things could and should be different. A major tenet of the Industrial Revolution was that a decent life begins with material advantages and the production of everyday items of good quality. As it took hold, the Industrial Revolution did achieve some of that betterment for many people. On the other hand, it also spawned small factory towns such as Coketown, a place that Charles Dickens wrote disparagingly about in the nineteenth century.

Coketown

Charles Dickens

Coketown was a town of red brick, or of brick that would have been red if the smoke and ashes had allowed it; but, as matters stood, it was a town of unnatural red and black, like the painted face of a savage. It was a town of machinery and tall chimneys, out of which interminable serpents of smoke trailed themselves for ever and ever, and never got uncoiled. It had a black canal in it, and a river that ran purple with ill-smelling dye, and vast piles of building full of windows where there was a rattling and a trembling all day long, and where the piston of the steam-engine worked monotonously up and down, like the head of an elephant in a state of melancholy madness. It contained several large streets all very like one another, and many small streets still more like one another, inhabited by people equally like one another, who all went in and out at the same hours, with the same sound upon the same pavements, to do the same work, and to whom every day was the same as yesterday and tomorrow, and every year the counterpart of the last and the next.

These attributes of Coketown were in the main inseparable from the work by which it was sustained; against them were to be set off comforts of life which found their way all over the world, and elegances of life which made, we will not ask how much of the fine lady, who could scarcely bear to hear the place mentioned. The rest of its features were voluntary, and they were these.

You saw nothing in Coketown but what was severely workful. If the members of a religious persuasion built a chapel there—as the members of eighteen religious persuasions had done—they made it a pious warehouse of red brick, with sometimes (but this is only in highly ornamented examples) a bell in a bird-cage on the top of it. The solitary exception was the New Church; a stuccoed edifice with a square steeple over the door, terminating in four short pinnacles like florid wooden legs. All the public inscriptions in the town were painted alike, in severe characters of black and white. The jail might have been the infirmary, the infirmary might have been the jail, the Town-hall might have been either, or both, or anything else, for anything that appeared to the contrary in the graces of their construction. Fact, fact, fact, everywhere in the material aspect of the town; fact, fact, fact, everywhere in the immaterial. The M'Choakumchild school was all fact, and the school of design was all fact, and the relations between master and man were all fact, and everything was fact between the lying-in hospital and the cemetery, and what you couldn't state in figures, or show to be purchasable in the cheapest market and salable in the dearest, was not, and never should be, world without end, Amen.

A town so sacred to fact, and so triumphant in its assertion, of course got on well? Why, no, not quite well. No? Dear me!

No less dismal is this reminder by Robert Owen that while the Industrial Revolution producing new luxuries for the world, it was trampling rather heavily on the rights and lives of the men, women, and children who ran the machines that made it possible. Here is Owen's letter to the Earl of Liverpool, with some acerbic comments on what then passed for reform.

178 Letter to the Earl of Liverpool

Robert Owen

Generally speaking, the occupation which manufactures under the existing arrangements afford, is more or less unhealthy to those employed in them; who are called upon to sacrifice their strength and substantial comforts for the advantage of others, and not unfrequently for the benefit of their enemies.

The numbers thus employed now constitute an important part of the population—so great indeed, as necessarily to disseminate the good or evil effects of the practices which obtain among them, over every corner of the empire.

These practices are:—

First—The employment of children before they possess sufficient strength for their work; before they can be initiated in their necessary domestic duties; and before they can acquire any fixed moral habits or knowledge that may render them useful, or not injurious, members of the community.

Second—The employment of adults, both male and female, in situations unfavourable to health, for an excessive and unreasonable number of hours per day.

The Bill before the House of Commons is intended to apply to the former evil only, and to remedy which it is surely very inadequate.

It permits children to be employed at nine years of age for twelve hours and a half per day, allowing them only an hour and a half during this period for meals and for exercise in the open air.

Experience proves that men may be habituated by early practice to consider the most inhuman and barbarous customs, not only with indifference, but as amusements from which they derive exquisite enjoyment. Even the best natural dispositions may be trained with ease to delight in cannibalism. It would be clearly unjust, therefore, to blame manufacturers for practices with which they have been familiar from childhood; or to suppose that they have less humanity than any other class of men.

Those practices, however, to which they have been familiarised from early life, are nevertheless unjust in the extreme to their fellow-creatures, and peculiarly injurious to the best interests of the country. . . .

To advance one step beyond a mere pretence to ameliorate the wretched condition of the working classes,—the sole instruments by which our necessaries, our comforts, and our luxuries are produced,—they must be protected from that oppression to which they are now exposed, and their children must be placed under circumstances in which they may be trained in habits useful to themselves and to society.

If the Bill now before the Legislature secure not these objects, it will not effect anything that had not much better be left undone.

Instead of confining the provisions of the Act to cotton mills only, they should extend to all manufactures whatever not carried on in private houses.

Instead of children being admitted at nine years of age to work in cotton mills twelve and a half hours per day, with only an hour and a half for meals and recreation, no child should be admitted to work in any manufactory before ten years of age and not for more than six hours per day until he is twelve years old.

And no manufacturer should be permitted to employ either young or old for a longer period than twelve hours per day, allowing them out of that time one hour for breakfast and another for dinner, leaving ten hours for full and constant work, which is one more than our ancestors thought beneficial; and I doubt whether nine hours of regular and active employment, established as the measure of daily labour to be required from the working classes, would not be still more economical and profitable for the country.

I am fully aware of the clamour which these propositions will at first call forth from the blind avarice of commerce; for commerce, my Lord, trains her children to see only their immediate or apparent interest; their ideas are too contracted to carry them beyond the passing week, month, or year at the utmost.

They have been taught, my Lord, to consider it to be the essence of wisdom to expend millions of capital, and years of extraordinary scientific application, as well as to sacrifice the health, morals, and comforts of the great mass of the subjects of a mighty empire, that they may uselessly improve the manufacture of, and increase the demand for, pins, needles, and threads;—that they may have the singular satisfaction, after immense care, labour, and anxiety on their own parts, to destroy the real wealth and strength of their country by gradually undermining the morals and physical vigour of its inhabitants, for the sole end of relieving other nations of their due share of this enviable process of pin, needle, and thread making.

I trust, my Lord, it is not by such men that our great national concerns are henceforward to be directed.

If their voice is to predominate in our councils, the vital interests of the country must soon be utterly sacrificed to the grossest political errors.

It is the creed of this class, that no effort or expense should be spared to improve these trifling baubles and luxuries which when perfected are of no intrinsic value whatever, which cannot add a particle of strength or comfort to the empire (no real advantage has occurred from enabling our fashionable females to purchase fine lace and muslins at one-fourth of the former prices; but, to produce them at this price, many thousands of our population have existed amidst disease and wretchedness, and have been carried prematurely to their graves); while any attempts to ameliorate the condition of human beings are decried as unnecessary and visionary, as travelling out of the proper business of life, which to them is solely and exclusively to accumulate wealth—wealth which is not only acquired at the expense of every thing that is truly great or valuable in the character of a nation, but which, when acquired, is useless, nay, in the highest degree injurious, to themselves and others.

In spite of those dismal contemporary viewpoints, Dr. Bronowski maintains that "it is silly to treat the Industrial Revolution as a major disaster." He contends that science and technology—as utilized in the Industrial Revolution—was used for the first time to improve the lives of ordinary people. He expands that point in these selections from *William Blake and the Age of Revolution.*

William Blake and the Age of Revolution

J. Bronowski

The turbulent age from about 1760 to 1815 was shaped, of course, by the American and the French Revolutions; and it was also shaped by the Industrial Revolution. There is a unity between all three of these: the movement of industry from the village to the factory was pushed on by the same forces that made the political movements. A common restlessness runs under that time, a discontent with the traditional ways of doing and thinking, and an urge to band men together in new alliances. The age of the revolutions was powered by common aspirations in its poetry and its politics, its ideals and its inventions, and as far apart as the organization of Napoleon's armies and Richard Arkwright's factories.

The massive expansion of manufacture by machines about the year 1800 obviously changed the economy of England. And, more subtly, it changed the way in which men looked at the economy and at their own place in it. So long as men worked in villages, they thought of their livelihood as a gift: a gift from God or from nature, which God or nature might bless with plenty in one year and blast with famine in another. The cottage weaver at his loom thought in this way, even though the loom made the greater part of his living for him; he thought in this way because (as Daniel Defoe insists) he still turned to his patch of garden and his rights on the common land for a critical part of his living. A slump seemed as much an Act of God to him as a drought. The village worker did not think of himself as the master of his destiny, because his craft did not make up his whole possession.

The move to the factory and the enclosure of common land changed all this. They put the worker into a different place in his economy, to which he now had only one thing to give: the labour of his hands. No longer did a superhuman dispensation stand between his work and its reward. If the society of masters rejected his handiwork, and found no use for his skill and diligence, there was something wrong with society, and not with God or with nature.

It is strange to see that the spread of machines made the men who worked them conscious, first of their own work, and then of themselves as men. Yet this is what the Industrial Revolution did. It forced men in the long run to seek their destiny, and to find their station, not in the hand of God but in their own hands. In 1720, the village workers had accepted the slump that followed the South Sea Bubble as an accident of nature. In the slumps a hundred years later, the cloth-shearers and the stocking-knitters drilled on the dark moors in open mutiny against the economy that starved them.

The machines changed the organization of society, and shifted the centre of a man's life from his cottage home to the daily factory. In that shift, the man ceased to be a member of his family and his village, and in the long run became simply himself: a person. Because the machine in the factory changed the order in his life, it slowly changed the status of the worker who served it. It regimented and brutalized and starved him, it exploited him and (for a long time) his family, and it robbed him of everything but his skill. And yet, by these acts in the end it made him a man—a man alone.

This is the change of conscience which links the political revolutions to the rise of industry, and the American smallholder in 1775 to the Luddite of 1811. Here the citizens who proclaimed the Rights of Man in America and in France were at one with the tanners and the nailmakers who quoted from *The Wealth of Nations* by Adam Smith. They spoke with different voices, but they spoke for the same cast of mind, in the homely pamphlets of Benjamin Franklin, in the educational schemes of Rousseau and Jefferson, and in the romantic poetry of Robert Burns and William Blake and Wordsworth. We catch in all of them a new sense that the dignity of man is the simple dignity of labour, and a new image of man the maker. . . .

The steps which took industry to its own revolution are now plainer. The village workmen could remain masters only so long as they could buy their raw stuffs. They could buy their raw stuffs only so long as they could sell their manufactures. When the price of manufactures slumped, the small masters had to borrow. And the price of manufactures often slumped: it slumped whenever war narrowed the export market: and England was at war for more than half of the eighteenth century. The small masters borrowed raw stuffs, and became workers for a putting-out merchant. They borrowed on their tools, and came to rent them from an owning merchant. Before Blake was born, most weavers, stocking-knitters, nailmakers, and others had already lost their small capital. Many had lost their tools, and were selling only their labour. The threshold to ownership was growing too high, even before the coming of the factory.

The merchants and middlemen who had bought out these men had wanted to run industry: they had not set out to own it. But now that they had sunk capital in raw stuffs and tools, they could keep up profits on this capital only by using it fully. They must rationalize industry. The worker at home did not give the best return. He was careless with stuffs and tools which he did not own; he was idle when he had earned his keep; having little hope of becoming a master, he was thriftless. He was still suffered to stay at home, because there he had the help of his family. As soon as greater help was offered by drawing industry together, by breaking its processes into unskilled linked steps, and by working its machines by power, he was moved to the factory. With these means Arkwright made his spinneries great, and Boulton his metalworks. The means of rationalization: the machine and the factory; the ring, the cartel, and the trust: these were not new. But in Blake's lifetime they became the norms of industry; and Blake knew their purpose.

Wheel without wheel,
To perplex youth in their outgoings and to bind to labours
in Albion
Of day and night the myriads of eternity: that they may
grind
And polish brass and iron hour after hour, laborious task,
Kept ignorant of its use: that they might spend the
days of wisdom
In sorrow drudgery to obtain a scanty pittance of bread.

The industries best fitted to gain by factory methods were those making textiles. Of these, the woollen industry was heavily protected; and rationalization was therefore pressed in the rival cotton industry. Both industries had unbounded markets in Europe, in America, and in Africa. In both, the skilled and well-paid weaver had outpaced the casual spinster, who spun only to eke out farm wages. For twenty-five years before Blake's birth, weaving was being speeded up: for example, by the use of a mechanical shuttle, at least in the cotton industry; and there had been a growing shortage of yarn. Two inventions of the 1760s righted the balance. One was the many-spindled jenny of James Hargreaves, in which the spindles themselves stretched and twisted the thread which they were winding. The other was the frame patented by Richard Arkwright, which spun the thread as it were with mechanical fingers, by drawing it through rollers each pair of which turned faster than the pair before it. By 1779, Samuel Crompton had made an invention which in effect married these two together, in a mule which is largely the spinning machine of to-day.

Arkwright had probably stolen his invention from the mechanic Thomas Highs; and it had been made by others before Highs, about 1736. But Arkwright turned this and kindred inventions into real power machines; he organized the machines in factories; and he did what the mechanics did not know how to do, he made the factories pay. Hitherto, English cotton cloths had had linen warps; and Parliament had protected these fustians against the pure cotton cloths. It was Arkwright's gift, not merely to spin a cotton thread strong enough to make warps, but to get Parliament to withdraw its law against it. He got bankers and hosiers to back him; he bought his way into rival firms; when his patents were overthrown, he beat down his rivals without them. Vigorous, ambitious, ruthless, self-made, and with no past knowledge of his industry, he was the model of the new captains of the textile industry. . . .

Like the woollen and the cotton industries, the English iron industry had long been held back by the shortage of one of its raw stuffs: charcoal to smelt and fine the plentiful iron ore. Once the Surrey ironmasters, who had enraged Queen Elizabeth by selling cannon to Navy and pirates alike, had made their charcoal in the oak scrub there.

The Surrey hills glow like the clinkers of the furnace;
Lambeth's Vale
Where Jerusalem's foundations began, where they were
laid in ruins,
Where they were laid in ruins from every Nation, and Oak
Groves rooted,
Dark gleams before the Furnace-mouth a heap of burning
ashes.

Since then, the furnaces had moved west and north in search of charcoal. But charcoal probably remained scarce, and was certainly dear and awkward in use: for it is so easily crushed that iron ore could be smelted with it only a little at a time. In 1749 pig iron was so scarce that it had to be imported from America, in spite of the petition of the tanners of Sheffield and elsewhere.

If the Bill should pass the English iron would be undersold; consequently a great number of furnaces and forges would be discontinued; in that case the woods used for fuel would stand uncut, and the tanners be deprived of oak bark sufficient for the continuation of their occupation.

A stronger and more plentiful stuff was needed to smelt iron ore. The stuff was coke, that is coal from which have been burnt off some of the impurities which make iron brittle and unworkable. It is not easy to use coke in smelting, for it must be done in a large furnace and with a strong blast. The process was tried by William Wood in 1726, after Swift had driven his halfpence out of Ireland. Swedenborg, then an expert in mining and metals, reported its failure. But the process had in fact been worked secretly since about 1709 by a family of Shropshire ironmasters, the Darbys. It had since been learnt by a neighbouring family, the Wilkinsons. Iron cast straight from the blast furnace began to take over many of the uses of fined wrought iron. It began to make new uses for itself, for example in the founding of engines. By 1762 the Darbys and the Wilkinsons were forming a ring to charge the same prices for their castings. The Darbys were already a vertical trust, feeding their furnaces from their own coal- and iron-mines. Ways were found by 1784 of stirring and rolling the dross out of iron also fined with coke, so that it could be wrought. But the stuff of the Industrial Revolution remained cast iron, the pride of John Wilkinson. By the 1790s, Matthew Boulton was minting Wilkinson's wage tokens, and they were good coin in every Shropshire iron town. The iron-masters were masters of the West Country. It is pleasant to know that the artless tanners were also saved. In 1765, the *Society for the Encouragement of Arts, Manufactures, and Commerce* gave a prize of £100 to a poor country tanner, who had discovered how to tan with oak sawdust in place of bark.

The Industrial Revolution was many things: a blend of science, engineering, and social consciousness, another stepping stone in man's ascent, aided and abetted by machines and newly plumbed sources of energy. The polluted skies and oceans of today bear some of the unanticipated effects of that continuing revolution. Nevertheless, it has been a great boon to many men, as Marc Sequin, a French railway pioneer realized as early as 1839. Here is his tribute to the Industrial Revolution.

184 On the Industrial Revolution

Marc Sequin

To increase the well-being and the enjoyment of material life is today the dominant idea of civilized nations. All their efforts are turned to industry because it is from that alone that one can expect progress. It is industry that gives birth to and develops in mankind new needs and gives them at the same time the means to satisfy them. Industry has become the life of the peoples. It is thus to this development that all hopes and wishes should tend, all our talents and knowledge be devoted. It is around this powerful lever that should be united all superior spirits that aspire to the honour of contributing to our social regeneration.

Where are the limits before which human power will come to a stop? Commonplace individuals can never imagine them beyond their own horizon but nevertheless every day that horizon is widened. Every day its limits are put back. Look around us, everywhere in the last twenty years the elements of the old civilization have been modified, perfected, renewed; everywhere wonders have been worked. The enjoyment, the commodity of life which had been reserved only for men of fortune are now enjoyed by artisans. In a few more steps they will be shared equally with all classes. A thousand inventions are simultaneously born and lead to other discoveries and these in turn will become the starting-point for new progress; all these changes concur to the profit of the whole public and tend to make well-being common property. This is a new era based on the love of the good and the beautiful which is being built on the ruins of class prejudices and the monopolies of the wealthy. In all its creation, in all its innovations the same character is shown. Low prices and general utility are the essential conditions of the vitality of the industrial arts. Governments, national as well as local, drawn into this irresistible mass movement have followed the same drive and it is only by granting reforms to meet the needs of the times that they have maintained themselves in office, submitting these reforms to the point of view of modern ideas that they have been able to secure their acceptance. The old world has shaken off the yoke of its old habits. It is refreshing and remaking itself. So look, everything is changed around us—the towns, the face of the countryside, the course of rivers, the work of the peoples, the production of the soil and industry, the distribution of property—everything has taken a new face. And just where the direct force of the material power of men has shown itself insufficient to accomplish its work and to persevere in progress; where his will seems to be broken against insurmountable obstacles, just there a drop of water turned into steam acts to supplement his weakness, to create for him a power of which we cannot now, not yet for a long time to come, measure the extent.

From now on with the help of this agent these prodigies have been accomplished, and the wonders which our fathers would not have thought realizable with the united efforts of all their magicians have become the ordinary run of things. Machines which only need from man an idle supervision now spin and weave by themselves linen, cotton, wool and silk and return to us in a variety of stuffs the material which we supply to them in the raw state. Then after having been submitted to a chemical preparation for some hours these cloths dipped into a vat come out as if by enchantment with the most lovely colours, with the most graceful designs. These make the pretty prints, which on holiday, the working class itself wears, and which in the country as much as the town, light up with their sparkle and freshness groups of young girls and spread all around them an atmosphere of joy, comfort and happiness. Besides, the old rags that you throw into the masher soon come back transformed into paper of the purest whiteness ready to receive and spread and make eternal your thoughts; some minutes only are sufficient to achieve this metamorphosis. Everywhere the most delicate objects of use and of luxury are poured into consumption at prices which are always decreasing. By means of that same steam, rivers, seas are navigated. It transports us with inconceivable speed to all the extremities of the world in floating palaces which shelter the poor man, the rich man, offering them alike luxury and conveniences which they often lack in their own houses. Finally, in our own valleys and across our hills wind and spread long ribbons of iron, along which rush, rapid as thought, those formidable machines which seem to eat up space with a spontaneous impatience and which seem almost alive in their breathing and their movement. When we consider the majestic elegance of these lines which speed with grace and level themselves out to cross plains, valleys, precipices and the granite mountains; which one hears the noise of these trains which carry thousands of individuals so fast that you cannot distinguish them; when one remembers that these results are the work of an industry which is only a few years old, of an agent which we have as yet studied only very imperfectly, an art which is in its infancy, one asks what will be the last prodigies to be realized by the perfectioning of this art, and one feels a noble desire to contribute to the realization of these incalculable blessings.

As the Industrial Revolution advanced, a romantic movement among contemporary writers and poets moved in a parallel column. Power was the common key. Just as a man sought to tap the power of nature to drive machines, he also sought its strength in uplifting and ennobling his consciousness. A leading poet of the romantic movement was William Wordsworth, who in the following poems describes the city of London as seen from Westminster Bridge and evokes other sentiments in "Lines Written Above Tintern Abbey."

Composed upon Westminster Bridge, September 3, 1802

Earth has not anything to show more fair:
Dull would be he of soul who could pass by
A sight so touching in its majesty:
This city now doth, like a garment, wear
The beauty of the morning; silent, bare,
Ships, towers, theatres, and temples lie
Open unto the fields, and to the sky;
All bright and glittering in the smokeless air.
Never did sun more beautifully steep
In his first splendour, valley, rock, or hill!
Ne'er saw I, never felt, a calm so deep!
The river glideth at his own sweet will:
Dear God! the very houses seem asleep;
And all that mighty heart is lying still.

William Wordsworth

Composed a Few Miles Above Tintern Abbey, on Revisiting the Banks of the Wye During a Tour, July 13, 1798

These beauteous forms,
Through a long absence, have not been to me
As is a landscape to a blind man's eye:
But oft, in lonely rooms, and 'mid the din
Of towns and cities, I have owed to them,
In hours of weariness, sensations sweet,
Felt in the blood, and felt along the heart;
And passing even into my purer mind,
With tranquil restoration: —feelings too
Of unremembered pleasure: such, perhaps,
As have no slight or trivial influence
On that best portion of a good man's life,
His little, nameless, unremembered, acts
Of kindness and of love. Nor less, I trust,
To them I may have owed another gift,
Of aspect more sublime; that blessed mood,
In which the burthen of the mystery,
In which the heavy and the weary weight
Of all this unintelligible world,

Is lightened: —that serene and blessed mood,
In which the affections gently lead us on,—
Until, the breath of this corporeal frame
And even the motion of our human blood
Almost suspended, we are laid asleep
In body, and become a living soul:
While with an eye made quiet by the power
Of harmony, and the deep power of joy,
We see into the life of things.
        For nature then
(The coarser pleasures of my boyish days,
And their glad animal movements all gone by)
To me was all in all. —I cannot paint
What then I was. The sounding cataract
Haunted me like a passion: the tall rock
The mountain, and the deep and gloomy wood,
Their colours and their forms, were then to me
An appetite; a feeling and a love,
That had no need for a remoter charm,
By thought supplied, nor any interest
Unborrowed from the eye. —That time is past,
And all its aching joys are now no more,
And all its dizzy raptures. Not for this
Faint I, nor mourn nor murmur; other gifts
Have followed; for such loss, I would believe,
Abundant recompense. For I have learned
To look on nature, not as in the hour
Of thoughtless youth; but hearing oftentimes
The still, sad music of humanity,
Nor harsh nor grating, though of ample power
To chasten and subdue. And I have felt
A presence that disturbs me with the joy
Of elevated thoughts; a sense sublime
Of something far more deeply interfused
Whose dwelling is the light of setting suns,
And the round ocean and the living air,
And the blue sky, and in the mind of man:
A motion and a spirit, that impels
All thinking things, all objects of all thought,
And rolls through all things. Therefore am I still
A lover of the meadows and the woods,
And mountains; and of all that we behold
From this green earth: of all the mighty world
Of eye, and ear, —both what they half create,
And what perceive; well pleased to recognise
In nature and the language of the sense
The anchor of my purest thoughts, the nurse,

The guide, the guardian of my heart, and soul
Of all my moral being.
William Wordsworth

Mary Shelley's *Frankenstein*, illustrates a macabre twist taken by the romantic movement. It underscored what might happen if nature's power is unleashed without adequate concern for its long-range consequences. Frankenstein's creature, instead of being beautiful, is hideously ugly, and although he was intended to be benevolent, he later became a raging monster. From our own perspective, it is worth asking whether or not the Industrial Revolution has gone beyond what its original creators intended. Here is a selection from *Frankenstein* that touches on the problem.

Frankenstein—an Excerpt

Mary Shelley

One of the phenomena which had peculiarly attracted my attention was the structure of the human frame, and, indeed, any animal endued with life. Whence, I often asked myself, did the principle of life proceed? It was a bold question, and one which has ever been considered as a mystery; yet with how many things are we upon the brink of becoming acquainted, if cowardice or carelessness did not restrain our enquiries. I revolved these circumstances in my mind, and determined thenceforth to apply myself more particularly to those branches of natural philosophy which relate to physiology. Unless I had been animated by an almost supernatural enthusiasm, my application to this study would have been irksome, and almost intolerable. To examine the causes of life, we must first have recourse to death. I became acquainted with the science of anatomy: but this was not sufficient; I must also observe the natural decay and corruption of the human body. In my education my father had taken the greatest precautions that my mind should be impressed with no supernatural horrors. I do not ever remember to have trembled at a tale of superstition, or to have feared the apparition of a spirit. Darkness had no effect upon my fancy; and a churchyard was to me merely the receptacle of bodies deprived of life, which, from being the seat of beauty and strength, had become food for the worm. Now I was led to examine the cause and progress of this decay, and forced to spend days and nights in vaults and charnel-houses. My attention was fixed upon every object the most insupportable to the delicacy of the human feelings. I saw how the fine form of man was degraded and wasted; I beheld the corruption of death succeed to the blooming cheek of life; I saw how the worm inherited the wonders of the eye and brain. I paused, examining and analysing all the minutiae of causation, as exemplified in the from life to death, and death to life, until from the midst of this darkness a sudden light broke in upon me—a light so brilliant and wondrous, yet so simple, that while I became dizzy with the immensity among so many men of genius who had directed their enquires towards the same science, that I alone should be reserved to discover so astonishing a secret.

Remember, I am not recording the vision of a madman. The sun does not more certainly shine in the heavens, than that which I now affirm is true. Some miracle might have produced it, yet the stages of the discovery were distinct and probable. After days and nights of incredible labour and fatigue, I succeeded in discovering the cause of generation and life; nay, more, I became myself capable of bestowing animation upon lifeless matter. . . .

It was on a dreary night of November, that I beheld the accomplishment of my toils. With an anxiety that almost amounted to agony, I collected the instruments of life around me, that I might infuse a spark of being into the lifeless things that lay at my feet. It was already one in the morning; the rain pattered dismally against the panes, and my candle was nearly burnt out, when, by the glimmer of the half-extinguished light, I saw the dull yellow eye of the creature open; it breathed hard, and a convulsive motion agitated its limbs.

How can I describe my emotions at this catastrophe, or how delineate the wretch whom with such infinite pains and care I had endeavoured to form? His limbs were in proportion, and I had selected his features as beautiful. Beautiful! Great God! His yellow skin scarcely covered the work of muscles and arteries beneath; his hair was of a lustrous black, and flowing; his teeth of a pearly whiteness; but these luxuriances only formed a more horrid contrast with his watery eyes, that seemed almost of the same colour as the dun white sockets in which they were set, his shrivelled complexion and straight black lips.

The different accidents of life are not so changeable as the feelings of human nature. I had worked hard for nearly two years, for the sole purpose of infusing life into an inanimate body. For this I had deprived myself of rest and health. I had desired it with an ardour that far exceeded moderation; but now that I had finished, the beauty of the dream vanished, and breathless horror and disgust filled my heart. Unable to endure the aspect of the being I had created, I rushed out of the room, and continued a long time traversing my bedchamber, unable to compose my mind to sleep. At length lassitude succeeded to the tumult I had before endured; and I threw myself on the bed in my clothes, endeavouring to seek a few moments of forgetfulness. But it was in vain: I slept, indeed, but I was disturbed by the wildest dreams. . . .

. . . I started from my sleep with horror; a cold dew covered my forehead, my teeth chattered, and every limb became convulsed: when, by the dim and yellow light of the moon, as it forced its way through the window shutters, I beheld the wretch—the miserable monster whom I had created. He held up the curtain of the bed; and his eyes, if eyes they may be called, were fixed on me. His jaws opened, and he muttered some inarticulate sounds, while a grin wrinkled his cheeks. He might have spoken, but I did not hear; one hand was stretched out, seemingly to detain me, but I escaped, and rushed down stairs. I took refuge in the courtyard belonging to the house which I inhabited; where I remained during the rest of the night, walking up and down in the greatest agitation, listening attentively, catching and fearing each sound as if it were to announce the approach of the demoniacal corpse to which I had so miserably given life.

Oh! no mortal could support the horror of that countenance. A mummy again endued with animation could not be so hideous as that wretch. I had gazed on him while unfinished; he was ugly then; but when those muscles and joints were rendered capable of motion, it became a thing such as even Dante could not have conceived.

I passed the night wretchedly. Sometimes my pulse beat so quickly and hardly, that I felt the palpitation of every artery; at others, I nearly sank to the ground through languor and extreme weakness. Mingled with this horror, I felt the bitterness of disappointment; dreams that had been my food and pleasant rest for so long a space were now become a hell to me; and the change was so rapid, the overthrow so complete!

The key to the Industrial Revolution, according to Dr. Bronowski, is man's new understanding of how to use the energy and power that nature provides in a variety of forms. There was heat energy and there was mechanical energy, but until the first half of the nineteenth century, there was little thought or hope that one could be converted to the other. The discovery that they could led to the development of the steam engine, and how it came about is described in this reading by George Gamow from his *Biography of Physics.*

Heat Is Motion

George Gamow

The idea that heat is some kind of internal motion of a material body, and not a special substance as Black and others thought, occurred first to a professional soldier and was substantiated by experiments carried on in a gun factory. Benjamin Thompson was born in Massachusetts, and during his youthful days participated in the Revolutionary War. Later on he changed his affiliation to England and soon became undersecretary of state in the ministry of the colonies. Still later he went to Bavaria as the ministry of war, and was given the title of Count Rumford for reorganization of the German army. Between all these military activities, he was deeply interested in the problems of science and in particular in the nature of heat. He was not satisfied with the contemporary view that heat is a certain substance, not unlike all other chemical substances, which unites with ice to produce water (ice + heat = water), or is liberated in various combustion processes. The reason for his doubts was the fact that heat is produced "from nothing" in friction processes which have apparently nothing to do with chemical transformation. Watching the boring of cannons at the munitions factory in Munich he was wondering why the casting became so hot, especially when the borer was blunt. He considered the possibility that material bodies may have a larger capacity for caloric fluid when they are in one solid block than when they are broken into small fragments; this would explain the liberation of heat during cannon boring when a large number of gun-metal turnings are produced. He carefully measured the heat capacity of a solid metal block and an equal weight of metal turnings and found them exactly the same. He tried to compare the weight of hot bodies with their weight when cold in an attempt to discover the weight of the escaping heat fluid but came to a negative result.

According to the figures given in his article in the *London Philosophical Transactions* (1799), a calorie cannot weigh more than 0.000013 mg. We know now that any form of energy possesses a ponderable mass which is obtained, according to the famous Einstein relation, by dividing it by the square of the velocity of light. The weight of a calorie of heat is actually: 0.00000000004 mg, which is well below the precision of any measurements. All this brought him to a conclusion that heat cannot be an ordinary substance and must be some kind of motion. He writes: "What is heat? It cannot be material substance. It appears difficult, if not quite impossible, for me to imagine heat to be anything else than that which in this experiment [cannon boring] was supplied continuously to the piece of metal, as the heat appeared, namely motion."

The ideas of Count Rumford were further developed several decades later by a German physician, Julius Robert Mayer, in his article "Remarks on the Forces of Inanimate Nature," published in 1842. Mayer arranged an experiment at a paper factory where the pulp contained in a large cauldron was stirred by a mechanism powered by a horse going around the circle. Measuring the rise of pulp temperature, he obtained a figure for the amount of heat produced by a given amount of mechanical work done by a horse. However, being too busy with his medical practice, he never pursued this line any further by carrying out more precise experiments, and the honor of measuring exactly the mechanical equivalent of heat went to an Englishman, James Prescott Joule. In his experiments Joule used an apparatus . . . which consisted of a water-filled vessel containing a rotating axis with several stirring

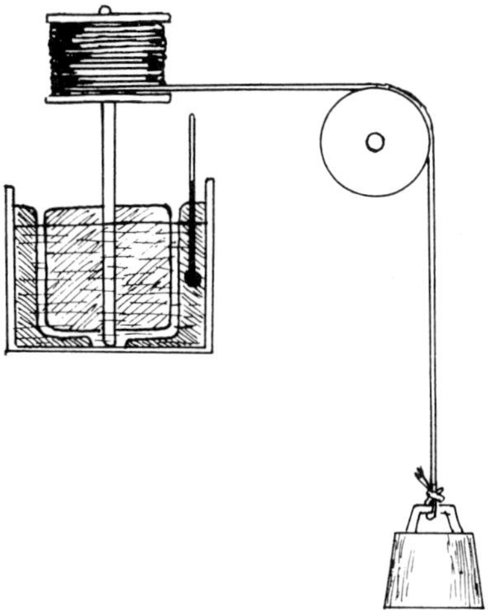

Joule's experiment on transformation of mechanical energy into heat. The descending weight rotates the paddles in a water-filled container and the temperature of water is increased by internal friction. Comparing the work done by descending weight with the increase of the heat content of water, Joule obtained the value of the mechanical equivalent of heat.

paddles attached to it. The water in the vessel was prevented from rotating freely along with the paddles by special vanes attached to the walls of the vessel which increase the internal friction. The axis with the paddles was driven by a weight suspended across a pulley, and the work done by the descending weight was transformed into friction-heat communicated to the water. Knowing the amount of water in the vessel, and measuring the rise of temperature, Joule could calculate the total amount of heat produced. On the other hand, the product of the driving weight and the distance of its descent provided the value of mechanical work. Repeating this experiment many times, and under different conditions, Joule established that there is a direct proportionality between the work done and the heat produced. Announcing in 1843 the result of his studies, he wrote: "The work done by the weight of one pound through 778 feet in Manchester will, if spent in producing heat by friction of water, raise the temperature of one pound of water by one degree Fahrenheit." This is the figure which, being expressed in these or other units, is now used universally whenever one has to translate heat energy into mechanical energy or vice versa.

The members of the Lunar Society could be considered the commissioned officers of the Industrial Revolution. They wore many uniforms and included Joseph Priestley, the minister-chemist; Josiah Wedgwood, the maker of tableware for the man in the street; and Matthew Boulton, rugged individualist and industrial tycoon. It was Boulton who made it possible for James Watt to build a workable steam engine, which provided a new form of power for the Industrial Revolution. Their meeting and collaboration is discussed here by J.D. Bernal.

James Watt and the Steam Engine

J.D. Bernal

[James] Watt . . . had an unfortunate life despite his work. He was . . . taken up by a great pioneer of the times called Roebuck, the founder of the modern chemical industry, who also started the first iron and steel works in this country, the Carron Iron Works near Falkirk in Scotland. Roebuck had the right technical ideas but he was not sufficiently bussinesslike, and the Carron Iron Works went bankrupt. Among his assets were Watt and his [steam] engine which, incidentally, had never worked. Roebuck was a Birmingham man and he brought it to the attention of his old friend Boulton, who was the greatest industrialist of his time. Boulton owned the largest button factory in Birmingham, but although his factory was suitable for its purpose up to a point when it was mostly a matter of fiddly little bits of metal work, when he got going in a big way, it needed some power and he acquired a millpond and a mill. But Birmingham is not a very hilly district or a very wet one, and the millpond always dried up in the summer time. That was sufficient for working in a small way because men could then be temporarily dismissed during the summer when they could work in the fields bringing in the harvest, and that was the way industry was run until 1870.

Excerpted from J.D. Bernal, *The Extension of Man*, M.I.T. Press, 1953; used with permission of George Weidenfeld and (Publishers) Limited and M.I.T. Press.

Roebuck suggested that Watt came along with his engine so that it could be used to pump the water up from the millpond into the mill race and so work the wheel. This he did and Boulton made a great deal of money out of it. This gave Boulton an idea, his ideas were very important scientifically, or, rather, the ideas of Boulton and Watt between them were. Boulton produced the money but he also produced an enormous amount of push, which was very necessary to get things done because Watt was not a very pushful person, but a rather melancholy Scot. Until he was eighty years old, he said he had never had a day free from anxiety about money.

Boulton's idea was that he would sell something that no one had ever sold before—power. He actually used those words; he wrote to Empress Catherine of Russia saying, "I am selling what the whole world wants: power." And this is how he did it. He sent his people down to Cornwall to say:

We are offering engines on these terms. Our firm, Boulton and Watt, will set up the engines, free, gratis and for nothing, at your mine. We will service them for the first five years and all we are asking in return is one-third of the difference between the cost of coals for our engines and the cost of hay for the horses that would have to do the same amount of work.

Well, the mine-owners thought he was obviously crazy but they accepted his offer.

Now, of course, came the disputed question of how much work a horse could do. First of all it is necessary to determine what is work? Galileo and Newton never discussed "work"; it is a very odd concept, you see, it is not this or that inertia, it has not a Latin name, it is just work. There were two terms used, one of which has gone out of use and that is "duty"—the turn of duty in a treadmill or in anything else, was work multiplied by time. Watt measured the amount of work a horse could do by making a horse pull something lifted over a pulley. He conceived the idea of work being the product of force and distance and of power being the rate of doing work. All these, as we think, elementary concepts of physics and thermodynamics, really came entirely from practice, they all came from Manchester or Birmingham people and they were not really recognised in London, as you will see later on. Having done this, Watt also very wisely secured a patent and, by various political pulls of one sort or another, had the patent extended for a fantastic period of forty years, therefore holding up anyone else's development, unless they were prepared to cheat—which, of course, a great many people did.

Meanwhile came the next problem. The mines were all right; the mine-owners complained, of course, after the first ten years, about having to keep on paying Mr. Boulton and Mr. Watt; it seemed to them they would go on paying for ever. They had quite forgotten what the situation was at the beginning. No one wanted horses now, everyone wanted engines. The trouble with an engine, however, was that though it could pump water, it could not actually make anything go round; and the reason for this was that the action of the cylinder was so irregular. Here, then, was where Watt introduced two developments which transformed the whole situation.

. . . One of the original rotative machines . . . actually combined the feature of an old beam machine with a rotating part. It is a very curious thing, a sun-and-planet gear. The part with the planet on it goes round and round and turns the wheel at half the speed. We might ask: why could he not use a crank? Well, the reason was, believe it or not, that the crank was patented by somebody else. Watt was so furious that someone had patented the crank

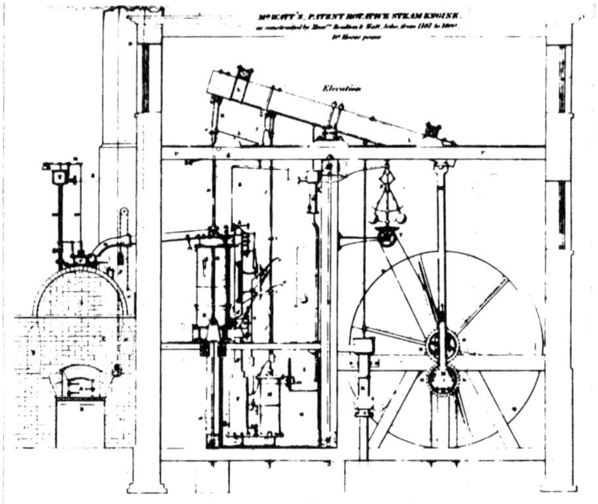

Watt's double-acting rotative steam engine of 1784; the sun-and-planet gear may be seen on the right.

that he determined not to use one and not to pay the royalties. In fact, they came round to cranks afterwards; I do not know what happened to the patent, it probably expired. So we now had the first modern engine which was a real rotary engine, one that could be used for the new factories, which did not so much want water pumping as turning of machinery.

There was, however, one element which for the future was to be even more important. You can see what made the rotary process possible; the centrifugal governor. If you translate "governor" into Greek, it becomes *kybernetes.* That is all cybernetics means—governorship, steersmanship—and this governor adjusted the steam inlet to the cylinders according to how rapidly the machinery was going. This is the first example of real feedback. The rest is all valve gear and so forth, Watt's rather ingenious devices to drive a piston straight up and down from a beam instead of hanging it from a beam. All the previous machines worked on tension, they could not be driven, you could just pull one side and then the other. To drive them needed one of these gears.

. . . [The figure] shows a steam-engine which was used in a big flour mill; it still has this sun-and-planet gear with the governor and by now it is double-acting, it has the cylinder which admits steam at both ends and it really is quite a remarkably efficient engine. From this time on things never looked back.

As science and technology came into full power in the Industrial Revolution, a few men asked whether these two manifestations of man's ascent had somehow removed themselves from the arena of human values. It is a question that is being asked even more urgently today, as is shown in this reading by Joseph B. Platt.

Science, Technology, and Values
Joseph B. Platt

What do we mean by science, by technology, and for that matter, value?

Science is knowledge. In particular, the science with which we are concerned is a body of knowledge which derives its facts from observation, connects these facts with theories and then tests or modifies these theories as they succeed or fail in predicting or explaining new observations. In this sense, science has a relatively recent history—perhaps four centuries. The test of science is never whether it is useful, but to the extent to which it is true.

Over these four centuries, science has led to our world view, much different from that of our ancestors. We were moved centuries ago from the center of the universe to a planet; we now find that universe is much larger than we knew, and it is inhabited also by quasars and neutron stars and black holes. Over recent years we have learned a great deal about this planet: the oceans, the Earth's crust, the many fascinating topics of this scientific meeting, and much more. We have been humbled in our own importance, but we have been moved into a universe of order, where we can learn that order; we are much less the playthings of inscrutable fates.

Science is international; we build on the observations and the theories of persons from many nations. Science is critical; a scientist is judged not on rank or position, but on the ability of others to repeat his observations or on whether his theories predict accurately. It is not surprising that the last several centuries have seen a growing faith in the dignity of the individual, an appreciation of the unity of mankind and criticism of arbitrary authority. These are not only the beliefs of scientists; they are now also the beliefs of many of our fellow citizens who are not scientists.

Technology has a much longer history than science—a history as long as the history of man. We have evolved together with our techniques and tools over millions of years. The major changes in human population are due to the technology we have developed—the domestication of grain, the development of irrigation, the invention of methods for storing and preserving food. We live on the bounty of the Earth, but how many of us live and how many starve depends on how well we can use that bounty. At the time of Columbus, perhaps 15 million people lived in the Americas, and most of them lived in the major civilizations of Mexico, Central America, and South America, where agriculture was relatively advanced. Most human labor was used to obtain food. We now have over half a billion people in the Americas. In my own country less than 5 per cent of our labor force is needed to produce food.

Excerpted from Joseph B. Platt, "The Value of Science and Technology to Human Welfare," *Bulletin of the Atomic Scientists*, October 1973. Reprinted by permission of the Bulletin of the Atomic Scientists.

The race between population and food is not over, as the titles of the symposia at this meeting attest: nutrition, food technology, uses of arid lands, the sea and its resources. Fifteen years ago there were many predictions of widespread famine in the world in the early 1970s. The various technologies now known as the Green Revolution have made these predictions premature, but the race continues. Any ultimate solution must depend on a limit on population, but what that limit can be depends on technology.

Technology is not science—technology is how we do things, not how we think of them. We have used many technologies we do not really understand. It is only recently that research—the increase of understanding—has been considered part of the development of new technologies. Each new technology has increased the choices—the opportunities—available to those who could use that technology. In this century those choices have increased enormously. Just as people breed people, ideas breed ideas, and technologies breed technologies. (The death rates do differ; all men die, but some ideas and some technologies have lasted through history.) For those societies which a generation ago had committed significant resources and numbers of people to developing new technologies, the result in this generation has been greater choice, greater opportunity—and most of these societies have chosen to have longer lives and a better standard of living for their peoples.

It is not easy to have many choices. It is the nature of a choice that to take advantage of one, a person must refuse another. As technology presents us with more choices, many of them interrelated, the tasks of planning and of selection become more complex. How does an individual or a government decide among the courses of action that technology makes possible?

Choices, for an individual or for a society, are guided by values. We tend to identify the values of a society—or a person—by the choices made. "Death before dishonor" calls to our minds a chivalrous code of conduct governing one's responsibility to his fellows, and a knight would choose death before breaking that code. The citizens of Sparta chose rigid discipline over freedom and comfort. We still speak of Spartan values: this one places the welfare of the family above that of the individual; that one places greatest value on military valor; still another places high value on artistic creativity. I leave it to the poets to tell and the social scientists to explain what mixture of religious heritage, folk history, shared experiences or national aspirations determine these differences. I simply note that the differences exist. Given the opportunity to earn a second income, one man will work longer and another will go fishing. Given a surplus in the treasury, one nation increases its educational expenditures and another reduces taxes. Values differ, and values determine choices—choices among technologies or among forms of government. Values also change with time. We do learn from experience, and from each other. Choices force us to examine our values: Do we really value clean air more than low-cost fuel? Would we rather go hungry than change our food habits?

It is widely believed by non-scientists and scientists alike that science and the methods of science are value-free, that they simply do not interact with other human systems of belief. I do not believe this. It is true—and one of the great strengths of science—that its observations and theories once made can be verified by other observers of different nationalities, different cultures, and different values. Any observations, any theories, that cannot stand this test do not remain part of science. It is also true that it is more difficult—and less certain—to apply the methods of science to human affairs than it is to the motion of the planets; one must identify more variables, events are less repeatable, and theories are less detailed in their prediction. Fortunately or unfortunately, people are not completely predictable.

But the choice to do science—and the elegance with which it is done—is a human choice, a value choice. Each new understanding added to science is an individual human creation, created by someone who found the effort worthwhile. The result of the science of the last four centuries has not simply been excitement and discovery for scientists; it has also been a growing belief on the part of non-scientists that we and our world are comprehensible, that we do live in an orderly universe that we can make our choices with increasing knowledge of the consequences. It is tempting to speculate on what lies ahead. There are many more people in the world than previously; there are still more to come. Through communication and travel, we are increasingly in touch with each other. What we do to our common environment influences each other as never before. In short, we are interdependent in time and space as never before, and these interdependencies can be expected to increase.

Do the newer methods of science influence our view of such a world? We are increasingly using statistical methods to understand complex phenomena, using intricate mathematical models to explore the behavior of physical and biological systems, tracing the ways in which the information content of a large molecule is used or withheld in the construction of other molecules. It is entirely conceivable that these methods of thought may in another generation also be part of the way in which we understand our world, ourselves and our interrelationships—again, for scientist and non-scientist alike. So the role of science is to understand; technology, to make new choices possible; of values, to determine which choices "ought" to be made. The choices which shape the lives of people are in some cases made by government, in some cases are made in the market place by consumers, in some cases by many individual farmers, merchants or fishermen.

Begun in England, the Industrial Revolution spread quickly throughout the world. In the United States, it blossomed profusely at the end of the nineteenth century. It produced men like Thomas Edison, Charles Eastman, and Henry Ford, tinkers all, who shared a common vision that anything is possible. The following reading gives a glimpse of the life and times of one of these visionaries, Henry Ford.

The Model T

Allan Nevins and Frank Ernest Hill

The Model T was the handiwork of a dozen men, among whom Henry Ford, Joseph Galamb, and C. Harold Wills were the most important. Ford must be named first, not only because he bore the principal responsibility, but because we have ample evidence that he contributed the guiding concepts and furnished the guiding hand. Of course, he knew all rival cars well, foreign as well as domestic (we hear of a Renault and later a Minerva imported for study), and borrowed ideas from them. In the main, however, the new model was an original product of the company's engineering group, with Ford as leader. Its basic merit lay in a hitherto unmatched combination of lightness, simplicity, and power; and its most striking features—the guiding mind; his was clearly the most powerful personality. We might be tempted to fall back on the usually accurate theory of social invention, but all the survivors of that era who have left reminiscences agree as to his leadership and his paramount contributions.

In some mechanical matters, men then—as later—found in Ford an extraordinary power of divination. He was quick to see precisely how an adjustment of parts could be bettered. When a problem of thickness of materials, distribution of weight, or tensile strength arose, he sometimes solved it by sheer inspiration. A university-trained engineer with precision instruments and slide-rule could make computations far beyond him, but he had an instinct which left expert engineers wondering at his superior insight. Here lay his genius. In most other matters he was a shrewd, skilled, but undisciplined and, in some respects, ignorant mechanic; his faults of taste and temper were so evident that a perceptive observer like Percival Perry wondered uneasily as to the use he would make of the great power gathering in his hands. But in dealing with the challenge presented in the production of Model T the tremendous challenge of designing cheap, efficient transport for the masses of America and the world—he had just the single gift required.

Once the feat was performed, he was exultant. The day the first Model T was brought out of the experimental room, and (after some last-minute tinkering with a stuck transmission) set running, was a day to be celebrated. Ford asked George Holley to take him downtown. They traversed all the main streets—Woodward Avenue, Michigan Avenue, Lafayette. . . . Throughout the ride Holley found Ford as pleased as a boy with an athletic prize. To another worker, George Brown, he seemed "tickled to death." He bubbled over with exuberant joy. "Every time he'd meet somebody, he'd give him a kick in the pants or a punch between the shoulders." Looking around a circle of his associates in the plant, he grinned happily and ejaculated: "Well, I guess we've got started."

They were started indeed. Once loosed, the cataract of orders never stopped. The wave of enthusiasm for the new model rolled across America and round the world. At the Olympia Show in London in 1909 the two sensations were the Model T from the Middle West and the new Darracq from France, both low priced and both conspicuously well made. Other striking cars were brought out in 1908: the large Buick Model D at $1750, for example, of which Buick manufactured 543 that year, and the little Maxwell two-cylinder fourteen-horsepower runabout at $825. None, however, combined anything like so many new features as the ungraceful, bouncing, noisy, tough-looking, and endlessly useful new Ford.

The faults of the early Model T's, as we have suggested, were numerous and painful. One weak point was the bearings on the rear axle, made during the first year of babbitt metal; pounding on the road elongated them, requiring frequent replacements. Roller bearings . . . were introduced the second year. The transmission bands gave users constant trouble, for the lining easily burned out until a better fabric was employed. Complaint arose over the fact that front and rear wheels were of different sizes, which necessitated carrying double sets of tires and tubes. The rear seat of the touring car was too narrow, an addition of a few inches presently sufficing to accommodate three instead of two persons. Riveting had to be bettered, and appropriate steps were quickly taken. By 1911 fore doors for the touring car, missing in those first made, were available for ten dollars a pair. Starting the Model T, especially in cold weather, was a source of innumerable broken arms and endless profanity; and we find the company suggesting that owners with cold garages attach an electric light to a long cord and keep it burning under the hood to keep the motor warm.

Many of the special features of which Ford dealers made much seem quaint today; the hand-lever employed to reverse the machine, for example. But wherever the Model T went curious groups gathered to gaze and confident owners paused to point out its advantages. The left-hand drive, for example, for both the Model N and Model K had kept the right-hand drive; the lever just under the steering-wheel which controlled the flow of gasoline; the double system of braking, hand and foot; the extra-large springs; the good leather upholstery. Early buyers were able to get touring cars painted red, and roadsters in gray, but beginning June 1, 1909, all model T's became a Brewster Green with black trimmings and red striping. The company was too busy turning them out in quantity to bother about different styles of painting.

The faults and merits, the graces and gaucheries of the early Model T have been hymned by many writers who fell into alternate moods of exasperation and rapture as they dealt with its temperamental, skittish ways, but who almost unanimously cherished an intimate affection for the unpredictable creature. No two cars were quite alike. Mastery of any one involved highly personal qualities of courage, skill, intuition, and luck. As of Cleopatra, it could be said that time could not wither nor custom stale the infinite variety of the flivver; with all its superior dependability and simplicity it combined an arch and mercurial eccentricity. It was more like a human being (of feminine gender) than any other car ever known to man.

9

Ladder of Creation

These, as successive generations bloom,
New powers acquire, and larger limbs assume.

Erasmus Darwin

Man now stands on the top rung of "the ladder of creation," a ladder built by the physical and biological forces that were set in motion more than four and one-half billion years ago, according to the most reliable scientific estimates. These forces converged when Charles Darwin formulated his theory about natural selection and the origin of the species. Both Darwin and Alfred Russel Wallace used the writings of Thomas Robert Malthus as the premise for their studies and theories of the evolutionary process. Among the topics discussed by Malthus in his "First Essay on Population" was his concern that the world's population would eventually reach the point that there would not be enough food to go around. Speculation on the consequences of such a crisis for the survival of man and animals led Darwin to develop the concept of natural selection. Here is a section of the influential Malthus essay.

First Essay on Population–an Excerpt

Thomas Robert Malthus

I said that population, when unchecked, increased in a geometrical ratio; and subsistence for man in an arithmetical ratio.

Let us examine whether this position be just.

I think it will be allowed, that no state has hitherto existed (at least that we have any account of) where the manners were so pure and simple, and the means of subsistence so abundant, that no check whatever has existed to early marriages; among the lower classes, from a fear of not providing well for their families; or among the higher classes, from a fear of lowering their condition in life. Consequently in no state that we have yet known, has the power of population been left to exert itself with perfect freedom.

Whether the law of marriage be instituted, or not, the dictate of nature and virtue, seems to be an early attachment to one woman. Supposing a liberty of changing in the case of an unfortunate choice, this liberty would not affect population till it arose to a height greatly vicious; and we are now supposing the existence of a society where vice is scarcely known.

In a state therefore of great equality and virtue, where pure and simple manners prevailed, and where the means of subsistence were so abundant, that no part of the society could have any fears about providing amply for a family, the power of population being left to exert itself unchecked, the increase of the human species would evidently be much greater than any increase that has been hitherto known.

In the United States of America, where the means of subsistence have been more ample, the manners of the people more pure, and consequently the checks to early marriages fewer, than in any of the modern states of Europe, the population has been found to double itself in twenty-five years.

This ratio of increase, though short of the utmost power of population, yet as the result of actual experience, we will take as our rule; and say,

That population, when unchecked, goes on doubling itself every twenty-five years, or increases in a geometrical ratio.

Let us now take any spot of earth, this Island for instance, and see in what ratio the subsistence it affords can be supposed to increase. We will begin with it under its present state of cultivation.

If I allow that by the best possible policy, by breaking up more land, and by great encouragements to agriculture, the produce of this Island may be doubled in the first twenty-five years, I think it will be allowing as much as any person can well demand.

In the next twenty-five years, it is impossible to suppose that the produce could be quadrupled. It would be contrary to all our knowledge of the qualities of land. The very utmost that we can conceive, is, that the increase in the second twenty-five years might equal the present produce. Let us then take this for our rule, though certainly far beyond the truth; and allow that by great exertion, the whole produce of the Island might be increased every twenty-five years, by a quantity of subsistence equal to what it at present produces. The most enthusiastic speculator cannot suppose a greater increase than this. In a few centuries it would make every acre of land in the Island like a garden.

Yet this ratio of increase is evidently arithmetical.

It may be fairly said, therefore, that the means of subsistence increase in an arithmetical ratio.

202 Let us now bring the effects of these two ratios together.

The population of the Island is computed to be about seven millions; and we will suppose the present produce equal to the support of such a number. In the first twenty-five years the population would be fourteen millions; and the food being also doubled, the means of subsistence would be equal to this increase. In the next twenty-five years the population would be twenty-eight millions; and the means of subsistence only equal to the support of twenty-one millions. In the next period, the population would be fifty-six millions, and the means of subsistence just sufficient for half that number. And at the conclusion of the first century, the population would be one hundred and twelve millions, and the means of subsistence only equal to the support of thirty-five millions; which would leave a population of seventy-seven millions totally unprovided for.

A great emigration necessarily implies unhappiness of some kind or other in the country that is deserted. For few persons will leave their families, connections, friends, and native land, to seek a settlement in untried foreign climes, without some strong subsisting causes of uneasiness where they are, or the hope of some great advantages in the place to which they are going.

But to make the argument more general, and less interrupted by the partial views of emigration, let us take the whole earth, instead of one spot, and suppose that the restraints to population were universally removed. If the subsistence for man that the earth affords was to be increased every twenty-five years by a quantity equal to what the whole world at present produces; this would allow the power of production in the earth to be absolutely unlimited, and its ratio of increase much greater than we can conceive that any possible exertions of mankind could make it.

Taking the population of the world at any number, a thousand millions, for instance, the human species would increase in the ratio of—1, 2, 4, 8, 16, 32, 64, 128, 256, 512, etc. and subsistence as—1, 2, 3, 4, 5, 6, 7, 8, 9, 10, etc. In two centuries and a quarter, the population would be to the means of subsistence as 512 to 10; in three centuries as 4096 to 13; and in two thousand years the difference would be almost incalculable, though the produce in that time would have increased to an immense extent.

No limits whatever are placed to the productions of the earth; they may increase for ever and be greater than any assignable quantity; yet still the power of population being a power of a superior order, the increase of the human species can only be kept commensurate to the increase of the means of subsistence, by the constant operation of the strong law of necessity acting as a check upon the greater power.

The effects of this check remain now to be considered.

Among plants and animals the view of the subject is simple. They are all impelled by a powerful instinct to the increase of their species; and this instinct is interrupted by no reasoning, or doubts about providing for their offspring. Wherever therefore there is liberty, the power of increase is exerted; and the superabundant effects are repressed afterwards by want of room and nourishment, which is common to animals and plants; and among animals, by becoming the prey of others.

The effects of this check on man are more complicated.

Impelled to the increase of his species by an equally powerful instinct, reason interrupts his career, and asks him whether he may not bring beings into the world, for whom he cannot provide the means of subsistence. In a state of equality, this would be the simple question. In the present state of society, other considerations occur. Will he not lower his rank in life? Will he not subject himself to greater difficulties than he at present feels? Will he not be obliged to labour harder? and if he has a large family, will his utmost exertions enable him to support them? May he not see his offspring in rags and misery, and clamouring for bread that he cannot give them? And may he not be reduced to the grating necessity of forfeiting his independence, and of being obliged to the sparing hand of charity for support?

In his introduction to *The Origin of the Species*, which was first published in 1859, Darwin acknowledges his debt to Malthusian doctrine. In addition, he gives some insight into how he arrived at the conclusions he so boldly presents.

On *The Origin of the Species*

Charles Darwin

When on board H.M.S. 'Beagle', as naturalist, I was much struck with certain facts in the distribution of the organic beings inhabiting South America, and in the geological relations of the present to the past inhabitants of that continent. These facts, as will be seen in the latter chapters of this volume, seemed to throw some light on the origin of the species—that mystery of mysteries, as it has been called by one of our greatest philosophers. On my return home, it occurred to me, in 1837, that something might perhaps be made out on this question by patiently accumulating and reflecting on all sorts of facts which could possibly have any bearing on it. After five years work I allowed myself to speculate on the subject, and drew up some short notes; these I enlarged in 1844 into a sketch of the conclusions which then seemed to me probable; from that period to the present day I have steadily pursued the same object. I hope that I may be excused for entering on these personal details, as I give them to show that I have not been hasty in coming to a decision.

My work is now (1859) nearly finished; but as it will take me many more years to complete it, and as my health is far from strong, I have been urged to publish this Abstract. I have more especially been induced to do this, as Mr. Wallace, who is now studying the natural history of the Malay archipelago, has arrived at almost exactly the same general conclusions that I have on the origin of species. In 1858 he sent me a memoir on this subject, with a request that I would forward it to Sir Charles Lyell, who sent it to the Linnean Society, and it is published in the third volume of the Journal of that Society. Sir C. Lyell and Dr. Hooker, who both knew of my work—the latter having read my sketch of 1844—honoured me by thinking it advisable to publish, with Mr. Wallace's excellent memoir, some brief extracts from my manuscripts.

This Abstract, which I now publish, must necessarily be imperfect. I cannot here give references and authorities for my several statements; and I must trust to the reader reposing some confidence in my accuracy. No doubt errors will have crept in, though I hope I have always been cautious in trusting to good authorities alone. I can here give only the general conclusions at which I have arrived, with a few facts in illustration, but which, I hope, in most cases will suffice. No one can feel more sensible than I do of the necessity of hereafter publishing in detail all the facts, with references, on which my conclusions have been grounded; and I hope in a future work to do this. For I am well aware that scarcely a single point is discussed in this volume on which facts cannot be adduced, often apparently leading to conclusions directly opposite to those at which I have arrived. A fair result can be obtained only by fully stating and balancing the facts and arguments on both sides of each question; and this is here impossible.

I much regret that want of space prevents my having the satisfaction of acknowledging the generous assistance which I have received from very many naturalists, some of them personally unknown to me. I cannot, however, let this opportunity pass without expressing my deep obligations to Dr. Hooker, who, for the last fifteen years, has aided me in every possible way by his large stores of knowledge and his excellent judgment.

In considering the Origin of Species, it is quite conceivable that a naturalist, reflecting on the mutual affinities of organic beings, on their embryological relations, their geographical distribution, geological succession, and other such facts, might come to the conclusion that species have not been independently created, but had descended, like varieties, from other species. Nevertheless, such a conclusion, even if well founded, would be unsatisfactory, until it could be shown how the innumerable species inhabiting this world have been modified, so as to acquire that perfection of structure and coadaptation which justly excites our admiration. Naturalists continually refer to external conditions, such as climate, food, etc., as the only possible cause of variation. In one limited sense, as we shall hereafter see, this may be true; but it is preposterous to attribute to mere external conditions, the structure, for instance, of the woodpecker, with its feet, tail, beak, and tongue, so admirably adapted to catch insects under the bark of trees. In the case of the mistletoe, which draws its nourishment from certain trees, which has seeds that must be transported by certain birds, and which has flowers with separate sexes absolutely requiring the agency of certain insects to bring pollen from one flower to the other, it is equally preposterous to account for the structure of this parasite, with its relations to several distinct organic beings, by the effects of external conditions, or of habit, or of the volition of the plant itself.

It is, therefore, of the highest importance to gain a clear insight into the means of modification and coadaptation. At the commencement of my observations it seemed to me probable that a careful study of domesticated animals and of cultivated plants would offer the best chance of making out this obscure problem. Nor have I been disappointed; in this and in all other perplexing cases I have invariably found that our knowledge, imperfect though it be, of variation under domestication, afforded the best and safest clue. I may venture to express my conviction of the high value of such studies, although they have been very commonly neglected by naturalists.

From these considerations, I shall devote the first chapter of this Abstract to Variation under Domestication. We shall thus see that a large amount of hereditary modification is at least possible; and, what is equally or more important, we shall see how great is the power of man in accumulating by his Selection successive slight variations. I will then pass on the variability of the species in a state of nature; but I shall, unfortunately, be compelled to treat this subject far too briefly, as it can be treated properly only by giving long catalogues of facts. We shall, however, be enabled to discuss what circumstances are most favorable to variation. In the next chapter the Struggle for Existence amongst all organic beings throughout the world, which inevitably follows from the high geometrical ratio of their increase, will be considered. This is the doctrine of Malthus, applied to the whole animal and vegetable kingdoms. As many more individuals of each species are born than can possibly survive; and as, consequently, there is a frequently recurrent struggle for existence, it follows that any being, if it vary however slightly in any manner profitable to itself, under the complex and sometimes varying conditions of life, will have a better chance of surviving, and thus be *naturally selected.* From the strong principle of inheritance, any selected variety will tend to propagate its new and modified form.

This fundamental subject of Natural Selection will be treated at some length in the fourth chapter; and we shall then see how Natural Selection almost inevitably causes much Extinction of the less improved forms of life, and leads to what I have called Divergence of Character. In the next chapter I shall discuss the complex and little known laws of variation. In the five succeeding chapters, the most apparent and gravest difficulties in accepting the theory will be given; namely, first, the difficulties of transitions, or how a simple being or a simple organ can be changed and perfected into a highly developed being or into an elaborately constructed organ; secondly, the subject of Instinct, or the mental powers of animals; thirdly, Hybridism, or the infertility of species and the fertility of varieties when intercrossed; and fourthly, the imperfection of the Geological Record. In the next chapter I shall consider the geological succession of organic beings throughout time; in the twelfth and thirteenth, their geographical distribution throughout space; in the fourteenth, their classification or mutual affinities, both when mature and in an embryonic condition. In the last chapter I shall give a brief recapitulation of the whole work, and a few concluding remarks.

No one ought to feel surprise at much remaining as yet unexplained in regard to the origin of species and varieties, if he make due allowance for our profound ignorance in regard to the mutual relations of the many beings which live around us. Who can explain why one species ranges widely and is very numerous, and why another allied species has a narrow range and is rare? Yet these relations are of the highest importance, for they determine

the present welfare, and, as I believe, the future success and modification of every inhabitant of this world. Still less do we know of the mutual relations of the innumerable inhabitants of the world during the many past geological epochs in its history. Although much remains obscure, and will long remain obscure, I can entertain no doubt, after the most deliberate study and dispassionate judgment of which I am capable, that the view which most naturalists until recently entertained, and which I formerly entertained—namely, that each species has been independently created—is erroneous. I am fully convinced that species are not immutable; but that those belonging to what are called the same genera are lineal descendants of some other and generally extinct species, in the same manner as the acknowledged varieties of any one species are the descendants of that species. Furthermore, I am convinced that Natural Selection has been the most important, but not the exclusive, means of modification.

When Darwin first proposed his theory that the human race was the product of natural selection processes and evolutionary changes that originated in the lower animals, he was the immediate target of critics from every quarter. Some of the criticism was sharp, but good-natured, as evident in this verse from the May 1861 issue of *Punch* magazine:

Let pigeons and doves
Select their own loves,
And grant them a million of ages,
Then doubtless you'll find,
They've altered their kind,
And changed into prophets and sages.

Much of the most scathing criticism of Darwin's theory came from churchmen who argued that Darwin was attempting to overturn the biblical account of creation, in which

God is said to have made man in his own image, in one miraculous instant. In a letter to Huxley, Darwin reveals that he was very sensitive to the critical barrage that greeted the publication of his book.

I have got fairly sick of hostile reviews. Nevertheless, they have been of use in showing me when to expatiate a little and to introduce a few new discussions.

I entirely agree with you, that the difficulties on my notions are terrific, yet having seen what all the Reviews have said against me, I have far more confidence in the general *truth of the doctrine than I formerly had. Another thing gives me confidence, viz. that some who went half an inch with me now go further, and some who were bitterly opposed are now less bitterly opposed. . . . I can pretty plainly see that, if my view is ever to be generally adopted, it will be by young men growing up and replacing the old workers, and then young ones finding that they can group facts and search out new lines of investigation better on the notion of descent than on that of creation.*

Some who read Darwin's book, including some churchmen, insisted that the naturalist had not been outrageously heretical and that "creative evolution" may indeed explain how man came to his presently superior place among the animals. They argued that straight-line evolution, or *orthogenesis*, was an acceptable middle ground between Darwinian extremists, on the one hand, who said that evolution was purely accidental and the most fundamental adherents of the Bible, on the other hand. The viewpoint of the creative evolutionists is presented in this reading by Canon C.E. Raven.

Science, Religion, and the Future

Canon C.E. Raven

In . . . *The Origin of Species* . . . Darwin explained his doctrine of evolution. May I remind you of it?

(1) All living things show slight variations—no two members of a family are exactly alike
(2) These variations are inherited—John takes after his father
(3) There is a fierce struggle for existence, so that a very slight variation in strength or cunning or camouflage will give its possessor an advantage and determine its survival
(4) Over the long periods of geological time these successful variations have produced the vast differences between living plants and animals that we now know.

Darwin called his doctrine Natural Selection; and collected a vast amount of evidence to prove that Nature was doing slowly what human dog-fanciers did more quickly—produce totally distinct types.

Two or three times in the history of science there have come sudden solutions to problems that have been perplexing people for generations. Isaac Newton's discovery of the law of gravity and Charles Darwin's doctrine of natural selection each produced an epoch-making change. Each of them altered man's whole idea of the universe. In each case men had long been aware that there was a whole realm of experience waiting to be appreciated and understood; they had been groping and guessing; and suddenly the key was found, the doors were opened, and the new realm could be explored. In Newton's case the effect was to reveal the universe as a scene of ordered law, to suggest that it could be interpreted in terms of mathematics and physics, and so to give an immense impetus to mechanical and afterwards to industrial development. In Darwin's case the consequences were still more revolutionary; for his doctrine dealt not with the laws governing inanimate objects, not with physics and chemistry, but with living creatures, with man and man's position in the scheme of things, and with religion.

Newton's work . . . did not challenge religion directly. Men had long been realising that God must be a God of order and of law, not arbitrary or self-contradictory; so to think of Him as the Great Mechanic and of the Universe as a perfect piece of machinery did not seem irreligious or even inadequate. Moreover, Newton, like the other great scientists of the seventeenth century,

Robert Boyle and John Ray, the naturalist, was a profoundly religious man; and the Christian leaders, great thinkers like Ralph Cudworth and great bishops like Wilkins and Tillotson, were wholly in sympathy with the scientific movement. In any case the scientists were relatively few and their ideas had little popular influence.

With Darwin the position was entirely different. People were already accustomed to the doctrine of *laissez-faire*, the refusal, that is, to interfere with the ruthless competition of "free" industry, or to allow any measure of social reform. The Chartists and Lord Shaftesbury, Maurice and the Christian Socialists had protested against it in vain. *Laissez-faire* was in fact the doctrine of struggle for existence applied to industry. Darwin's extension of *laissez-faire* as the law of all evolutionary progress was certain of a general hearing. Moreover, since Tom Paine's time, that is, since the French Revolution, the historical value of the book of Genesis and of the stories of miracles and of the Old Testament pictures of God had been very widely challenged. Darwin's new evidence against them was eagerly accepted and employed. So both conservatives and reformers were ready to support him. His doctrine of struggle won the approval of the hard-headed and hard-hearted employers who profited by child-labour and the sweated industries, and his challenge to the literal accuracy of the Bible delighted the opponents of tradition and of prestige.

How, then, were his ideas received by thoughtful Christians? Well, the Christians of the time were in no mood to examine his teaching carefully or to ask themselves how far it really upset their convictions. It is probably true . . . that most of them still believed that the earth was made in six days and that man was a special creation totally separate from everything else. But there were very many who had long come to feel that the book of Genesis could not be taken literally, and that creation was not a work done once for all, but a continuing and still uncompleted process. Such people were ready, like Charles Kingsley, to welcome Darwin and to say: "Thank goodness we have at last got rid of the idea that God is a sort of master-magician: we must now choose, if we can, between belief in a world of mere accident and a world which is in the care of an ever-active and indwelling Deity." Unfortunately (and it is a tragic event, whatever one's views of Darwinism may be), unfortunately the Church leaders as a whole were already scared by the development of biblical criticism and by the growth of liberal theology, and so were quick to feel the danger of Darwin's book as an attack upon their tradition. So they made common cause with the scientists like Richard Owen who had at once denounced Darwinism, and allowed themselves to be represented by the Bishop of Oxford, Samuel Wilberforce, in his violent assault upon *The Origin of Species* at the famous meeting of the British Association at Oxford in the year 1860. There the bishop, a clever and popular speaker but with no first-hand knowledge of the subject, attacked the whole idea of evolution as involving a denial of man's special creation and place in the universe, and concluded by asking Huxley, Darwin's young and enthusiastic supporter who was sitting in front of him, whether he claimed descent from an ape on his father's or his mother's side. Huxley's reply and retort were so crushing as to discredit once for all both the arguments and the taste of the bishop, and to make it appear that in a conflict between eminent representatives of religion and of science the victory was wholly with science. It is hardly too much to say that we are still suffering from the controversy thus begun.

In order to understand the issues at stake it is important to realise the essential point of controversy. This was not the matter of Adam and Eve or of the Creation Stories; nor was it in fact the question of man's ancestry whether from ape or angel. A more deep-seated conviction was involved—nothing less than the oldest and most familiar of the arguments for the existence of God, the argument from design. It seemed to Darwin himself and to many, if not most, of his readers that he was destroying all belief in there being any element of purpose in the evolution of life. Now this is an ancient and universal belief, indeed one of the most immediate and powerful of all our human reactions to the world of nature. Let us look at it for a moment.

As soon as men began to reflect and to work upon their environment—that is as soon as they attained any human status—they realised that the earth was so constituted that growth took place in it. There were a thousand difficulties to be overcome: life, whether human or animal, was always precarious: but in spite of all the hazards and adventures living creatures somehow managed to survive. The more men learnt about the world, about the plants and animals, about themselves and their needs, the more evident was the extraordinary intricacy of their existence. It was unthinkable, even if some philosophers might assert it, that so complex a design could be the result of blind chance, of a fortuitous coincidence of atoms. If by spilling a million letters at random you could produce a play like *Hamlet*, that would be a trifle in comparison with the continuous marvel of an accidental world. So the argument that the design implies the designer, and that the intelligible order of the cosmos cannot be explained in terms of a less than intelligent and purposive cause, became the familiar commonplace of all human belief.

Cicero formulated it in the century before Christ, but it was an ancient conviction long before his day. It has always been man's favourite argument for the existence of God; and in the scientific movement, which had so immensely enlarged man's knowledge of the complexity and the interdependence of every part of the universe, there had until Darwin seemed to be a plain reinforcement of the truth of that argument.

Now Darwin's contention seemed to destroy the whole idea. Random variations acted upon by the selective influence of the struggle for existence were represented as responsible for all those beautiful and ingenious adaptations of the organism to its environment which John Ray had collected as evidence of "the Wisdom of God in the Works of Creation"—and which, in spite of Darwin, continue to thrill you and me with a sense of wonder and of awe. Here, then, was no proof of an ever-vigilant master-designer, but merely the operation of a purely automatic process. Here, indeed, was a vast mass of evidence to show that no intrusions of the supernatural, no succession of miraculous acts of creation, need be postulated or could be found in the orderly course of nature. Granted a world like this, its evolution could be explained without recourse to divine interventions and special creations. Scientists could not yet set out the whole course of events from beginning to end. There were still gaps in the record. But it would be unwise for the religious to try to fit God into the gaps.

Now it is, of course, obvious that the old idea of special creative acts, as if every orchid were specially manufactured for its insect, could no longer be accepted. In that sense Darwinism was contradictory of the traditional picture of creation. That is, I think, indisputable. But it is, or ought to be, equally obvious that the real problem is unaffected. It remains as true as ever that the world is so ordered that life develops and the evolution of higher types takes place upon it; and it is as absurd as ever it was to suppose that it is "merely accidental". New achievements like the flight of birds, or complex chains of behaviour like the parasitism of the cuckoo, depend upon the simultaneous change of a vast number of factors involving in many cases the creature's whole structure; and the odds against such changes happening all at once by accident are astronomical. Several of Darwin's strongest supporters, like Huxley and Asa Gray, insisted that the belief in an element of purpose in the evolution of life was unaffected. Darwin himself was no philosopher and became on his own confession hopelessly muddled. But for the majority of the combatants on both sides the issue seemed to be that between a godless and automatic mechanism and the traditional imagery of Genesis, and with it the whole Christian religion.

If this was the basic issue raised by *The Origin of Species*, there were two other points on which it came into sharp conflict with the outlook of most Christians. The first of these is as to the place of struggle and destruction in evolution; the second concerns the status and spirit of man.

Darwin himself, though he regarded natural selection as a chief agent in the evolution of new species, never claimed that it operated alone, and as he grew older insisted strongly that he "had not allowed sufficient weight to the direct action of the environment, i.e. food, climate, and so on". But Weismann and the most influential of Darwin's successors repudiated every other factor except natural selection. In any case, if elimination played a large part in evolution, could "Nature, red in tooth and claw with ravine", be the appointed agent of a God of love? We can see from Tennyson and a multitude of others how violently the pain and suffering of the animal world shocked the sensibility and the optimism of the great Victorians.

Similarly, they were convinced that even if Descartes was wrong in treating all animals as mere automata, he was quite right in claiming that man alone possessed a soul. God had breathed into Adam His own breath; separating him at once thereby from all the rest of His creatures. How could the lord of creation be descended from an ape? Did not the whole doctrine of immortality and indeed of the Incarnation depend upon denying to him so lowly a pedigree?

Now neither of these objections is from the Christian standpoint as serious as was then supposed. If Christ crucified is a true symbol of our faith, then it need not surprise us to find that what Fabre, the great French naturalist, called "the sublime law of sacrifice" holds true for all creation. And if God is both Creator and Redeemer, then it is appropriate that His creation should be "all of a piece" and man the end-product of an age-old succession. But here, as in the first issue, though some few Christians valued the new knowledge, it was presented in too challenging a shape to escape arousing hostility.

Looking at the whole question now that it has become a matter of history, we can see that at the time Christians might be excused for feeling grave anxiety. That creation was a process, not an act, continuous, not intermittent, operating through the orderly sequence of natural events, upset the whole idea of a God outside the world who set it going, and then on special occasions intervened by miracle to alter its course. We can now see that such an idea was never satisfactory; and in fact it was not the belief of St. Paul or of the first great Christian theologians. But at the time the only alternative to the traditional view seemed to be a purely mechanical and automatic system in which everything could be weighed and measured so that there would be no room for human freedom and responsibility, no sanctions for morality, and no place for religion. Such fears were not in fact groundless. There was in my own young days a period of sheer materialism when it looked as if the scientific and the religious outlooks were wholly contradictory, and when some of us had to say how hard it is to be a Christian not only on moral and spiritual but on intellectual grounds. Indeed, it is still true that we have not yet realised the full consequences of Darwin's work; or seen all its effect upon our traditional ideas.

But this could, I think, be said by way of summary:

(1) The fact of evolution is certain, whatever the precise means involved. Life advances from simple to complex in response to, and in relationship with, its environment. The process of creation is, as St. Paul said in the Epistle to the Romans, still incomplete.

(2) In spite of many set-backs, there is plain evidence of real progress—a travail as St. Paul puts it. Nature and history give the record of fuller

and fuller life attained by effort and sacrifice. "Whoso loveth his life loseth it."

(3) Religion has gained from Darwinism a new sense of the scale, continuity and significance of the creative process, and a new stress upon the perpetual energy of the living God. If His Spirit shares in the process we can have confidence in its result.

(4) There is room for hope that the goal to which creation moves is the "manifestation of the sons of God", the emergence of a fellowship or family of free persons, by which His will can be done, as in heaven so on earth.

At the same time that Darwin's pronouncements on man's origins burst upon the public, Louis Pasteur was busily engaged in disproving the theory of spontaneous generation—the view that various life forms, such as mice, frogs, and worms, grew fully formed from apparently nonliving substances. Some insight into the long history of the notion of spontaneous generation is provided in this reading by A.I. Oparin.

Spontaneous Generation

A.I. Oparin

The writings of scholars of the Middle Ages abound in the most fantastic descriptions and sometimes even pictures of the spontaneous generation of various insects, worms, and fish from the slime and moist earth, frogs from the May dew and even lions from the stones of the deserts. Especially characteristic of the methods of perceiving nature prevalent in the Middle Ages were the legends at that time of the goose tree, of the plant lamb, and the homunculus.

We find mention of the goose tree at the beginning of the eleventh century in the writings of Cardinal Peitro Damiani (1007–1072). The English encyclopedist Alexander Neckam (1157–1217) develops the doctrine of the formation of birds from the resin of pine trees on contact with sea salt. Subsequently this doctrine of the plant origin of geese and ducks became so generally accepted that their meat was eaten on fast days, which then had to be prohibited by special dispensation of Pope Innocent III. But in spite of this, even at the end of the fifteenth century (that is, almost three centuries later) Ritter Leo von Rotsmithal describes a dinner given in his honor in London by Herzog Klarenskii, at which ducks were served as a hot dish called "fish" (as a fasting food), the ducks having been spontaneously generated from the sea. However, Rotsmithal notes that these "fish" tasted very much like duck.

The famous traveler Odorico di Pordenona (died in 1331) first mentions the vegetable lamb. He told about a "reliable" people, of the Tatar Khanate of Khadli who cultivated tremendous gourds, which were split open on ripening; they contained lambs covered with white wool which had a very tasty meat. Mandeville (1300–1372), describing his journey to the eastern countries, also tells of a whole tree, in the melonlike fruit of which live lambs

Excerpted from A.I. Oparin, *Genesis and Evolutionary Development of Life*, Academic Press, 1968.

were growing. This tale has been handed down from century to century and even in the middle of the seventeenth century was again repeated by Adam Olearius in his description of his travels to Moscow and Persia. "They told us," he wrote, "that in Samara, between the rivers Volga and Don, there grows a rare form of melon, or rather gourd, which is huge and of a variety resembling the usual melon, in external appearance looking like a lamb, its members having that outline, and this is why the Russians call it 'baranets' (for "sheep"). This vegetable lamb ate the grass around it, and often is taken by wolves, which are great hunters of it." In addition, Olearius wrote, he had himself seen the wool of this "baranet."

The legend of the homunculus grew up in the wake of alchemical experiments. Apparently it appeared as early as the first century of our era. This legend states that by mixing a passive maternal source and an active paternal source, one could artificially produce the phenomenon of birth and obtain the embryo of a small man—the homunculus.

Like the legend of the goose tree and the vegetable lamb, stories about the homunculus occur throughout all the Middle Ages and are found in a number of alchemical writings. A typical early natural philosopher of the sixteenth century, Theophrastus von Hohenheim, known under the name of Paracelsus (1498–1541),[1] even gives in his writings an "accurate receipt" for the preparation of an homunculus.

Paracelsus in general was a convinced adherent of the theory of spontaneous generation of living beings. He believed that an active life force—Archaei—predominates in the body of animals and man; this force can be controlled by means of certain magic rules. It determines the formation of the organism and its future behavior. In accordance with these philosophical views, Paracelsus evolved an hypothesis on the spontaneous generation of life. He even carried out a series of his own observations on the sudden generation of mice, frogs, toads, and turtles from water, air, straw, rotting wood, and every form of waste product.

In the second half of the sixteenth century and especially in the seventeenth century, observations on the phenomena of nature became more accurate. Copernicus (1473–1543), Bruno (1548–1600), Galileo (1564–1642) demolished the old Ptolemaic system and formed a correct representation of the world of stars and planets surrounding us. But this flowering of accurate knowledge still did not apply to biological problems. The idea of the primordial spontaneous generation of living things remained in full strength in the minds of the investigators of the day.

The well-known Brussels doctor Van Helmont (1577–1644) can here be cited as an example. He was so completely a master of the methods of accurate experiment that he could closely approach a solution to the complicated problem of plant nutrition, but in spite of this, he considered completely indisputable the theory of spontaneous generation of living things. Moreover, he based his position on a series of observations and experiments. Van Helmont himself is the author of a widely known recipe for obtaining mice from grains of wheat. Since human vapors, in his view, were a breeding

[1] See also Chapter 4.—Ed.

source, it was only necessary to put a dirty shirt in some sort of vessel containing wheat grains.[2] After 21 days the "fermentation" is broken off and the emanations of the shirt, together with the emanations of the grain, form living mice. Van Helmont was very much surprised that these artificial mice which he obtained were entirely the same as natural mice produced from paternal sperm. The creator of the theory of blood circulation, Harvey (1578–1657), also did not deny arbitrary spontaneous generation. Although he is the author of the famous phrase, "Everything living comes from the egg" *(Omne vivum ex ovo)*, he attributed to the word "egg" here a very broad interpretation and considered entirely possible the arbitrary spontaneous generation *(generatio aequivoca)* of worms, insects, etc. as a result of the action of special forces developing during putrefaction and analogous processes.

This same point of view was held by a contemporary of Harvey's, the founder of seventeenth century English materialism, Francis Bacon (1561–1626). In his writings he expressed the opinion that different plants and animals, (flies, ants and frogs, for example) could emerge spontaneously during the rotting of various materials. But he approached this phenomenon from the materialistic viewpoint and saw in it only an indication of the absence of an impenetrable barrier between the inorganic and organic worlds.

An especially clear expression of the materialistic interpretation of spontaneous generation is found in the teachings of Descartes (1596–1650).

This great French philosopher, although he did not doubt the spontaneous generation of living things, categorically rejected the position that this generation occurs under the influence of the "anima vegetative" of the scholasticists, the "Archaei" of Paracelsus, or the "spirit of life" of Van Helmont or any sort of spiritual source. In sharp contrast with the religious scholars and with the anthropocentric tendencies of medieval natural philosophy, the physicist Descartes attempted to turn all the qualitative multitude of natural phenomena into a question of matter and its movement.

From Descartes' point of view, the living organism does not need to be explained by any special obedience to a "vital force"; it is nothing other than a machine, extremely complex in its structure, but completely understandable; its movement depends exclusively on pressure and the impact of particles on it similar to the movement of a wheel in a tower clock. This is why various living things can be spontaneously generated from the nonliving matter surrounding them. Specifically, every possible plant and animal such as worms, flies, and other insects emerge from the exposure of moist earth to the sun's light or from the phenomena of putrefaction. But the intervention of some sort of "spiritual source" is not required. Spontaneous generation is only the natural process of autoformation of those complex machines—living things—a process which always occurs under predetermined conditions, although it is true, they are as yet only partly understood.

Thus by the middle of the seventeenth century the theory of spontaneous generation of living things had not been doubted by anyone. The dispute between the mystical doctrines of the Middle Ages and briskly developing

[2] Van Helmont specified that a woman's shirt worked best.—Ed.

materialism applied only to the theoretical interpretation of this phenomenon: whether to regard spontaneous generation as the manifestation of a "spiritual genesis" or as the natural process of autoformation of living things.

However the more and more extensive and accurate investigations of living nature began to shake peoples' faith in the authenticity of the "fact" of spontaneous generation.

The turning point in this respect may fairly be considered the experiments of the Tuscan doctor Francesco Redi (1626–1698). . . . Redi had the honor of being the first to come forward with experimental refutation of spontaneous generation so firmly established for many centuries. In his paper "Esperienze intorno alla generazione degl'inetti" (1668) this scholar describes a series of experiments he carried out which indicate that white maggots in meat are non [*sic*] other than the larvae of flies. He placed meat or fish in a large vessel covered with a very fine Neapolitan muslin, and for yet more careful protection covered the vessel with a muslin cloth on a frame. Although many flies lit on the cloth, worms did not appear in the meat itself. Redi stated that he had observed that the flies laid their eggs on the cloth and that only when the eggs fell through onto the meat did they develop into maggots. He concluded from this that the decaying substance was only a place or nest for the development of the insects, and that the necessary prerequisite for their appearance was the laying of eggs, without which the flies never would emerge.

"Life has a specific chemical character," wrote Dr. Bronowski. This concept was first illuminated by Louis Pasteur, whom René Dubos has called a "Freelance of Science." His interests were many, his accomplishments great. He saved the French wine industry, lay the groundwork for the "pasteurization" of milk, and as mentioned earlier, killed the classical theory of spontaneous generation. But one of his most dazzling findings was that there were right-handed crystals and left-handed crystals in living systems and that one type could be separated from another. In the following reading, Louis Pasteur himself recounts the details of that experiment.

Louis Pasteur

René Dubos

Pasteur retained throughout his life a vivid memory of this first scientific triumph and never tired of referring to it in conversations and in lectures. As late as 1883, almost forty years after the event, he described it again in the course of a lecture delivered before the Société Chimique de Paris.

I was a student at the Ecole Normale Supérieure, from 1843 to 1846. Chance made me read in the school library a note of the learned crystallographer, Mitscherlich, related to two salts: the tartrate and the paratartrate of sodium and ammonium. I meditated for a long time upon this note; it disturbed my schoolboy thoughts. I could not understand that two substances could be as similar as claimed by Mitscherlich, without being completely identical. To know how to wonder and question is the first step of the mind toward discovery.

Hardly graduated from the Ecole Normale, I planned to prepare a long series of crystals, with the purpose of studying their shapes. I selected tartaric acid and its salts, as well as paratartaric acid, for the following reasons. The crystals of all these substances are as beautiful as they are easy to prepare. On the other hand, I could constantly control the accuracy of my determinations by referring to the memoir of an able and very precise physicist, M. de la Provostaye, who had published an extensive crystallographic study of tartaric and paratartaric acid and of their salts.

I soon recognized that . . . tartaric acid and all its combinations exhibit asymmetric forms. Individually, each of these forms of tartaric acid gave a mirror image which was not superposable upon the substance itself. On the contrary, I could not find anything of the sort in paratartaric acid or its salts.

Suddenly, I was seized by a great emotion. I had always kept in mind the profound surprise caused in me by Mitscherlich's note on the tartrate and paratartrate of sodium and ammonium. Despite the extreme thoroughness of their study, I thought, Mitscherlich, as well as M. de la Provostaye, will have failed to notice that the tartrate is asymmetric, as it must be; nor will they have seen that the paratartrate is not asymmetric, which is also very likely. Immediately, and with a feverish ardor, I prepared the double tartrate of sodium and ammonium, as well as the corresponding paratartrate, and proceeded to compare their crystalline forms, with the preconceived notion that I would find asymmetry in the tartrate and not in the paratartrate. Thus, I thought, everything will become clear; the mystery of Mitscherlich's note will be solved, the asymmetry in the form of the tartrate crystal will correspond to its optical asymmetry, and the absence of asymmetry in the form of the paratartrate will correspond to the inability of this salt to deviate the plane of polarized light. . . . And indeed, I saw that the crystals of the tartrates of sodium and ammonium exhibited the small facets revealing asymmetry; but when I turned to examine the shape of the crystals of paratartrate, for an instant my heart stopped beating: all the crystals exhibited the facets of asymmetry!

The fortunate idea came to me to orient my crystals with reference to a plane perpendicular to the observer, and then I noticed that the confused mass of crystals of paratartrate could be divided into two groups according to the orientation of their facets of asymmetry. In one group, the facet of asymmetry nearer my body was inclined to my right with reference to the plane of orientation which I just mentioned, whereas the facet of asymmetry was inclined to my left in the other. The paratartrate appeared as a mixture of two kinds of crystals, some asymmetric to the right, some asymmetric to the left.

A new and obvious idea soon occurred to me. These crystals asymmetric to the right, which I could separate manually from the others, exhibited an absolute identity of shape with those of the classical right tartrate. Pursuing my preconceived idea, in the logic of its deductions, I separated these right crystals from the crystallized paratartrate; I made the lead salt and isolated the acid; this acid appeared absolutely identical with the tartaric acid of grape, identical also in its action on polarized light. My happiness was even greater the day when, separating now from the paratartrate the crystals with asymmetry at their left, and making their acid, I obtained a tartaric acid absolutely similar to the tartaric acid of grape, but with an opposite asymmetry, and also with an opposite action on light. Its shape was identical to that of the mirror image of the right tartaric acid and, other things being equal, it rotated light to the left as much in absolute amount as the other acid did it to the right.

Finally, when I mixed solutions containing equal weights of these two acids, the mixture gave rise to a crystalline mass of paratartaric acid identical with the known paratartaric acid.·

The major impact of Darwinism was that it introduced a "natural" explanation for a process that was believed to be supernatural in origin. The creation myths were swept away by the new tide of mechanistic evolution; the metaphysical was, in a sense, replaced by the physical. Among those who accepted the Darwinian gospel on the evolution of man were many who proposed that the origin of life itself was nothing more than a series of physical and chemical accidents, the process of abiogenesis, the foundation of life from non-life. Abiogenesis, in another sense, is a more sophisticated and more realistic statement of spontaneous generation. Within the last twenty years, since the pioneer experiment was carried out by chemist Stanley L. Miller, the theory has gained prominence. Miller explained in the original paper that he had succeeded in duplicating in the laboratory the synthesis of some of the chemical ingredients of life, under conditions that may have existed on the primitive earth four billion years ago.

The Origin of Life

Stanley M. Miller

The idea that the organic compounds that serve as the basis of life were formed when the earth had an atmosphere of methane, ammonia, water, and hydrogen instead of carbon dioxide, nitrogen, oxygen, and water was suggested by Oparin and has been given emphasis recently by Urey and Bernal.

In order to test this hypothesis, an apparatus was built to circulate CH_4, NH_3, H_2O, and H_2 past an electric discharge. The resulting mixture has been tested for amino acids by paper chromatography. Electrical discharge was used to form free radicals instead of ultraviolet light, because quartz absorbs wavelengths short enough to cause photo-dissociation of the gases. Electrical discharge may have played a significant role in the formation of compounds in the primitive atmosphere.

The apparatus used is shown. Water is boiled in the flask, mixes with the gases in the 5–1 flask, circulates past the electrodes, condenses and empties back into the boiling flask. The U-tube prevents circulation in the opposite direction. The acids and amino acids formed in the discharge, not being volatile, accumulate in the water phase. The circulation of the gases is quite

From Stanley Miller, "A Production of Amino Acids Under Possible Primitive Earth Conditions," *Science*, May 15, 1953, pp. 528–529.

slow, but this seems to be an asset, because production was less in a different apparatus with an aspirator arrangement to promote circulation. The discharge, a small corona, was provided by an induction coil designed for detection of leaks in vacuum apparatus.

The experimental procedure was to seal off the opening in the boiling flask after adding 200 ml of water, evacuate the air, add 10 cm pressure of H_2, 20 cm of CH_4, and 20 cm of NH_3. The water in the flask was boiled, and the discharge was run continuously, for a week.

During the run the water in the flask became noticeably pink after the first day, and by the end of the week the solution was deep red and turbid. Most of the turbidity was due to colloidal silica from the glass. The red color is due to organic compounds adsorbed on the silica. Also present are yellow organic compounds, of which only a small fraction can be extracted with ether, and which form a continuous streak tapering off at the bottom on a one-dimensional chromatogram run in butanol-acetic acid. These substances are being investigated further.

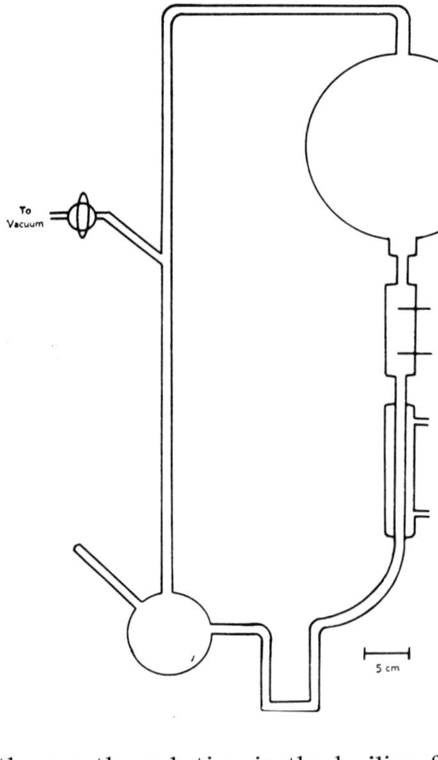

At the end of the run the solution in the boiling flask was removed and 1 ml of saturated $HgCl_2$ was added to prevent the growth of living organisms. The ampholytes were separated from the rest of the constituents by adding $Ba(OH)_2$ and evaporating *in vacuo* to remove amines, adding H_2SO_4 and evaporating to remove the acids, neutralizing with $Ba(OH)_2$, filtering and concentrating *in vacuo*.

The amino acids are not due to living organisms because their growth would be prevented by the boiling water during the run, and by the $HgCl_2$, $Ba(OH)_2$, H_2SO_4 during the analysis.

A paper chromatogram was prepared using *n*-butanol-acetic acid-water mixture followed by water-saturated phenol, and spraying with ninhydrin. Identification of an amino acid was made when the R_f value (the ratio of the distance travelled by the amino acid to the distance travelled by the solvent front), the shape and the color of the spot were the same on a known, unknown, and mixture of the known and unknown, and when consistent results were obtained with chromatograms using phenol and 77% ethanol.

On this basis glycine, α-alanine and β-alanine are identified. The identification of the aspartic acid and α-amino-*n*-butyric acid is less certain because the spots are quite weak. There are other spots, unidentified as yet, but may be beta and gamma amino acids. These are the main amino acids present, and others are undoubtedly present but in smaller amounts. It is estimated that the total yield of amino acids was in the milligram range.

In this apparatus an attempt was made to duplicate a primitive atmosphere of the earth, and not to obtain the optimum conditions for the formation of amino acids. Although in this case the total yield was small for the energy expended, it is possible that, with more efficient apparatus (such as mixing of the free radicals in a flow system, use of higher hydrocarbons from natural gas or petroleum, carbon dioxide, etc., and optimum ratios of gases), this type of process would be a way of commercially producing amino acids.

Since Miller completed his historic experiment in 1953, it has been repeated with many variations to create other components of living systems. In those cases, the energy that must have been responsible for these transformations on the primitive earth were simulated in the laboratory with ultraviolet light, radiation, and heat. Similarly, the hypothetical compositions of the atmosphere and ponds of the primitive earth were recreated and used as a source of the building blocks for man-made proteins, nucleic acids, and other important life-sustaining chemicals.

Darwin himself had such experiments in mind when he wrote:

But if (and oh! what a big if!) we could conceive in some warm little pond, with all sorts of ammonia and phosphoric salts, light, heat, electricity, etc. present, that a protein compound was formed ready to undergo complex changes. . . .

Some who criticize experiments aimed at simulating the terrestrial environment in which life may have begun maintain that the scientists involved, without access to a time machine, have no way to prove that their experiments really mimic conditions on the prebiotic earth. The field of abiogenesis *is* rife with speculation. It is possible, for example, that life began in a frozen ocean, rather than in Darwin's "warm little pond," as shown by Leslie Orgel. Recently, in collaboration with Francis Crick, codiscoverer of the double helix (Chapter 12), Orgel wrote an interesting and admittedly speculative paper giving an alternative explanation for the appearance of life on earth.

Directed Panspermia

F.H.C. Crick and L.E. Orgel

It was not until the middle of the nineteenth century that Pasteur and Tyndall completed the demonstration that spontaneous generation is not occurring on the Earth nowadays. Darwin and a number of other biologists concluded that life must have evolved here long ago when conditions were more favourable. A number of scientists, however, drew a quite different conclusion. They supposed that if life does not evolve from terrestrial nonliving matter nowadays, it may never have done so. Hence, they argued, life reached the earth as an "infection" from another planet.

Arrhenius (1908) proposed that spores had been driven here by the pressure of the light from the central star of another planetary system. His theory is known as Panspermia. Kelvin suggested that the first organisms reached the Earth in a meteorite. Neither of these theories is absurd, but both can be subjected to severe criticism. [Carl] Sagan has shown that any known type of radiation-resistant spore would receive so large a dose of radiation during its journey to the Earth from another Solar System that it would be extremely unlikely to remain viable. The probability that sufficiently massive objects escape from a Solar System and arrive on the planet of another one is considered to be so small that it is unlikely that a single meteorite of extrasolar origin has ever reached the surface of the Earth. These arguments may not be conclusive, but they argue against the "infective" theories of the origins of life that were proposed in the nineteenth century.

It has also been argued that "infective" theories of the origins of terrestrial life should be rejected because they do no more than transfer the problem of origins to another planet. This view is mistaken; the historical facts are important in their own right. For all we know there may be other types of planet on which the origin of life *ab initio* is greatly more probable than on our own. For example, such a planet may possess a mineral, or compound, of crucial catalytic importance, which is rare on Earth. It is thus important to know whether primitive organisms evolved here or whether they arrived here from somewhere else. Here we reexamine this problem in the light of more recent biological and astronomical information.

Our Present Knowledge of the Galaxy

The local galactic system is estimated to be about 13×10^9 yr old. The first generation of stars, because they were formed from light elements, are unlikely to have been accompanied by planets. However, some second generation stars not unlike the Sun must have formed within 2×10^9 yr of the origin of the galaxy. Thus it is quite probable that planets not unlike the Earth existed as much as 6.5×10^9 yr before the formation of our own Solar System.

We know that not much more than 4×10^9 yr elapsed between the appearance of life on the Earth (wherever it came from) and the development of our own technological society. The time available makes it possible, therefore, that technological societies existed elsewhere in the galaxy *even*

From F.H.C. Crick and L.E. Orgel, "Directed Panspermia," *Icarus*, Vol. 19, pp. 341–346, Academic Press, 1973.

before the formation of the Earth. We should, therefore, consider a new "infective" theory, namely that a primitive form of life was deliberately planted on the Earth by a technologically advanced society on another planet.

Are there many planets which could be infected with some chance of success? It is believed, though the evidence is weak and indirect, that in the galaxy many stars, of a size not dissimilar to our Sun, have planets, on a fair fraction of which temperatures are suitable for a form of life based on carbon chemistry and liquid water, as ours is. Experimental studies of the production of organic chemicals under prebiotic conditions make it seem likely that a rich prebiotic soup accumulates on a high proportion of such Earthlike planets. Unfortunately, we know next to nothing about the probability that life evolves within a few billion years in such a soup, either on our own special Earth, or still less on other Earthlike planets.

If the probability that life evolves in a suitable environment is low we may be able to prove that we are likely to be alone in the galaxy (Universe). If it is high the galaxy may be pullulating with life of many different forms. At the moment we have no means at all of knowing which of these alternatives is correct. We are thus free to postulate that there have been (and still are) many places in the galaxy where life could exist but that, in at least a fraction of them, after several billion years the chemical systems had not evolved to the point of self-replication and natural selection. Such planets, if they do exist, would form an excellent breeding ground for external microorganisms. Note that because many if not all such planets would have a reducing atmosphere they would not be very hospitable to the higher forms of life as we know them on Earth.

Our Proposal

The possibility that terrestrial life derives from the deliberate activity of an extraterrestrial society has often been considered in science fiction and more or less light-heartedly in a number of scientific papers. For example, [Tommy] Gold (1960) has suggested that we might have evolved from the microorganisms inadvertently left behind by some previous visitors from another planet (for example, in their garbage). Here we wish to examine a very specific form of Directed Panspermia. Could life have started on Earth as a result of infection by microorganisms sent here deliberately by a technological society on another planet, by means of a special long-range unmanned spaceship? To show that this is not totally implausible we shall use the theorem of detailed cosmic reversibility; if we are capable of infecting an as yet lifeless extrasolar planet, then, given that the time was available, another technological society might well have infected our planet when it was still lifeless.

The Proposed Spaceship

The spaceship would carry large samples of a number of microorganisms, each having different but simple nutritional requirements, for example blue-green algae, which could grow on CO_2 and water in "sunlight." A payload of 1000 kg might be made up of 10 samples each containing 10^{16} microorganisms, or 100 samples each of 10^{15} microorganisms.

It would not be necessary to accelerate the spaceship to extremely high velocities, since its time of arrival would not be important. The radius of our galaxy is about 10^5 light years, so we could infect most planets in the galaxy within 10^8 yr by means of a spaceship traveling at only one-thousandth of the velocity of light. Several thousand stars are within a hundred light years of the Earth and could be reached within as little as a million years by a spaceship travelling at only 60,000 mph, or within 10,000 yr if a speed of one-hundredth of that of light were possible.

The technology required to carry out such an act of interstellar pollution is not available at the present time. However, it seems likely that the improvements in astronomical techniques will permit the location of extrasolar planets within the next few decades. Similarly, the problem of sending spaceships to other stars, at velocities low compared with that of light, should not prove insoluble once workable nuclear engines are available. This again is likely to be within a few decades. The most difficult problem would be presented by the long flight times; it is not clear how long it will be before we can build components that would survive in space for periods of thousands or millions of years.

Although there are some technological problems associated with the distribution of the microorganisms in viable form after a long journey through space, none of them seems insuperable. Some radiation protection could be provided during the journey. Suitable packaging should guarantee that small samples, including some viable organisms, would be widely distributed. The question of how long microorganisms, and in particular bacterial spores, could survive in a spaceship has been considered in a preliminary way by P.H.A. Sneath (1962). He concludes "that life could probably be preserved for periods of more than a million years if suitably protected and maintained at temperatures close to absolute zero." Sagan (1960) has given a comparable estimate of the effects of radiation damage. We conclude that within the foreseeable future we could, if we wished, infect another planet, and hence that it is not out of the question that our planet was infected.

We can in fact go further than this. It may be possible in the future to send either mice or men or elaborate instruments to the planets of other Solar Systems (as so often described in science fiction) but a rocket carrying microorganisms will always have a much greater effective range and so be advantageous if the sole aim is to spread life. This is true for several reasons. The conditions on many planets are likely to favour microorganisms rather than higher organisms. Because of their extremely small size vast numbers of microorganisms can be carried, so much more wastage can be accepted. The ability of microorganisms to survive, without special equipment, both storage for very long periods at low temperatures and also an abrupt change back to room temperatures is also a great advantage. Whatever the potential range for infection by other organisms, microorganisms can almost certainly be sent further and probably much further.

It should be noted that most of the earliest "fossils" so far recognized are somewhat similar to our present bacteria or blue-green algae. They occur in cherts of various kinds and are estimated to be up to 3×10^9 yr old. This makes it improbable that the Earth was ever infected merely by higher organisms.

Motivation

Next we must ask what motive we might have for polluting other planets. Since we would not derive any direct advantage from such a programme, presumably it would be carried through either as a demonstration of technological capability or, more probably, through some form of missionary zeal.

It seems unlikely that we would deliberately send terrestrial organisms to planets that we believed might already be inhabited. However, in view of the precarious situation on Earth, we might well be tempted to infect other planets if we became convinced that we were alone in the galaxy (Universe).* As we have already explained we cannot at the moment estimate the probability of this. The hypothetical senders on another planet may have been able to prove that they were likely to be alone, and to remain so, or they may have reached this conclusion mistakenly. In either case, if they resembled us psychologically, their motivation for polluting the galaxy would be strong, if they believed that all or even the great majority of inhabitable planets could be given life by Directed Panspermia.

The psychology of extraterrestrial societies is no better understood than terrestrial psychology. It is entirely possible that extraterrestrial societies might infect other planets for quite different reasons than those we have suggested. Alternatively, they might be less tempted than we would be, even if they thought that they were alone. The arguments given above, together with the principle of cosmic reversibility, demonstrate the possibility that we have been infected, but do not enable us to estimate the probability.

Possible Biological Evidence

Infective theories of the origins of terrestrial life could be taken more seriously if they explained aspects of biochemistry or biology that are otherwise difficult to understand. We do not have any strong arguments of this kind, but there are two weak facts that could be relevant.

The chemical composition of living organisms must reflect to some extent the composition of the environment in which they evolved. Thus the presence in living organisms of elements that are extremely rare on the Earth might indicate that life is extraterrestrial in origin. Molybdenum is an essential trace element that plays an important role in many enzymatic reactions, while chromium and nickel are relatively unimportant in biochemistry. The abundance of chromium, nickel, and molybdenum on the Earth are 0.20, 3.16, and 0.02%, respectively. We cannot conclude anything from this single example, since molybdenum may be irreplaceable in some essential reaction—nitrogen fixation, for example. However, if it could be shown that the elements represented in terrestrial living organisms correlate closely with those

*In a somewhat different context the seeding of Venus and other solar planets has been suggested by C. Sagan (1961) and T. Gold. . . .

that are abundant in some class of star—molybdenum stars, for example—we might look more sympathetically at "infective" theories.

Our second example is the genetic code. Several orthodox explanations of the universality of the genetic code can be suggested, but none is generally accepted to be completely convincing. It is a little surprising that organisms with somewhat different codes do not coexist. The universality of the code follows naturally from an "infective" theory of the origins of life. Life on Earth would represent a clone derived from a single extraterrestrial organism. Even if many codes were represented at the primary site where life began, only a single one might have operated in the organisms used to infect the Earth.

Conclusion

In summary, there is adequate time for technological society to have evolved twice in succession. The places in the galaxy where life could start, if seeded, are probably very numerous. We can foresee that we ourselves will be able to construct rockets with sufficient range, delivery ability, and surviving payload if microorganisms are used. Thus the idea of Directed Panspermia cannot at the moment be rejected by any simple argument. It is radically different from the idea that life started here *ab inito* [*sic*] without infection from elsewhere. We have thus two sharply different theories of the origin of life on Earth. Can we choose between them?

At the moment it seems that the experimental evidence is too feeble to make this discrimination. It is difficult to avoid a personal prejudice, one way or the other, but such prejudices find no scientific support of any weight. It is thus important that both theories should be followed up. Work on the supposed terrestrial origin of life is in progress in many laboratories. As far as Directed Panspermia is concerned we can suggest several rather diverse lines of research.

The arguments we have employed here are, of necessity, somewhat sketchy. Thus the detailed design of a long-range spaceship would be worth a careful feasibility study. The spaceship must clearly be able to home on a star, for an object with any appreciable velocity, if dispatched in a random direction, would in almost all cases pass right through the galaxy and out the other side. It must probably have to decelerate as it approached the star, in order to allow the safe delivery of the payload. The packets of microorganisms must be made and dispersed in such a way that they can survive the entry at high velocity into the atmosphere of the planet, and yet be able to dissolve in the oceans. Many useful feasibility studies could be carried out on the engineering points involved.

On the biological side we lack precise information concerning the life-time of microorganisms held at very low temperatures while traveling through space at relatively high velocities. The rocket would presumably be coasting most of the time so the convenient temperature might approximate to that of space. How serious is radiation damage, given a certain degree of shielding? How many distinct types of organism should be sent and which should they be? Should they collectively be capable of nitrogen fixation, oxidative phosphorylation and photosynthesis? Although many "soups" have been produced artifically [*sic*] in the laboratory, following the pioneer experiments of Miller, as far as we know no careful study has been made to determine which present-day organisms would grow well in them under primitive Earth conditions.

At the same time present-day organisms should be carefully scrutinized to see if they still bear any vestigial traces of extraterrestrial origin. We have already mentioned the uniformity of the genetic code and the anomalous abundance of molybdenum. These facts amount to very little by themselves but as already stated there may be other as yet unsuspected features which, taken together, might point to a special type of planet as the home of our ancestors.

These enquiries are not trivial, for if successful they could lead to others which would touch us more closely. Are the senders or their descendants still alive? Or have the hazards of 4 billion years been too much for them? Has their star inexorably warmed up and frizzled them, or were they able to colonise a different Solar System with a short-range spaceship? Have they perhaps destroyed themselves, either by too much aggression or too little? The difficulties of placing any form of life on another planetary system are so great that we are unlikely to be their sole descendants. Presumably they would have made many attempts to infect the galaxy. If the range of their rockets were small this might suggest that we have cousins on planets which are not too distant. Perhaps the galaxy is lifeless except for a local village, of which we are one member.

One further point deserves emphasis. We feel strongly that under no circumstances should we risk infecting other planets at the present time. It would be wise to wait until we know far more about the probability of the development of life on extrasolar planets before causing terrestrial organisms to escape from the solar system.

10

World Within World

Mock on, Mock on, Voltaire, Rousseau:
Mock on, Mock on: 'tis all in vain!
You throw the sand against the wind,
And the wind blows it back again.

And every sand becomes a Gem
Reflected in the beams divine;
Blown back they blind the mocking Eye,
But still in Israel's paths they shine.

The Atoms of Democritus
And Newton's Particles of light
Are sands upon the Red sea shore,
Where Israel's tents do shine so bright.

William Blake

"World Within World" is a theme that has been building steadily as man has traveled the road of progress and change. It is exemplified in the planetary wheels within wheels of Ptolemy's universe, in the atoms of Democritus and Dalton, and in the "grain in the stone" that the sculptor sees in the creation of his greatest works. Dimitri Ivanovich Mendeleevf showed that the elemental building blocks for everything in nature fit into less than one hundred orderly cubby holes on a paper chart. Then, at the turn of the century, a great scientific explosion ripped apart the inner core of matter, peeling away layer after layer of the structure of the atom like the layers of an onion.

The tools used by man in this phase of his ascent were themselves subatomic particles, including x rays and neutrons. For Harry Moseley, x rays were the clue that helped him to decipher the unique atomic configuration of different elements. They were a new probe into the nature of living things, including the human body. As such, they have proved their worth over and over in the field of diagnostic medicine. But for some people, including L.K. Russell, they revealed too much:

Lines on an X-ray Portrait of a Lady
She is so tall, so slender, and her bones–
Those frail phosphates, those carbonates of lime–
Are well produced by cathode rays sublime,
By oscillations, amperes and by ohms.
Her dorsal vertebrae are not concealed
By epidermis, but are well revealed.

Around her ribs, those beauteous twenty-four,
Her flesh a halo makes, misty in line,
Her noseless, eyeless face looks into mine,
And I but whisper, "Sweetheart, Je t'adore."
Her white and gleaming teeth at me do laugh.
Ah! Lovely, cruel, sweet cathodograph!

As for Harry Moseley, he died a "war hero" at the age of 27, before his capacity as a scientist could be fully realized. His biographer, J.L. Heilbron, opines that Moseley's early demise was a wrenching human loss to his friends and family. There was a boyishness about his approach to science and to life that comes through in this letter that young Harry wrote to his mother when his x-ray work was reaching its peak. His scientific contributions are discussed further in the next reading.

A Letter to My Mother

H.G.J. Moseley

The University,
Manchester
Sunday [2 Nov 13]

My dearest of Mothers,

When I last wrote I prophesied that my work was at last going to go well. Since Wednesday it has been astonishingly successful, which you will be glad to hear as it shortens the time that I will have to stay here. I can now get in five minutes a strong sharp photograph of the X rays spectrum, which would mean days work by the ionisation method. In the last four days I have got the spectrum given by Tantalum. Chromium. Manganese. Iron. Nickel. Cobalt and Copper and part of the Silver spectrum. The chief result is that all the elements give the same kind of spectrum, the result for any metal being quite easy to guess from the results for the others. This shews that the insides of all the atoms are very much alike, and from these results it will be possible to find out something of what the insides are made up of.

You are over captious to grumble about the quality of my letter writing. When I break off work in time to write for the midnight post you should feel grateful that I have anything to say at all. Anyhow when you choose to go in for a game (Ladies' Chess) which is run almost exclusively by fussies, it is hypercritical to refuse to accept their standards. I know it does not come natural to you, but it is part of the game as far as I gather from what you tell me, so you ought to see that you are not outmanouvred.

I want to know the result of your two games, as I suppose that £2 hung in the balance, and surely that is not to be despised. The poor Scotch lady must have [been] heartbroken at throwing away her chances of the booty. Anyhow I am sorry you missed your two games with her, as it would have been great joy to you to have beaten her. . . .

I have returned the official postcard of the British Ass* on the question of Australia telling them to put me provisionally on the list of acceptances, as the time limit expired on Nov 1st, and I thought it a pity to burn my boats. It is quite easy to cry off later. I fancy that there will be £100 grant going for anyone who wants it, so that anyone going second class would be self supporting. The fares are the same by the Canadian Pacific as by the Red Sea I am surprised to see, though of course going across Canada takes longer. If our New Zealand scheme falls through I might like to go.

Y.I.H.M.

Moseley, X-Rays, and the Structure of the Atom

E.N. Da C. Andrade

Two men whose work was destined to establish the nuclear atom as of fundamental importance were closely associated with, and inspired by, [Ernest] Rutherford in his Manchester days, H.G.J. Moseley and Niels Bohr. Moseley, killed in 1915 while serving as an army officer at Suvla Bay, in the Dardanelles, had studied physics at Oxford, but, full of enthusiasm for the work being carried out at Manchester, applied to the laboratory and was appointed to a junior position on the staff at the beginning of the academic year in 1910. He was then aged twenty-three. With him came, from Cambridge, Charles Galton Darwin—grandson of the Charles Darwin of the *Origin of the Species*—

*The British Association for the Advancement of Science. . . .

who, as already noted, was appointed Schuster reader in mathematical physics. Darwin and Moseley became close friends and together they carried out some experiments on the reflections of X-rays from crystals, which convinced them that X-rays did actually behave like light of very short wave length.

Since Röntgen's first discovery, the nature of X-rays had been long in doubt, some considering that they were a kind of particle and others that they were a kind of wave, like light waves, but of immensely shorter wave length. In 1912 Max von Laue, with the assistance of W. Friedrich and P. Knipping, had succeeded in proving, by passing the rays through a crystal, that they were of a wave nature, a discovery of which the fiftieth anniversary was celebrated at Munich in 1962. In the year after Laue's discovery, W.H. and W.L. Bragg showed how the wave lengths of X-rays could be measured by reflecting them at the face of a suitable crystal. Darwin and Moseley had at once taken this up, as just mentioned, and Moseley then settled down to measure systematically the wave lengths of X-rays given out by different elements. The wave length, of course, gives the frequency, that is, the number of vibrations a second, if the velocity is known, and the velocity of X-rays is the same as that of light. The frequency, which for X-rays is in the region of millions of millions of millions of vibrations per second, is the fundamental quantity.

. . . The visible light which gaseous elements may be provoked to send out (by, for example, passing an electric discharge through a gas, as in a neon or mercury lamp, or by putting a salt in a flame and so producing metallic vapor) consists of monochromatic lights of certain characteristic frequencies, spoken of as spectral lines. The frequencies of the different lines that belong to a given element can be arranged in certain series, the frequencies in a given series being governed by a simple law. . . .

What Moseley first showed was that the X-rays sent out by a particular element comprised certain characteristic frequencies, called "lines," because, when the X radiation is spread out by reflection from a crystal, just as visible light is spread out by a prism or other device, these characteristic frequencies are made evident by lines on a photographic plate. To use Moseley's own words, "The present paper contains a description of a method of photographing these spectra, which makes the analysis of X-rays as simple as any other branch of spectroscopy." He then compared the frequencies of characteristic lines sent out by different elements, with results of the utmost importance. Let us look back at the way things were done in those days of great, simple discoveries, which will help us to recapture the spirit of Rutherford's lab.

For his experiments Moseley used a wide evacuated glass tube about a yard long, with toy railway lines on which was a trolley carrying a row of blocks of the elements to be examined. Each block in turn was brought into the path of the cathode rays, the impact of which produced the X-rays. These rays came out through a thin window and fell on the crystal used to measure the wave length. It is typical of the scarcity of apparatus, even in Manchester, in these times that the Gaede pump which Moseley used to evacuate his tube was borrowed from Balliol College, Oxford. The apparatus was a troublesome setup to handle, but Moseley was a man of great experimental skill and prodigious industry, who produced his results in surprisingly short time. He worked late at night: in fact, it was said of him that his specialized attainments included a knowledge of where in Manchester to get a meal at three o'clock in the morning. C.G. Darwin states that "he was, without exception, the hardest worker I have ever known." His first work on the subject, carried out in Manchester, was published in 1913. At the beginning of 1914 he went to Oxford, where his widowed mother dwelt, and there he completed the work, as described in a second fundamental paper in 1914.

His great discovery was that a quantity which involved in a simple way the frequency of a characteristic line of the X-ray spectrum of an element increased by a constant amount as one passed from one chemical element to the next one when the elements were arranged in order of atomic weight. What mattered was the number giving the order in the series, not the atomic weight itself. Thus when the elements are arranged in the order of atomic weight, hydrogen is the first, number 1; helium is number 2; lithium is number 3; and so on, gold being, for example, number 79. This number giving the place in the series is called the atomic number. As Moseley himself said of his results, "We have here a proof that there is in the atom a fundamental quantity, which increases by regular steps as we pass from one element to the next. This quantity can only be the charge on the central positive nucleus." That was the essence of his discovery. In an obituary notice which Rutherford wrote when Moseley was killed, he stated, "This proof of Moseley will, in my opinion, rank in importance with the discovery of the periodic law of the elements and of spectral analysis, and in some respects is more fundamental than either." G. Urbain, the great French chemist who was responsible for the discovery of the elements lutecium and dysprosium, wrote of Moseley . . . "His law substituted for Mendeleéff's somewhat romantic classification a completely scientific accuracy."

The essence, then, of Moseley's discovery was that if N gives the place number of a given element when all the elements are arranged in order of atomic weight, then the charge on the nucleus of that element is Ne, where e is the magnitude of the electronic charge. The charge is not half the atomic weight, as Rutherford originally thought, although this gives a very rough estimate of the value. Thus for gold half the atomic weight is 98, while the atomic number is 79.

Since the frequency of a characteristic line in the X-ray spectrum is connected simply with the atomic number—actually, for those who like exact laws, it is the square root of the frequency that increases by equal steps from one atomic number to the next—we can, from X-ray measurements, see if there is missing any element that ought to be in the series, since in this case there will be an unexpected jump in the characteristic X-ray quantity owing to a missing N. This Moseley clearly realized: in a letter to Rutherford he said, "I do not doubt that it will be possible to put every rare earth element into its right pigeon-hole and to settle whether any of them are complex and where to look for new ones." In this way it was found that there were four gaps where, to put the atomic numbers right, there ought to be an element. These four elements were, of course, searched for. The first to be found was number 72, hafnium, discovered in Bohr's laboratory at Copenhagen by D. Coster and G. von Hevesy, the Hevesy who has been already mentioned as a worker in Rutherford's laboratory and subsequently a lifelong friend of his. Hafnia was the old name of Copenhagen. There followed rhenium, number 75, named after Rhenus, the Latin name for the Rhine, and technetium, number 43, named from the Greek word *technetos*, artificial, because it was the first element to become known by being made artificially. The discovery of the last missing element, number 47, has been claimed by several different men or groups, who have given it different names, the most favored of which seems to be promethium. [Frederick] Soddy put the matter strikingly when he wrote, "Moseley, as it were, called the roll of the elements, so that for the first time we could say definitely the number of possible elements between the beginning and the end, and the number that still remained to be found."

For Dr. Bronowski, there is no difference between science and art in the ascent of man. They are both expressions of the unique nature of man's creative faculties, as he continually shapes and reshapes the world in which he lives. It is a theme elaborated in Chapters 3 and 4 of this anthology and elsewhere throughout the book and film versions of *The Ascent of Man*.

Georges Seurat successfully meshed science and art in his painting, by breaking light into little dots of paint. He was, Dr. Bronowski says, "visibly taking the world to pieces and putting it together on the same canvas." The scientific basis of Seurat's work is discussed in the following reading by John Rewald, from his biography of the French impressionist.

Seurat and Science

John Rewald

In a conversation with Gustave Kahn, Seurat once defined painting as *the art of hollowing a surface.* At the same time he explained that while the science of aesthetics was not the whole story, since there were intimate elements in art, even in its technique, which only the painter himself could recognize and deal with, for his part he was convinced of the absolute need to base his theories on scientific truths.

While Seurat the painter was indivisible from Seurat the theorist, there are certain sayings of his that may be considered to spring from the one rather than the other. It was Seurat the painter who said that he could only paint what he saw before his eyes and declared that the drawing was the fundamental element in painting, and also that harmony of color should flow from harmony of line. But it was Seurat the theorist who attempted to generalize the laws of color and line.

The art of hollowing a surface became for Seurat the theorist a *science* to be studied by reading mathematical writings and to be developed by painting pictures. During the years which followed the completion of *La Grande Jatte* Seurat systematically attacked every one of the problems of painting. Only still lifes are lacking in his work, and even these can be located in sections of his large canvases. Having studied in *La Grande Jatte* the movements of people outdoors and reproduced in landscapes nature in repose, Seurat painted, in turn, immobile nudes in the studio (*Les Poseuses*), the portrait (*Jeune Femme se poudrant*), immobile figures outdoors under artificial lighting (*La Parade*) and moving figures indoors, also under artificial lighting (*Le Chahut* and *Le Cirque*). The figures in these paintings are dominated by monotony or joy (there is no sadness in the pictures of Seurat) and are, of course, governed by strict rules, being controlled by that play of line and color whose laws Seurat had studied. In these canvases Seurat, without yielding in any way to the literary or the picturesque, rehabilitated the *subject* which had been relinquished by the impressionists. His works are "exemplary specimens of a highly developed decorative art, which sacrifices the anecdote to the arabesque, nomenclature to synthesis, the fugitive to the permanent, and confers on nature—weary at last of its precarious reality—an authentic reality," wrote Fénéon.

If Seurat never let himself be caught in the "charming traps" set by nature, if he had the strength to renounce all details that would detract from the simplicity of the whole, it was because he had a cold reasonableness which, prompted by his lack of imagination, thrust him towards the essential. "When he wanted to feel in a certain way," says Paul Adams, "he modified nature in accordance with his will." And will, often literary with Gauguin, bent mainly on observation of the model with Cézanne, with Seurat was directed towards line and ornament, livened by tints and tones. Signac says: "It is by the harmony of lines and colors which he can manipulate in accordance with his needs and will, and not by his subject that the painter should move."

While Gauguin said disdainfully, "In art we have just passed through a period of bewildered wandering, caused by physics, chemistry, mechanics and the study of nature," and while Cézanne complained that "the ignorance of harmony is being revealed more and more in the discordance of the coloring and, what is even worse, in the deadness of the tone," Seurat continued to work for a new organization of painting based on science. Hence he kept up with all the new theories of color and perspective. He studied *The Scientific Theory of Color and Its Relation to Art and Industry* by N.O. Rood, he became extremely interested in the researches of Charles Henry (Signac collaborated with Henry for a time), who published in 1888 *Le cercle chromatique présentant tous les compléments et toutes les harmonies de couleurs*, and also *Le rapporteur aesthétique permettant l'étude et la rectification aesthétique de toute forme.* And Seurat collected David Sutter's articles on visual phenomena which appeared in the review *L'Art* in 1880.

Seurat was interested in the laws Sutter arrived at (Sutter summed them up in 167 rules) and also in the more general import of Sutter's conclusions. Indeed, Sutter's point of departure was a principle dear to Seurat:

One must look at nature with the eyes of the mind and not merely with the eyes of the body, like beings without reason. . . . There are colorists' eyes as there are tenor voices, but these gifts of nature have to be stimulated by sci-

ence to develop to the full. . . . Despite their absolute character, rules do not hamper the spontaneity of invention or execution. Science delivers us from every form of uncertainty and enables us to move freely within a wide circle; it would be an injury both to art and science to believe that one necessarily excludes the other. Since all rules are derived from the laws of nature, nothing is easier to learn or more necessary to know. In art everything should be willed.

Seurat also found in Sutter's writings the answer to a problem which had always occupied him: the relation between the laws of optics and of music. Many times he had asked his friends whether they thought the innovations in painting had any technical similarity to, or at least any intuitive kinship with, the new music of Wagner, and he found in Sutter this categorical assertion which could not but confirm his own convictions: "The laws of the aesthetic harmony of colors can be taught as the rules of musical harmony are taught."

The 167 rules formulated by Sutter contained many propositions, or rather, precepts, which also confirmed the results of Seurat's experiments. These are some of Sutter's assertions:

Perspective is the primary element in the composition of a painting; for the gradation of light and colors is proportional to the perspective gradation of lines.

The unity of the picture is comprised by the unity of aesthetic lines, masses, disjunctions, chiaroscuro, colors and the moral traits of the given subject.

There are two characteristic dispositions in which the subject is held: the vertical and the horizontal.

The mass of light should exceed or be less than the mass of shadow.

A mass appears to be greater to the degree that it includes in itself fewer small details.

That plane is most luminous on which the light falls most perpendicularly.

The more intense the light, the more color the shadows will reflect.

The dominant color of the picture should be supported by analogous tones which constitute the unity of colors.

The complementary colors: red and green, orange and blue, yellow and violet, have the property of reproducing white light.

There are three warm colors: red, orange, yellow.

There are three cold colors: green, blue, violet.

When the light is warm, the shadows are cold.

When the light is cold, the shadows are warm.

The intensity of reflections is proportional to the quantity of light.

A color is always modified by the proximity of another color.

The best ordered form is that which most completely imprisons the intelligence.

Science can be learned, sentiment can perfect, genius comes from God.

And David Sutter concluded: "Instinct does not change, but reason progresses without pause, man takes advantage of his own experience and of the experiences of other men. Reason is one of the most beautiful faculties of the human mind, and he who does not passionately seek to extend his knowledge by that fact alone renounces his greatest privilege.

Max Planck almost got lost in the shuffle of scientific history as the field of atomic physics moved into the early part of the twentieth century. And even though a distinguished research institute in Germany bears his name, Planck's work does tend to be overshadowed by the achievements of Einstein, J.J. Thomson, and Niels Bohr. It was Max Planck's elucidation of quantum theory, however, that allowed Einstein to conclude that light was made up of individual packets, or quanta, of energy. In addition, Bohr gave credit to Planck for laying the groundwork for his own monumental discovery that the light spectra of various elements were merely reflections of different energy states—or quantum levels—of the electrons within a given atom. With the following reading from Barbara Lovett Cline's *The Questioners* we attempt to correct the historical oversight that forced even Planck to conclude, "I was not to have at all the satisfaction of seeing myself vindicated."

The Quantum Theory of Max Planck

Barbara Lovett Cline

A young physicist who was staying in Max Planck's home during a visit to Berlin gradually became aware of a certain regularity in the Planck household. The visitor's curiosity was aroused; he made an observation: stationing himself by the door of his room he waited to see what would happen as the clock in the hall struck the hour. Sure enough, while the clock was still sounding, Planck emerged from his room, proceeded down the stairs and then out the front door. Further observation confirmed the regularity of his comings and goings; as the big hall clock struck a certain hour, invariably there Max Planck would be, on his way down the stairs.

Systematically, part of Planck's day was allotted for a walk, just as thirty minutes regularly were devoted to playing the piano. And when this highly organized scientist worked in the study among his treasured collection of scientific volumes, he worked standing up. His desk was a high one like those used in Dicken's time by clerks, who sat on high stools. But Planck did not use a stool.

It was in 1897, nine years after he joined von Helmholtz in Berlin, that Planck began to work, hard and upright, on the problem later called "the ultraviolet catastrophe," the problem which led him to the idea that energy is not continuous, contrary to prior observations and to the science based on these.

As mentioned earlier, Planck was not the only person to become interested in the problem of accounting, with physical theory, for the light and heat sent out by the illumination standard called a "black body." Physicists were attracted to this ideal case where the radiation spectrum, uninfluenced by other factors, depended only on temperature; others besides Planck were trying to account for the ideal case and thus gain an understanding of the radiation process in all cases.

Radiation similar to that emitted by a black body may be observed when any solid piece of metal, such as an iron poker, is heated. At low temperatures long-wave radiation is emitted, that of the infrared region of the spectrum. As the temperature rises, radiation of shorter and shorter wavelengths also appears, and the poker glows red, then orange until with the addition of other colors it appears white to the eye. A further increase in temperature will produce even shorter wavelengths which are not visible, those of the ultraviolet end of the spectrum.

The spectrum of the black body (or of anything else) shows how energy is distributed among the different wavelengths, for some colors are brighter than others. In Planck's day it was already possible to measure in the laboratory the energies represented in a spectrum. Those of the black body, at various temperatures, were known by experiment. The problem was to account for this specific energy distribution on the basis of general knowledge. Broadly speaking, this is the way physicists attacked the problem: they began with a hypothesis as to the cause of radiation, a hypothesis which seemed reasonable on the basis of what already was known. This hypothesis took the form of a model illustrating how radiation might arise from matter. Then physicists deduced the radiation energies which would occur according to the model, checking these theoretical deductions against laboratory measurements of the actual energy spectrum, hoping to find confirmation of the hypothesis.

The models which were used in attempts to solve the black-body problem were not models of the atom. Evidence was just being found at this time for the existence of the electron, and speculations as to atomic structure lay in the future. Physicists worked with rough general assumptions as to the material structure responsible for radiation. They postulated that the black body was composed fundamentally of some sort of electrically charged particle, the motion of which—speeded up by heat—produced the radiation. Actually such a rough hypothesis was adequate for the problem at hand; these were not the assumptions which were responsible for the failure to account for black-body radiation. The trouble came when physicists took the next step: in order to arrive at the distribution of energy in the spectrum, one first had to assign energies to the particles responsible for the radiation; energy had to be distributed among them so that motion of such-and-such kind yielded radiation of such-and-such energy. But if energy is continuous, as was assumed, there can be no restrictions on it, for these would introduce jumps, abrupt change. Therefore the oscillating motion of the electrical particles could not be restricted; the oscillations had to be indefinitely small. But from this it follows inescapably that the radiation energy at the short-wave (high-frequency) end of the spectrum will be indefinitely large. Speaking more technically, there will be a continuous rise to infinity with declining wavelength. Thus physics predicted ultraviolet catastrophe, for we know that a glowing material body does not radiate an infinite amount of energy. Moreover, experiment shows most of the energy appearing as waves of medium length. The solution is obvious: to avoid the ultraviolet catastrophe and reach an accurate description of black-body radiation one must limit the possible energy in such a way that it is not infinite and does not go chiefly to the shortest waves.

This is easy enough to see now—afterward. When Planck worked on the problem it was far more tangled. For one thing, experimental results were not as clear-cut then as they became later; one would seem to be on the right track only to be proved wrong by some new and more accurate measurements of the radiation. What is more, it was virtually impossible at that time to recognize the assumption of continuous energy as the root of the trouble. Black-body measurements were the first recognizable evidence that this assumption might be wrong; there was no background of unsolved problems to make one suspicious. Indeed the idea of continuity in nature, and therefore unbroken energy, did not appear to *be* an assumption. It was taken for granted. If, for example, one were trying to work out laws governing the growth rate of a plant variety, one would take certain things for granted, such as the fact that the sun will rise every day.

That was the way physicists "assumed" that energy was continuous. The fact that they were unable to solve the problem on their assumption did not lead them to question it. They had made numerous others as well, and it was much more reasonable to question these. Only much later did they understand that the laws and theories they were using rested on a questionable idea and that whether one treated the problem in terms of heat, mechanics, or electricity, one must come to the same conclusion, the ultraviolet catastrophe. All routes led to it.*

And Max Planck continued to trust the physics of his time even as he solved the black-body problem with an idea which contradicted that physics. That this was possible was due chiefly to two things: in the first place, Planck solved the problem backward, so to speak; he put together a formula which correctly described the energies of black-body radiation without understanding the full implications of the mathematical expression he had written. Secondly, in finding the meaning of his formula, he used a mathematical procedure which was new to him and which, without realizing it at the time, he misapplied. Let us see how this happened.

Planck arrived at his first radiation formula by doing some juggling. Among the different formulas which had been worked out in connection with the problem, there were two which were partly sound: one gave the energy distribution of the short-wave region of the spectrum correctly; the other accounted exactly for the long-wave region. Planck saw a simple, logical way to combine the best features of the two mathematical expressions. Whether this new formula represented the exact distribution of energy throughout the entire spectrum at every temperature he did not as yet know. But he did not have to wait very long to find out.

His formula was announced almost at once at a meeting of the Berlin Physical Society on October 19, 1900. Among those present was Heinrich Rubens, who had been working on black-body experiments. As soon as the

*Einstein was the first to point this out, five years after Planck's work.

meeting was over, Rubens went to his home and stayed up most of the night, comparing his measurements with Planck's formula. The agreement was excellent. Rubens appeared at Planck's home the next morning to tell him the good news.

Now Max Planck knew that he had found the solution to the black-body problem, but neither he nor anyone else knew what that solution meant. He had gotten it by guessing, by putting parts of formulas together, and in doing so he had automatically changed the assumptions on which these mathematical summations had been based. How had they been changed? What new assumptions had he, without knowing it, introduced? The only way to find out was to try different hypotheses and see if one of them led to his formula. Only then would he understand what the successful combination of symbols was saying about the process of radiation.

Planck began his attempt to draw a logical line from theory to fact so that fact not only was described but understood. For two months he tried to do this, working harder, he said later, than he ever had worked before or would again. Although the problem as we have defined it concerned electrical particles and their motion, Planck did not base his attempt to solve it on electrical theory. Instead, he used thermodynamics. As mentioned earlier, the range of this science is very broad and the black-body problem could be expressed in the vocabulary of thermodynamics as well as in the vocabulary of electrical theory.

Planck had a reason for using thermodynamics. Like other physicists, he wanted to solve the black-body problem, but he wanted to do something else as well: he wanted to demonstrate, by solving it, the fundamental nature of the second law of thermodynamics. He wanted to show, once again, the increase of entropy taken in an absolute sense. That is why he had chosen the black-body problem in 1897. That is why for four years he had been trying to solve it. He believed that if he could demonstrate the increase of entropy in the case of black-body radiation this would lead automatically to the correct solution to the problem. For four years he had been unsuccessful. When, in 1900, he found the correct solution, but not on the basis of theory, he returned at once to his earlier work and tried to deduce that right answer in terms of the increase of entropy, the increase of entropy in every single case.

It could not be done. This meant that either he must give up his four-year attempt to solve the problem on the basis of the second law of thermodynamics or try one last possibility: Boltzmann's statistical interpretation of that law. Planck never had used this interpretation. Suppose he turned to it now and defined the problem anew, as Boltzmann would, in terms of molecular motions and their statistical probabilities, and then, on *this* basis, was able to deduce his own correct formula? Automatically he also would be demonstrating that the second law was not an absolute, black-and-white certainty.

Thus Max Planck's systematic, patient, disciplined pursuit of this law brought him to the point where if he were to continue he might have to deny that it was an absolute truth. He did go on: he looked up the papers in which Boltzmann had presented his statistical method, applied it to the problem at hand and found that, indeed, he was led to his own formula. In this way he demonstrated that the increase of entropy, while overwhelmingly probable, was not absolutely certain, as for several decades he had conceived it to be.

If this were all Max Planck had done in physics, it would have been a great deal. But at the same time he did it he also stumbled on a new idea. In order to deduce the correct black-body formula it was necessary to follow a statistical route, but something else also was necessary: a hypothesis which broke drastically with nineteenth-century physics. By not applying Boltzmann's method as Boltzmann had intended it to be applied, Planck in effect based his work on the hypothesis that energy is not continuous. Let us see how this happened.

At a certain stage in solving a problem, Boltzmann's method required one to treat energy as if it were divided into separate portions, because in order to determine a degree of probability it is necessary to have something to count up. Boltzmann and others familiar with his method understood that this dividing up of the energy was merely a calculating technique. At a later stage in the calculations one always, by means of another technique, got rid of the portions, making the energy continuous again.

A like procedure sometimes is followed in solving problems which involve a curved figure such as, for example, a circle. Rather than calculating the exact circumference of the circle, it often is easier in terms of arithmetic to use a rough estimate instead. In effect, the mathematician gets rid of the unbroken curved line, which is difficult to handle arithmetically, and represents the circle as closely as possible by means of tiny straight lines equal to each other in length. (Such a figure if drawn would look almost exactly like a circle.) The total length of these straight sides may be taken for purposes of calculation as the circumference. Then later on, when it is more convenient, the mathematician obtains the exact circumference figure by letting the number of straight sides increase without limit. In this way he restores the curved unbroken line.

An unbroken flow of energy, like a curved line, is infinitely divisible. Had Planck applied Boltzmann's method as its originator had intended and, after calculating with portions of energy, put the pieces together, allowing them to increase without limit, an entirely different conclusion would have been reached. Assuming that his calculations were correct in other respects, he would have obtained the formula containing the ultraviolet catastrophe. As mentioned earlier, one had to come to this regardless of the route taken. During all the years Planck had worked on the black-body problem, he had chanced to avoid it; his methods never indicated the disastrous consequence of classical physics. When, finally, he came to the crucial step, he was using a procedure which was new to him. He did not join up the energy portions. At just that point he recognized that there was a way to reach the right answer: the formula he had before him. This meant that he avoided the ultraviolet catastrophe. For energy treated in portions is not infinitely divisible and therefore the radiation energy also is not infinite in quantity. What is more, by making the portions unequal, energy can be distributed in such a way that it does not go chiefly to the waves of shortest length.

That was how Planck found the simple but strange rule which lay behind his first formula and which forms the basis of the quantum theory. The rule states a relationship between a portion of energy, which Planck named a "quantum," using the Latin word for "how much," and the frequency of a wave, which is related to its length. To find the energy of a quantum E, one multiplies a wave frequency f by a fixed number, a constant, represented by h, or

$$E = hf$$

Planck's rule was strange because it announced an equality between energy conceived as discontinuous (E) and (since frequency pertains to wavelike phenomena) energy conceived as continuous. The full impact of this equation would not be felt until later, when it was shown that light could be understood as a stream of energy grains, or quanta. The black-body problem concerned the emission and absorption of light, not the structure of the radiation itself. Planck's work set a limit only on the energy of the electrical particles which compose matter: they might move in certain ways only so that a whole quantum hf was emitted or in turn absorbed. If light comes out of matter in portions and goes back into matter in portions, it would seem to follow that light must exist in particlelike form. But in 1900 there was no evidence whatsoever that light might have such a structure while its wave properties had been established beyond question. The broader extension of $E = hf$ would come later when there was some evidence, just a little, for the discontinuous structure of light. Until then the discontinuities announced by Planck's equation were ascribed only to the emitter, not to what was emitted.

Planck's work did not directly challenge the wave theory of light; still it broke drastically with accepted ideas, ideas based on the observation of nature. A swinging pendulum appears to lose its energy gradually and smoothly and here the pendulum stands for almost anything which is large enough to be observed directly. In every case, energy changes appear to be continuous.

Yet Planck's work denies this: in the case of black-body radiation the energy changes in bursts or jumps. The jumps are exceedingly small. In the equation $E = hf$, the constant h indicates their size and (in terms of the unit erg-seconds) h stands for the number

.0000000000000000000000000066
or (6.6×10^{-27}).

Then was Planck's work so revolutionary after all? The jumps simply were too small to be detected by ordinary means of observation. Up to about 1900, physicists had been concerned mainly with objects and happenings on the pendulum scale, or larger. Now they were beginning to probe more deeply and had found evidence that on a smaller scale nature did not conform to expectations. But physics still could be used where it had been used before. One would not have to consider minute discontinuities in energy in order to understand the motion of the pendulum, or the planet. Planck, in showing something new, had not necessarily demolished the old—but he had shown that there were limitations to it.

And it was just that which made his work revolutionary. It suggested that the understanding which classical physics represents might not after all be the last word. It hinted that the physical world might just possibly not be a giant mechanism whose workings are in an ultimate sense foreseeable. Planck had planted a seed of doubt. From this very gradually a new physics would grow as more was learned about the atomic region of nature. From this beginning would come an understanding of why statistical rules, as in the case of radioactive change, must be used; of why the atom does not ordinarily glow with light and why its electrons do not bury themselves in the nucleus.

Two weeks before Christmas, in the year 1900, Max Planck announced the results of his theoretical work on black-body radiation. In his speech to the Berlin Physical Society he said nothing of the challenge to classical physics which this work represents, and other evidence strongly suggests that he did not at first recognize the full implications of his mathematics. He did not know that the route which led to $E = hf$ was not Boltzmann's route.

What is more, when Planck later on did recognize the revolutionary character of his work he tried to undo it. Returning to the problem which he already had solved, he tried to find a different solution, a different theoretical route which also led to the right answer and did not require the fatal equation $E = hf$. For years he tried, to no avail. But he did not give up hope. Perhaps future discoveries still would make it possible to mend the break with classical physics which he had made. Planck continued to hope for this even though it would diminish the importance of his own work.

Instead, the years brought more and more confirmations of this work, as well as more and more honors in recognition of it. In Germany, Planck's position would be that of a Hermann von Helmholtz. Everywhere in the world physicists would regard him as the father of a revolution in their science. Again, Max Planck did not have the satisfaction of seeing himself vindicated.

In the years after 1900, much of his research was devoted, as before, to the second law of thermodynamics. Now he used Boltzmann's statistical interpretation of that law but, as before, he studied the consequences of the law in case after case, problem after problem. Rarely, if at all, was the outcome of these labors dramatic. Some of Planck's colleagues wondered why he found such "thankless questions" worth pursuing. They thought he ground away like a machine, the victim of his own great energy, disciplined habits and precise, tidy methods.

Einstein disagreed. After 1913 he came to know Planck quite well and in spite of their differences in attitude and background, and in spite of Planck's formality which often made Einstein uncomfortable in his presence, he recognized something in the older man that united them. Planck did not work like a mechanical man, said Einstein. It was a longing to find harmony and order in nature which spurred him on, "a hunger of the soul." That is why Einstein, speaking to other physicists, referred to "our Planck." "And that," said Einstein, "is why we love him."

Enrico Fermi struck Dr. Bronowski as "a strange creature . . . the cleverest man I had ever set eyes on—well, perhaps the cleverest man with one exception." He was a "modern alchemist" who used the neutron to transmute one element to another. In fact, he was the father of the first nuclear reactor, which demonstrated that the immense energy of the atomic nucleus could be harnessed for man's welfare. As such, he was also the progenitor of the atomic bomb. On January 30, 1954, Enrico Fermi gave his last address before the American Physical Society, in which he recounted the genesis of the historic nuclear energy project at Columbia University. The address, presented here in its unedited form, gives us added insight into the character of Fermi and his contemporaries.

The Genesis of the Nuclear Energy Project

Enrico Fermi

It seems fitting to remember, on this 200th anniversary of Columbia University, the key role that the University played in the early experimentation and the organization of the early work that led to the development of atomic energy.

I had the good fortune to be associated with the Pupin Laboratories through the period of time when at least the first phase of this development took place. I had had some difficulties in Italy and I will always be very grateful to Columbia University for having offered me a position in the Department of Physics at the most opportune moment. And in addition this offer gave me, as I said, the rare opportunity of witnessing the series of events to which I have referred.

In fact I remember very vividly the first month, January, 1939, that I started working at the Pupin Laboratories because things began happening very fast. In that period, Niels Bohr was on a lecture engagement in Princeton and I remember one afternoon Willis Lamb came back very excited and said that Bohr had leaked out great news. The great news that had leaked out was the discovery of fission and at least an outline of its interpretation; the discovery as you well remember goes back to the work of Hahn and Strassmann and at least the first idea for interpretation came through the work of Lise Meitner and Frisch who were at that time in Sweden.

Then, somewhat later that same month, there was a meeting in Washington organized by the Carnegie Institution in conjunction with George Washington University where I took part with a number of people from Columbia University and where the possible importance of the new-discovered phenomenon of fission was first discussed in semi-jocular earnest as a possible source of nuclear power. Because it was conjectured, if there is fission with a very serious upset of the nuclear structure, it is not improbable that some neutrons will be evaporated. And if some neutrons are evaporated, then they might be more than one; let's say, for the sake of argument, two. And

From Enrico Fermi, "Last Address to the American Physical Society, *Physics Today*, November 1955, if they are more than one, it may be that the two of them, for example, may each one cause a fission and from that one sees of course a beginning of the chain reaction machinery.

So that was one of the things that was discussed at that conference and started a small ripple of excitement about the possibility of releasing nuclear energy. At the same time experimentation was started feverishly in many laboratories, including Pupin, and I remember before leaving Washington I had a telegram from Dunning announcing the success of an experiment directed to the discovery of the fission fragments. The same experiment apparently was at the same time carried out in half a dozen places in this country and in three or four, in fact I think slightly before, in three or four places in Europe.

Now a rather long and laborious work was started at Columbia University in order to firm up these vague suggestions that had been made as to the possibilities that neutrons were emitted and to try to see whether neutrons were in fact emitted when fission took place and if so how many they would be, because clearly a matter of numbers is in this case extremely important because a little bit greater or a little bit lesser probability might have made all the difference between possibility and impossibility of a chain reaction.

Now this work was carried on at Columbia simultaneously by Zinn and Szilard on one hand and by Anderson and myself on the other hand. We worked independently and with different methods, but of course we kept close contact and we kept each other informed of the results. At the same time the same work was being carried out in France by a group headed by Joliot and von Halban. And all the three groups arrived at the same conclusion—I believe Joliot may be a few weeks earlier than we did at Columbia—namely that neutrons are emitted and they were rather abundant, although the quantitative measurement was still very uncertain and not too reliable.

A curious circumstance related to this phase of the work was that here for the first time secrecy that has been plaguing us for a number of years started and, contrary to perhaps what is the most common belief about secrecy, secrecy was not started by generals, was not started by security officers, but was started by physicists. And the man who is most responsible for this certainly extremely novel idea for physicists was Szilard.

I don't know how many of you know Szilard; no doubt very many of you do. He is certainly a very peculiar man, extremely intelligent (*laughter*). I see that this is an understatement (*laughter*). He is extremely brilliant and he seems somewhat to enjoy, at least that is the impression that he gives to me, he seems to enjoy startling people.

So he proceeded to startle physicists by proposing to them that given the circumstances of the period—you see it was early 1939 and war was very much in the air—given the circumstances of that period, given the danger that atomic energy and possibly atomic weapons could become the chief tool for the Nazis to enslave the world, it was the duty of the physicists to depart from what had been the tradition of publishing significant results as soon as the *Physical Review* or other scientific journals might turn them out, and that instead one had to go easy, keep back some results until it was clear whether these results were potentially dangerous or potentially helpful to our side.

So Szilard talked to a number of people and convinced them that they had to join some sort of—I don't know whether it would be called a secret society, or what it would be called. Anyway to get together and circulate this information privately among a rather restricted group and not to publish it immediately. He sent in this vein a number of cables to Joliot in France, but he did not get a favorable response from him and Joliot published his results more or less like results in physics had been published until that day. So that the fact that neutrons are emitted in fission in some abundance—the order of magnitude of one or two or three—became a matter of general knowledge. And, of course, that made the possibility of a chain reaction appear to most physicists as a vastly more real possibility than it had until that time.

Another important phase of the work that took place at Columbia University is connected with the suggestion on purely theoretical arguments, by Bohr and Wheeler, that of the two isotopes of uranium it was not the most abundant uranium 238 but it was the least abundant uranium 235, present as you know in the natural uranium mixture to the tune of 0.7 of a percent, that was responsible at least for most of the thermal fission. The argument had to do with an even number of neutrons in uranium 238 and an odd number of neutrons in uranium 235 which, according to a discussion of the binding energies that was carried out by Bohr and Wheeler, made plausible that uranium 235 should be more fissionable.

Now it clearly was very important to know the facts also experimentally and work was started in conjunction by Dunning and Booth at Columbia University and by Nier. Nier took the mass spectrographic part of this work, attempting to separate a minute but as large as possible amount of uranium 235, and Dunning and Booth at Columbia took over the part of using this minute amount in order to test whether or not it would undergo fission with a much greater cross section than ordinary uranium.

Well, you know of course by now that this experiment confirmed the theoretical suggestion of Bohr and Wheeler, indicating that the key isotope of uranium, from the point of view of any attempt of—for example—constructing a machine that would develop nuclear energy, was in fact uranium 235. Now you see the matter is important for the following reasons that at the time were appreciated perhaps less definitely than at the present moment.

The fundamental point in fabricating a chain reacting machine is of course to see to it that each fission produces a certain number of neutrons and some of these neutrons will again produce fission. If an original fission causes more than one subsequent fission then of course the reaction goes. If an original fission causes less than one subsequent fission then the reaction does not go.

Now, if you take the isolated pure isotope U^{235}, you may expect that the unavoidable losses of neutrons will be minor, and therefore if in the fission somewhat more than one neutron is emitted then it will be merely a matter of piling up enough uranium 235 to obtain a chain reacting structure. But if to each gram of uranium 235 you add some 140 grams of uranium 238 that come naturally with it, then the competition will be greater, because there will be all this ballast ready to snatch away the not too abundant neutrons that come out in the fission and therefore it was clear at the time that one of the ways to make possible the production of a chain reaction was to isolate the isotope U^{235} from the much more abundant isotope U^{238}.

Now, at present we have in our laboratories a row of bottles labeled, more or less, isotope—what shall I say—iron 56, for example, or uranium 235 or uranium 238 and these bottles are not quite as common as would be a row of bottles of chemical elements, but they are perfectly easily obtainable by putting due pressure on the Oak Ridge Laboratory (*laughter*). But at that time isotopes were considered almost magically inseparable. There was to be sure one exception, namely deuterium, which was already at that time available in bottles. But of course deuterium is an isotope in which the two isotopes hydrogen one and hydrogen two have a ratio of mass one to two, which is a very great ratio. But in the case of uranium the ratio of mass is merely 235 to 238, so the difference is barely over one percent. And that, of course, makes the differences of these two objects so tiny that it was not very clear that the job of separating large amounts of uranium 235 was one that could be taken seriously.

Well, therefore, in those early years near the end of 1939 two lines of attack to the problem of atomic energy started to emerge. One was as follows. The first step should be to separate in large amounts, amounts of kilograms or maybe amounts of tens of kilograms or maybe of hundreds of kilograms, nobody really knew how much would be needed, but something perhaps in that order of magnitude, separate such at that time fantastically large-looking amounts of uranium 235 and then operate with them without the ballast of the associated much larger amounts of uranium 238. The other school of thought was predicated on the hope that perhaps using some little amount of ingenuity one might use them efficiently and one might perhaps be able to achieve a chain reaction without having to separate the isotopes, a task as I say that at that time looked almost beyond human possibilities.

Now I personally had worked many years with neutrons, and especially slow neutrons, so I associated myself with the second team that wanted to use nonseparated uranium and try to do the best with it. Early attempts and studies, discussions, on how to separate the isotopes of uranium were started by Dunning and Booth in close consultation with Professor Urey. On the other hand, Szilard, Zinn, Anderson, and myself started experimentation on the other line whose first step involved lots of measurements.

Now, I have never yet quite understood why our measurements in those days were so poor. I'm noticing now that the measurements that we are doing on pion physics are very poor, presumably just because we have not learned the tricks. And, of course, the facilities that we had at that time were not as powerful as they are now. It's much easier to carry out experimentation with neutrons using a pile as a source of neutrons than it was in those days using radium-beryllium sources when geometry was the essential item to control or using the cyclotron when intensity was the desired feature rather than good geometry.

Well, we soon reached the conclusion that in order to have any chance of success with natural uranium we had to use slow neutrons. So there had to be a moderator. And this moderator could have been water or other substances. Water was soon discarded; it's very effective in slowing down neutrons, but still absorbs a little bit too many of them and we could not afford that. Then it was thought that graphite might be perhaps the better bet. It's not as efficient as water in slowing down neutrons; on the other hand little enough was known of its absorption properties that the hope that the absorption might be very low was quite tenable.

This brings us to the fall of 1939 when Einstein wrote his now famous letter to President Roosevelt advising him of what was the situation in physics—what was brewing and that he thought that the government had the duty to take an interest and to help along this development. And in fact help came along to the tune of $6000 a few months after and the $6000 were used in order to buy huge amounts—or what seemed at that time when the eye of physicists had not yet been distorted—(*laughter*) what seemed at that time a huge amount of graphite.

So physicists on the seventh floor of Pupin Laboratories started looking like coal miners (*laughter*) and the wives to whom these physicists came back tired at night were wondering what was happening. We know that there is smoke in the air, but after all . . . (laughter).

Well, what was happening was that in those days we were trying to learn something about the absorption properties of graphite, because perhaps graphite was no good. So, we built columns of graphite, maybe four feet on the side or something like that, maybe ten feet high. It was the first time when apparatus in physics, and these graphite columns were apparatus, was so big that you could climb on top of it—and you had to climb on top of it. Well cyclotrons were the same way too, but anyway that was the first time when I started climbing on top of my equipment because it was just too tall—I'm not a tall man (*laughter*).

And the sources of neutrons were inserted at the bottom and we were studying how these neutrons were first slowed down and then diffused up the column and of course if there had been a strong absorption they would not have diffused very high. But because it turned out that the absorption was in fact small, they could diffuse quite readily up this column and by making a little bit of mathematical analysis of the situation it became possible to make the first guesses as to what was the absorption cross section of graphite, a key element in deciding the possibility or not of fabricating a chain reacting unit with graphite and natural uranium.

Well, I will not go into detail of this experimentation. That lasted really quite a number of years and required really quite many hours and many days and many weeks of extremely hard work. I may mention that very early our efforts were brought in connection with similar efforts that were taking place at Princeton University where a group with Wigner, Creutz and Bob Wilson set to work making some measurements that we had no possibility of carrying out at Columbia University.

Well, as time went on, we began to identify what had to be measured and how accurately these things that I shall call "eta," *f*, and *p*—I don't think I have time to define them for you—these three quantities "eta," *f*, and *p* had to be measured to establish what could be done and what could not be done. And, in fact, if I may say so, the product of "eta," *f*, and *p* had to be greater than one. It turns out, we now know, that if one does just about the best this product can be 1.1.

So, if we had been able to measure these three quantities to the accuracy of one percent we might have found that the product was for example 1.08 plus or minus 0.03 and if that had been the case we would have said let's go ahead, or if the product had turned out to be 0.95 plus or minus 0.03 perhaps we would have said just that this line of approach is not very promising, and we had better look for something else. However I've already commented on the extremely low quality of the measurements in neutron physics that could be done at the time—where the accuracy of measuring separately either "eta," or *f*, or *p* was perhaps with a plus or minus of 20 percent (*laughter*). If you compound, by the well-known rules of statistics, three errors of 20 percent you will find something around 35 percent. So if you should find, for example, 0.9 plus or minus 0.3—what do you know? Hardly anything at all (*laughter*). If you find 1.1 plus or minus 0.3—again, you don't know anything much. So that was the trouble and in fact if you look in our early work—what were the detailed values given by this or that experimenter to, for example, "eta" you find that it was off 20 percent and sometimes greater amounts. In fact I think it was strongly influenced by the temperament of the physicist. Shall we say optimistic physicists felt it unavoidable to push these quantities high and pessimistic physicists like myself tried to keep them somewhat on the low side (*laughter*).

Anyway, nobody really knew and we decided therefore that one had to do something else. One had to devise some kind of experiment that would give a complete over-all measurement directly of the product "eta," *f*, *p* without having to measure separately the three, because then perhaps the error would sort of drop down and permit us to reach conclusions.

Well, we went to Dean Pegram, who was then the man who could carry out magic around the University, and we explained to him that we needed a big room. And when we say big we meant a really big room, perhaps he made a crack about a church not being the most suited place for a physics laboratory in his talk, but I think a church would have been just precisely what we wanted (*laughter*). Well, he scouted around the campus and we went with him to dark corridors and under various heating pipes and so on to visit possible sites for this experiment and eventually a big room, not a church, but something that might have compared in size with a church was discovered in Schermerhorn.

And there we started to construct this structure that at that time looked again in order of magnitude larger than anything that we had seen before. Actually if anybody would look at that structure now he would probably extract his magnifying glass (*laughter*) and go close to see it. But for the ideas of the time it looked really big. It was a structure of graphite bricks and spread through these graphite bricks in some sort of pattern were big cans, cubic cans, containing uranium oxide.

Now, graphite is a black substance, as you probably know. So is uranium oxide. And to handle many tons of both makes people very black. In fact it requires even strong people. And so, well we were reasonably strong, but I mean we were, after all, thinkers (*laughter*). So Dean Pegram again looked around and said that seems to be a job a little bit beyond your feeble strength, but there is a football squad at Columbia (*laughter*) that contains a dozen or so of very husky boys who take jobs by the hour just to carry them through College. Why don't you hire them?

And it was a marvelous idea; it was really a pleasure for once to direct the work of these husky boys, canning uranium—just shoving it in—handling packs of 50 or 100 pounds with the same ease as another person would have handled three or four pounds. In passing these cans fumes of all sorts of colors, mostly black, would go in the air (*laughter*).

Well, so grew what was called at the time the exponential pile. It was an exponential pile, because in the theory an exponential function enters—which is not surprising. And it was a structure that was designed to test in an integral way, without going down to fine details, whether the reactivity of the pile, the reproduction factor, would be greater or less than one. Well, it turned out to be 0.87. Now that is by 0.13 less than one and it was bad. However, at the moment we had a firm point to start from, and we had essentially to see whether we could squeeze the extra 0.13 or preferably a little bit more. Now there were many obvious things that could be done. First of all, I told you these big cans were canned in tin cans, so what has the iron to do? Iron can do only harm, can absorb neutrons, and we don't want that. So, out go the cans. Then, what about the purity of the materials? We took samples of uranium, and with our physicists' lack of skill in chemical analysis, we sort of tried to find out the impurities and certainly there were impurities. We would not know what they were, but they looked impressive, at least in bulk (*laughter*). So, now, what do these impurities do?—clearly they can do only harm. Maybe they make harm to the tune of 13 percent. Finally, the graphite was quite pure for the standards of that time, when graphite manufacturers were not concerned with avoiding those special impurities that absorb neutrons. But still there was some considerable gain to be made out there, and especially Szilard at that time took extremely decisive and strong steps to try to organize the early phases of production of pure materials. Now, he did a marvelous job which later on was taken over by a more powerful organization than was Szilard himself. Although to match Szilard it takes a few able-bodied customers (*laughter*).

Well, this brings us to Pearl Harbor. At that time, in fact I believe a few days before by accident, the interest in carrying through the uranium work was spreading; work somewhat similar to what was going on at Columbia had been initiated in a number of different Universities throughout the country. And the government started taking decisive action in order to organize the work, and, of course, Pearl Harbor gave the final and very decisive impetus to this organization. And it was decided in the high councils of the government that the work on the chain reaction produced by nonseparated isotopes of uranium should go to Chicago.

That is the time when I left Columbia University, and after a few months of commuting between Chicago and New York eventually moved to Chicago to keep up the work there, and from then on, with a few notable exceptions, the work at Columbia was concentrated on the isotope-separation phase of the atomic energy project.

As I've indicated this work was initiated by Booth, Dunning, and Urey about 1940, 1939, and 1940, and with this reorganization a large laboratory was started at Columbia under the direction of Professor Urey. The work there was extremely successful and rapidly expanded into the build-up of a huge research laboratory which cooperated with the Union Carbide Company in establishing some of the separation plants at Oak Ridge. This was one of the three horses on which the directors of the atomic energy project had placed their bets, and as you know the three horses arrived almost simultaneously to the goal in the summer of 1945. I thank you.

If man has evolved from simpler life forms, as Darwin has taught us to believe, it is equally true, Dr. Bronowski points out, that "matter itself evolves." Within the furnace of the stars, small atoms became larger atoms; so all the atoms that make up everything—living and nonliving—in our universe were formed. In that sense we are the starfolk of which Carl Sagan writes.

Starfolk

Carl Sagan

Once upon a time, about ten or fifteen billion years ago, the universe was without form. There were no galaxies. There were no stars. There were no planets. And there was no life. Darkness was upon the face of the deep. The universe was hydrogen and helium. The explosion of the Big Bang had passed, and the fires of that titanic event—either the creation of the universe or the ashes of a previous incarnation of the universe—were rumbling feebly down the corridors of space.

But the gas of hydrogen and helium was not smoothly distributed. Here and there in the great dark, by accident, somewhat more than the ordinary amount of gas was collected. Such clumps grew imperceptibly at the expense of their surroundings, gravitationally attracting larger and larger amounts of neighboring gas. As such clumps grew in mass, their denser parts, governed by the inexorable laws of gravitation and conservation of angular momentum, contracted and compacted, spinning faster and faster. Within these great rotating balls and pinwheels of gas, smaller fragments of greater density condensed out; these shattered into billions of smaller shrinking gas balls.

Compaction led to violent collisions of the atoms at the centers of the gas balls. The temperatures became so great that electrons were stripped from protons in the constituent hydrogen atoms. Because protons have like positive charges, they ordinarily electrically repel one another. But after a while the temperatures at the centers of the gas balls became so great that the protons collided with extraordinary energy—an energy so great that the barrier of electrical repulsion that surrounds the proton was penetrated. Once penetration occurred, nuclear forces—the forces that hold the nuclei of atoms together—came into play. From the simple hydrogen gas the next atom in complexity, helium, was formed. In the synthesis of one helium atom from four hydrogen atoms there is a small amount of excess energy left over. This energy, trickling out through the gas ball, reached the surface and was radiated into space. The gas ball had turned on. The first star was formed. There was light on the face of the heavens.

The stars evolved over billions of years, slowly turning hydrogen into helium in their deep interiors, converting the slight mass difference into energy, and flooding the skies with light. There were in these times no planets to receive the light, and no life forms to admire the radiance of the heavens.

The conversion of hydrogen into helium could not continue indefinitely. Eventually, in the hot interiors of the stars, where the temperatures were high enough to overcome the forces of electrical repulsion, all the hydrogen was consumed. The fires of the stars were stoked. The pressures in the interiors could no longer support the immense weight of the overlying layers of star. The stars then continued their process of collapse, which had been interrupted by the nuclear fires of a billion years before.

In contracting further, higher temperatures were reached, temperatures so high that helium atoms—the ash of the previous epoch of nuclear reaction—became usable as stellar fuel. More complex nuclear reactions occurred in the insides of the stars—now swollen, distended red giant stars. Helium was converted to carbon, carbon to oxygen and magnesium, oxygen to neon, magnesium to silicon, silicon to sulfur, and upward through the litany of the periodic table of the elements—a massive stellar alchemy. Vast and intricate mazes of nuclear reactions built up some nuclei. Others coalesced to form much more complex nuclei. Still others fragmented or combined with protons to build only slightly more complex nuclei.

But the gravity on the surfaces of red giants is low, because the surfaces have expanded outward from the interiors. The outer layers of red giants are slowly dissipated into interstellar space, enriching the space between the stars in carbon and oxygen and magnesium and iron and all the elements heavier than hydrogen and helium. In some cases, the outer layers of the star were slowly stripped off, like the successive skins of an onion. In other cases, a colossal nuclear explosion rocked the star, propelling at immence velocity into interstellar space most of the outside of the star. Either by leakage or explosion, by dissipation slow or dissipation fast, star-stuff was spewed back to the dark, thin gas from which the stars had come.

But here, later generations of stars were aborning. Again the condensations of gas spun their slow gravitational pirouettes, slowly transmogrifying gas cloud into star. But these new second- and third-generation stars were enriched in heavy elements, the patrimony of their stellar antecedents. Now, as stars were formed, smaller condensations formed near them, condensations far too small to produce nuclear fires and become stars. They were little dense, cold clots of matter, slowly forming out of the rotating cloud, later to be illuminated by the nuclear fires that they themselves could not generate. These unprepossessing clots became the planets: Some giant and gaseous, composed mostly of hydrogen and helium, cold and far from their parent star; others, smaller and warmer, losing the bulk of their hydrogen and helium by a slow trickling away to space, formed a different sort of planet—rocky, metallic, hard-surfaced.

These smaller cosmic debris, congealing and warming, released small quantities of hydrogen-rich gases, trapped in their interiors during the processes of formation. Some gases condensed on the surface, forming the first oceans; other gases remained above the surface, forming the first atmospheres—atmospheres different from the present atmosphere of Earth, atmospheres composed of methane, ammonia, hydrogen sulfide, water, and hydrogen—an unpleasant and unbreathable atmosphere for humans. But this is not yet a story about humans.

Starlight fell on this atmosphere. Storms were driven by the Sun, producing thunder and lightning. Volcanoes erupted, hot lava heating the atmosphere near the surface. These processes broke apart molecules of the primitive atmosphere. But the fragments reassorted into more and more complex molecules, falling into the early oceans, there interacting with each other, falling by chance upon clays, a dizzying process of breakdown, resynthesis, transformation—slowly moving toward molecules of greater and greater complexity, driven by the laws of physics and chemistry. After a time, the oceans achieved the constituency of a warm dilute broth.

Among the innumerable species of complex organic molecules forming and dissipating in this broth there one day arose a molecule able crudely to make copies of itself—a molecule which weakly guided the chemical processes in its vicinity to produce molecules like itself—a template molecule, a blueprint molecule, a self-replicating molecule. This molecule was not very efficient. Its copies were inexact. But soon it gained a significant advantage over the other molecules in the early waters. The molecules that could not copy themselves did not. Those that could, did. The number of copying molecules greatly increased.

As time passed, the copying process became more exact. Other molecules in the waters were reprocessed to form the jigsaw puzzle pieces to fit the copying molecules. A minute and imperceptible statistical advantage of the molecules that could copy themselves was soon transformed by the arithmetic of geometrical progression into the dominant process in the oceans.

More and more elaborate reproductive systems arose. Those systems that copied better produced more copies. Those that copied poorly produced fewer copies. Soon most of the molecules were organized into molecular collectives, into self-replicating systems. It was not that any molecules had the glimmering of an idea or the ghostly passage of a need or want or aspiration; merely, those molecules that copied did, and soon the face of the planet became transformed by the copying process. In time, the seas became full of these molecular collectives, forming, metabolizing, replicating . . . forming, metabolizing, replicating . . . forming, metabolizing, mutating, replicating. . . . Elaborate systems arose, molecular collectives exhibiting behavior, moving to where the replication building blocks were more abundant, avoiding molecular collectives that incorporated their neighbors. Natural selection became a molecular sieve, selecting out those combinations of molecules best suited by chance to further replication.

All the while the building blocks, the foodstuffs, the parts for later copies, were being produced, mainly by sunlight and lightning and thunder—all driven by the nearby star. The nuclear processes in the insides of the stars drove the planetary processes, which led to and sustained life.

As the supply of foodstuffs gradually was exhausted, a new kind of molecular collective arose, one able to produce molecular building blocks internally out of air and water and sunlight. The first animals were joined by the first plants. The animals became parasites upon the plants, as they had been earlier on the stellar manna falling from the skies. The plants slowly changed the composition of the atmosphere; hydrogen was lost to space, ammonia transformed to nitrogen, methane to carbon dioxide. For the first time, oxygen was produced in significant quantities in the atmosphere—oxygen, a deadly poisonous gas able to convert all the self-replicating organic molecules back into simple gases like carbon dioxide and water.

But life met this supreme challenge: In some cases by burrowing into environments where oxygen was absent, but—in the most successful variants—by evolving not only to survive the oxygen but to use it in the more efficient metabolism of foodstuffs.

Sex and death evolved—processes that vastly increased the rate of natural selection. Some organisms evolved hard parts, climbed onto, and survived on the land. The pace of production of more complex forms accelerated. Flight evolved. Enormous four-legged beasts thundered across the steaming jungles. Small beasts emerged, born live, instead of in hardshelled containers filled with replicas of the early oceans. They survived through swiftness and cunning—and increasingly long periods in which their knowledge was not so much preprogrammed in self-replicating molecules as learned from parents and experiences.

All the while, the climate was variable. Slight variations in the output of sunlight, the orbital motion of the planet, clouds, oceans, and polar icecaps produced climatic changes—wiping out whole groups of organisms and causing the exuberant proliferation of other, once insignificant, groups.

And then . . . the Earth grew somewhat cold. The forests retreated. Small arboreal animals climbed down from the trees to seek a livelihood on the savannas. They became upright and tool-using. They communicated by producing compressional waves in the air with their eating and breathing organs. They discovered that organic material would, at a high enough temperature, combine with atmospheric oxygen to produce the stable hot plasma called fire. Postpartum learning was greatly accelerated by social interaction. Communal hunting developed, writing was invented, political structures evolved, superstition and science, religion and technology.

And then one day there came to be a creature whose genetic material was in no major way different from the self-replicating molecular collectives of any of the other organisms on his planet, which he called Earth. But he was able to ponder the mystery of his origins, the strange and tortuous path by which he had emerged from star-stuff. He was the matter of the cosmos, contemplating itself. He considered the problematical and enigmatic question of his future. He called himself Man. He was one of the starfolk. And he longed to return to the stars.

The universe then is the product of a great energy machine in which matter as we know it is continually altered. If the universe is such a machine, will it run perpetually, or will it, like all other machines, run down? That is a question that tantalized Rudolf Clausius in 1850, and in an attempt to answer it, he formulated the principle of *entropy*. With it, he developed what physics texts call "The Second Law of Thermodynamics." How that law and the prospects for a run-down universe relate to each other are discussed in three excerpts from the works of Isaac Asimov, chemist, novelist, essayist, and biblical scholar, who is equally at home in science fact and science fiction.

What Is Entropy?

Isaac Asimov

How does entropy relate to energy?

Energy can be turned into work only if, within the particular system you are using, there is an unevenness in the concentration of energy. The energy then tends to flow from the point of higher concentration to that of lower concentration until everything is evened out. It is by taking advantage of this flow that you can get work out of energy.

The water at the source of a river is higher and has more gravitational energy than the water at the mouth of the river. For that reason water flows down the river into the ocean. (If it weren't for rainfall, all the water on the continents would flow downhill to the ocean and the ocean level would rise slightly. The total gravitational energy would remain the same, but it would be more evenly distributed.)

It is while the water flows downhill that a waterwheel can be made to turn and work can be done by the water. Water on a single level cannot do work; not even if the water is on a high plateau and has an unusually high gravitational energy. It is the difference in energy concentration and the flow toward evenness that is crucial.

This is true of any kind of energy. In steam engines there is a heat reservoir which turns water into steam and a cold reservoir which condenses the steam to water again. It is this difference in temperature which is crucial. At any single undifferentiated temperature, however high, no work can be obtained.

"Entropy" is a term introduced in 1850 by a German physicist, Rudolf J.E. Clausius, to represent the evenness with which energy of any form is spread out. The more evenly it is spread out, the higher the entropy. If the energy is spread out perfectly evenly, entropy is at a maximum for the system under consideration.

It seemed to Clausius that, left to itself, the energy differences within a system always tended to even out. If a hot object is placed in contact with a cold object, heat flows in such a way that the hot object cools and the cold object warms until both are at the same temperature. If two water reservoirs are connected and one has a higher water level than the other, gravitational pull will make one water level sink and the other rise, till both levels are even and gravitational energy is smoothed out.

Clausius therefore said that it was a general rule in nature that differences in concentrations of energy tended to equalize. In other words, "Entropy increases with time."

The study of energy flow from point of higher concentration to points of lower concentration was carried out most thoroughly in connection with energy in the form of heat. The study of energy flow and of interchanges of energy and work has therefore come to be known as "thermodynamics," from Greek words meaning "heat-motion."

It had already been decided that energy could neither be created nor destroyed. This was so fundamental a rule that it is called "The First Law of Thermodynamics."

Clausius' suggesting that entropy increased with time seemed almost as fundamental a general rule and it is called the "Second Law of Thermodynamics."

Natural Changes Create Disorder

Isaac Asimov

Imagine nine people arranged in a square; three rows of three, all the rows and columns evenly spaced. We can call this an orderly arrangement because it is neat and symmetrical and is easily described.

If every one of the nine men takes one step forward simultaneously, they will remain in formation and the arrangement will still be orderly. The same is true if each takes a step backward, or each a step to the left, or each a step to the right.

Suppose, however, each man is told to take one step, either forward, backward, right or left, and each man is completely free to choose his direction. It may be that each man will independently just happen to decide to step forward one step. In that case, order will be maintained.

However, the chance that a particular man will step forward is only one out of four, since he is free to move in any of four different directions. The chance that all nine men will independently choose to move forward is one out of $4 \times 4 \times 4 \times 4 \times 4 \times 4 \times 4 \times 4 \times 4$, or one out of 262,144.

If all move right, or all left, or all back, they also keep in order so that the total chance of doing so is four out of 262,144, or one out of 65,536. So you can see what a tiny chance order has, and you know that if the men are given freedom to move, then a single step will be enough to break the square and decrease the amount of order. Even if, by some freak chance, they do all move together then the next step will break the square.

This is the case where only nine men are involved and where only four different directions of movement are allowed. In most natural processes, we deal with uncounted trillions of atoms free to move in very many different ways. If, by some chance, there were some sort of order imposed on the arrangement of atoms to begin with, then any free random motion, any spontaneous change, would be bound to decrease that order or to put it another way, increase disorder.

According to the Second Law of Thermodynamics, the entropy of the universe is always increasing; that is, the energy distribution in the universe is constantly evening out. It can be shown that any process which evens out energy concentration also increases disorder. Therefore, this tendency to increase disorder in the universe with the free random motions of the particles making it up is but another aspect of the Second Law, and entropy can be considered a measure of the disorder present throughout the entire Universe.

If we look at it this way, we can see the workings of the Second Law all about us, since natural changes clearly work in the direction of disorder and it is only with special effort, at a cost to ourselves, that we restore order. Our belongings get out of place, our houses get messy, our clothes get dirty, and we must constantly straighten and dust and clean to remain in one place. It may make us feel better to think that this is the result of the working out of a great cosmic law—(but somehow it doesn't in my case.)

Is the Universe Running Down?

Isaac Asimov

According to the "Second Law of Thermodynamics" entropy is always increasing. That means that differences in energy concentration are always evening out. When all differences in energy concentration have evened out completely, no more work can be extracted from energy; no more changes can take place.

Consider a watch. A watch goes because there is a concentration of energy in its mainspring or its battery. As the spring unwinds or the chemical reaction of the battery proceeds, there is a flow of energy from the point of high concentration to the point of low and as a result of this flow the watch goes. Once the spring is entirely unwound or the battery has completed its chemical change, the energy level has evened out throughout the watch, there is no more energy flow, and the watch no longer goes. It has "run down." It is by analogy with this that we say the universe will "run down" when all the energy is completely evened out.

Of course, we can wind up the watch again or buy a new battery. To wind up the watch, we use our own muscle power and we ourselves "run down" a little bit. We can buy a new battery, but that must be manufactured and, to make it, man's industrial establishment must "run down" a little bit.

We can renew our own muscle power by eating, but the food is formed originally by plants that make use of the sun's energy. Man's industrial establishment runs chiefly on coal and oil formed by plants in eons past from the sun's energy. As things "run down" on earth, we can always "wind them up" again by using something that usually stretches back to the sun's energy and the way it is "running down."

The sun is made up largely of hydrogen which contains much more energy per particle than do more complicated atoms such as helium, oxygen and iron. Inside the sun, there is a gradual evening out of energy concentration as hydrogen turns into more complicated atoms. (In atomic power plants on earth, the same thing happens as uranium atoms turn into less complicated atoms. If we ever develop hydrogen fusion plants we will be duplicating, in a way, what is going on in the sun.)

Eventually, as far as we can tell now, energy concentrations will even out in the sun and leave it with intermediate-sized atoms only. This will be true also of all the other stars in the universe and of everything in the universe.

If the Second Law of Thermodynamics is true then all energy concentrations everywhere in the universe are evening out and the universe is, in that sense, running down. If it is true, then entropy will reach a maximum when all the energy of the universe is perfectly evened out and then nothing more will happen because though the energy will still be all there, there will no longer be any flow to make things happen.

Depressing (IF the Second Law is actually true under all conditions) but there is no need for immediate alarm. The process will take many trillions of years to reach its end, and the universe as it now exists will not only last our time but, in all probability, it will last mankind's time and even the earth's time.

11

Knowledge or Certainty

Thus sciences, by the diverse motions of this globe of the brain of man, are become opinions, nay errors, and leave the imagination in a thousand labryinths. What is all we know, compared with what we know not?

William Drummond

Among the overriding myths about what science is and what science is not, perhaps one of the most widely accepted is that science—based as it is on numbers and apparently immutable natural laws—can provide absolute truth. Yet that is not so, as Dr. Bronowski explains. Despite each refinement of scientific instrumentation, there remains a hidden, forbidden zone, where some fraction of truth remains unknown.

The ultimate enigma of nature is what Werner Heisenberg tried to quantify by means of the uncertainty principle. He and other scientists concluded that it is impossible to locate the exact position of the electron in its quantum jumps around the nucleus. How that came about and what its implications are for developing instruments that will ultimately "tell all" are explored in the first selection, by Stanley L. Jaki. The second selection, by Paul G. Hewitt, is a more technical explanation of Heisenberg's principle and its ramifications.

The Edge of Precision

Stanley L. Jaki

Statistical considerations apart, once it was recognized that light consists of photons, the energy of which is equal to the product of their frequency and of Planck's quantum, it became impossible to localize particles and to register their interaction in time with absolute precision. For to obtain information about the position of an electron, a photon must be reflected from it; in this process, the electron must receive an impact that will change either its state of rest or its velocity. The magnitude of this impact is uncertain since the photon can follow any path within the angular opening of the objective lens. This situation obviously pointed to some deeper principle, one that was formulated by Heisenberg in 1927. Known as the "uncertainty principle," it states that magnitudes represented by canonically conjugate variables, or operators, cannot be measured simultaneously with unlimited accuracy. Such paired variables are position and momentum, angular momentum and angular position, moment of inertia and angular velocity. Since the limit of resolution in optics is of the order of the wavelength used, it would be possible to fix the position of the electron with a probable error of about one hundred thousandth of a centimeter by using ultraviolet light. The corresponding minimum uncertainty in the velocity of the electron would then be of the order of one kilometer per second. Conversely, if the velocity of the electron is given within an accuracy of one hundred thousandth of a centimeter per second, the electron could be anywhere within one kilometer. One could of course theoretically use a radiation of much shorter wavelength to achieve greater precision in measuring say the position of the electron, but this would necessarily increase the minimum error involved in measuring its velocity by the same factor as that by which the wavelength of radiation was reduced.

If the uncertainty relation connecting position and momentum, $\Delta x \Delta p \geqslant h$ is written as $\Delta x \Delta v \geqslant h/m$, then the situation can be grasped even more directly. The ratio h/m for an electron is about unity. Suppose therefore that one

From Stanley L. Jaki, *The Relevance of Physics,* The University of Chicago Press, © 1965.

desires to measure the classically computed value of the radius of the orbit of an electron in a hydrogen atom with an error of about 10^{-8} cm. This would be of the same magnitude as the radius itself. A larger error would evidently make the measurement meaningless. The corresponding minimum uncertainty in the value of the orbital velocity of the electron would be of the same order of magnitude as the value of the velocity computed classically. A greater precision in measuring the radius could only increase the already large uncertainty regarding the value of the orbital velocity. This essentially means that even with minimum uncertainty achieved, the position of the electron around the nucleus cannot be located. The uncertainty, in fact, is so great that one has to assume that the electron is "spread" around the nucleus. Thus the electron cannot be pictured as moving in a well-defined continuous orbit, or spinning on its axis as a perfect rotating solid. For the same reason such basic constants as the spin and magnetic moment of an electron cannot be directly measured by a magnetometer experiment. Such an experiment has to work with a moving electron, the position and velocity of which cannot be measured with perfect accuracy at the same time. The uncertainty in the velocity gives rise to an uncertainty in the magnetic field induced by the moving charge but the uncertainty thus produced is greater than the magnetic field owing to the spin magnetic moment that was supposed to be measured.

Another area in which precision has reached the ultimate limits set by the indeterminacy principle is the measurement of the width of a spectral line. Three main causes contribute to the broadening of spectral lines. Two of these, the translational motion of the atom and the interaction of atoms due to pressure, can be eliminated to almost any degree, at least theoretically. The third one, however, which produces the so-called natural width, is permanently inherent in any measurement of a spectral line. The energy width produced will be of the order of magnitude given by the ratio of Planck's constant divided by the time needed for the measurement. Since an ordinary excited state of an atom has a lifetime of the order of 10^{-8} seconds, it follows that lines in the visible spectrum will show a widening of about one ten-thousandth of an Angstrom. This is too small to measure, but in the X-ray region the natural width is easily observable. Thus the edge of precision actually opened up for physics what P.W. Bridgman called once in a rather pessimistic vein "the portals of a shadowy domain beyond which no refinement in measurement can carry the physicist."

Still, the limitations set by the uncertainty principle should not be construed as too narrow confines within which the possibilities will soon be exhausted by advances in instrumental precision. What the uncertainty principle essentially means is that the determination of the "state of the world at an instant" is not possible in terms of mechanistic physics. And this limitation holds also for that proverbial "superior spirit" to which Laplace and others liked to refer. It is, however, well to remember that while classical physicists recognized their inferiority to that superior spirit, it was not assumed to be superior to the principles of classical physics. At any rate, modern physics in general, and the indeterminacy principle in particular, made it abundantly clear that the mechanical intelligibility does not exhaust the whole range of intelligibility. Therefore one should not conclude on the basis of the indeterminacy principle that "the world is not a world of reason, understandable by the intellect of man." Those modern physicists who do so fall back unawares on the definition of intelligibility as formulated by classical mechanism. If, however, one is willing to part with this long discredited notion of intelligibility, cherished as it may be even today by some, the future of physics will appear promising not in the least because of the great potentialities of its precision instruments.

There is indeed nothing to indicate that the rate of progress in devising more precise instruments would hit a snag in the near future. To play the prophet in scientific matters is of course rather hazardous, but in the field of precision instruments the future usually vindicates those who keep expecting the almost impossible. The course of events has not justified even the slightest touch of resignation concerning the perfectibility of measurements in physics. How little did a Rowland (a true wizard of precision himself) suspect the degree of progress that was to take place in the field of measurements in the first half of the twentieth century! Not that he was pleased with everything physical science could offer in this respect as the nineteenth century came to its close. In 1899 he noted with some dissatisfaction that the attainment of precision has in some respect left much to be desired. As he surveyed the field, he saw that mechanical rotations in his day were limited to a few hundred revolutions per second, that the greatest pressure achieved did not surpass one hundred tons per square inch, that the sun's temperature was the upper limit in temperature measurements, and that the vacua, to use his own words, were very imperfect indeed.

Subsequent advances in all of these fields have been such that one who demanded them in 1899 would have seemed to be asking for the moon. Actually nothing short of the moon itself was reached in little more than half a century. At the same time, of course, spectacular successes also meant spectacular surprises, some of which were embarrassing. Photographs taken of the far side of the moon threw lunar theories into turmoil. The pictures obtained by Ranger IX, the last of which were taken just half a second before impact, provided ground for widely differing speculations about the structure of the moon's surface. The vigor of those speculations did not diminish after Surveyor I began to transmit thousands of pictures from the moon's surface. Although Mariner II, observing Venus from a vicinity of 21,000 miles in late 1962, did indeed confirm some of the findings of earthbound telescopes, Mariner IV accomplished what usually happens when the resolving power of the instruments of physics is considerably increased. By showing objects as small as two miles in diameter on the surface of Mars, Mariner IV not only surpassed anything achieved along these lines by astronomy, but also left in disorder a firmly established scientific consensus that expected anything but the revelation of a crater-pocked, moonlike Martian land. To be sure, the degree of precision built into the devices of space exploration is very great. Yet, no less astonishing are the results which the edge of precision provides in peeling off layer after layer of the mysteries enveloping the physical universe.

If the present-day search into space brings out sharply the fundamental role of precision in physics, the same kind of enlightenment is being achieved by efforts that aim at penetrating ever further into the structure of matter. In this connection it is well to remember that the chief probing instruments of matter, the accelerators, have not yet reached their ultimate possibilities. The greater the energy a particle acquires through acceleration, the smaller is its De Broglie wavelength and the smaller are the details it can resolve in its target. To "see" the rough details of a molecule, an electron microscope using electrons with an energy of several thousand electron volts is needed. To "see" the nucleus, a sort of "proton microscope," using protons of 2 Mev, is required. To resolve the details of the nucleus, accelerators operating in the Bev energy range have to be used. This is, however, only one aspect of the story behind the huge accelerators that somebody has called the cathedrals of the twentieth century. At sufficiently short wavelengths, particles are not only "seen" but also "created. . . ." Reflecting on the successive steps, that have improved acceleration machines, one ought to be struck by the fact that a major advance in peak energy led invariably to the unfolding of qualitatively new aspects of the structure of matter. This feature of the development of accelerators was stressed recently with special emphasis by C.N. Yang, the codiscoverer of the non-conservation of parity in weak interactions, who had this to say about the need for more powerful accelerators:

Sometimes people tend to think that to have greater intensity is only to increase experimental accuracies from ±10% to ±1%, or from ±1% to ±.1%. Let me emphasize that that is not the reason why greater intensity will be badly needed in high energy physics in the near future: greater intensity will be needed because one will have to understand qualitative *features of elementary particles. . . . At the present state of our knowledge of high energy physics, the danger that all these features will be qualitatively understood to such an extent that a high intensity machine will be left only with a job of adding decimal points to known results is very small indeed.*

That many startling discoveries will follow in studies of the basic structure of matter can be confidently expected from the fact that there is hardly any evidence in sight for a fixed ceiling for peak energy attainable in future accelerators. Plans for a 200 Bev [billion electron volts] accelerator have already reached the stage of selecting its location. Yet, while this enormous machine with a diameter of 4,528 feet is still about ten years away from becoming operational, more and more consideration is already given to accelerators with a peak energy of 600 to 1,000 Bev. Their construction might of course necessitate an international pooling of material and scientific resources. Clearly, to make such progress in little over three decades from Lawrence's first homemade cyclotron to the present-day plans, the execution of which may necessitate a cooperation on an international scale, is a cause for no small satisfaction. At the same time one should not forget that research here runs the full circle of scientific discovery. Bigger machines make new particles that in turn create problems that motivate further increases in the peak energy of accelerators. Still if a survey of past steps, triumphs, hopes, and dreams is any indication, one might say with confidence that the future belongs to those who like to think of possibilities in very bold terms. The late E. Fermi once suggested that the earth be ringed with a vacuum tube; by using the earth's magnetic field, one could obtain in such a way energies three thousand times greater than the peak energy of the largest accelerators now in use. This suggestion, truly fantastic only a few years ago, should appear more realistic since artificial satellites have come of age. A set of satellites, each equipped with injectors and focusing apparatus might make available even one million Bev without a single vacuum pump.

Physics has indeed reached the stage where the demands of precision have made inevitable the construction of machines that are nothing short of gigantic and fantastic. A case in point is the apparatus used for the detection of neutrinos. First devised by F. Reines and C. Cowan in the mid-1950's, these enormous installations provided in 1965 the first evidence of the existence of natural neutrinos. Here, bulkiness represents extraordinary sensitivity. Clearly, only a sensitivity that defies imagination is capable of detecting natural neutrinos, although billions of them bombard each square centimeter of the earth's surface every second. This feat took place under the direction of F. Reines near Johannesburg in a cave 10,492 feet deep. A similar experiment was devised to trap each day three to nine neutrinos out of trillions, and then after a month's time to extract their accumulated by-products, some 270 radioactive argon atoms, from 100,000 gallons of target fluid. All this represents a classic illustration that the more elusive a particle, the more elaborate are the instruments needed for its detection. At the same time instrumental layouts, such as the ones used for the detection of neutrinos, evidence not only an outstanding feat in reasoning but also a fantastic degree of precision in measurement. Thus if neutrinos are indeed carrying as much energy in the universe as do heat, light, and all sorts of cosmic particles, the apparatus described above might prove both one of the sharpest instruments man has yet devised and the most portentous as well in theoretical results. In all likelihood once again the edge of precision will unfold startling new aspects that might give entirely new turns and impetus to science, and might lead once more to a thorough overhaul of scientific concepts.

The quest for ever greater precision keeps science relentlessly on the move. Science is, in a sense, as De Sitter noted, "the struggle for the last decimal place." In his words, "the great triumphs of science are gained, when, by new methods or new instruments, the last decimal place is made into the penultimate." It is, however, precisely there, in the ever-continuing emergence of new digits that the paradox of the quest for precision lies. It is a quest that not only brings triumphs but also necessitates revolutionary reorientations. By turns it confirms and discards theories and points to aspects hitherto unsuspected. The moments when new, more precise instruments become available are usually turning points in science. The future is hidden, but precision will probably affect the future course of science as it has affected science in the past: scientific history should permit one to make that tentative conjecture.

The Uncertainty Principle

Paul G. Hewitt

Before the advent of quantum mechanics, it was believed that we could obtain better and better accuracy in measurement if we simply improved our measuring instruments. This is true to an extent, but *perfect* accuracy cannot be achieved no matter how fine the measuring instrument. There exists some uncertainty in every measurement we make, not only as a result of imperfect measuring instruments, but because of the unavoidable interaction between the observer and the observed.

To observe an object requires that at least one photon bounce off it. A photon bouncing off a baseball in flight alters its motion imperceptibly, and its motion is not altered by the presence or absence of a witness. To "observe" an electron, on the other hand, is a different story. A single photon bouncing off an electron appreciably alters its motion—and in an unpredictable way. If, for example, we wish to determine the position of the electron very accurately, the wavelength of light used to detect it must be short. A short wavelength, however, corresponds to a large quantum of energy, which greatly alters the electron's state of motion. If, on the other hand, a long wavelength corresponding to a small quantum of energy is used, the change in the electron's state of motion will be smaller, but the determination of its position by the coarser wave will be less accurate. The act of observing something as tiny as an electron produces a great uncertainty in either the position or the motion being measured. Although this uncertainty is completely negligible for measurements of position and motion for everyday objects, in the atomic domain it is a predominant fact of life.

The uncertainty of measurements in the atomic domain was first stated mathematically by Werner Heisenberg of Germany, and is called the *uncertainty principle*. It is a fundamental principle in quantum mechanics. Heisenberg found that when the uncertainties in the measurement of momentum and position for a particle are multiplied together, the product is Planck's constant h. Stated mathematically, this can be written:

$$\Delta M \Delta x \sim h$$

The Δ symbols are used here to mean "uncertainty of": ΔM is the uncertainty of momentum, Δx is the uncertainty of position. The product of these two uncertainties is approximately equal to the size of h. Of course if the measured uncertainties of position and momentum are large, their product may be many times greater than h. The significance of the uncertainty principle is that even in the best of conditions, the lower limit of uncertainty is h. This means, for example, if we wish to know the momentum of an electron with great accuracy, the corresponding uncertainty in position will be large. Or if we wish to know the position with great accuracy, the corresponding uncertainty in momentum will be large. The price we pay to know one of these quantities with great accuracy is great uncertainty about the other quantity. We cannot in principle know both the position and the momentum of a particle with absolute certainty.

The uncertainty principle operates similarly with *energy* and *time*. We cannot measure a particle's kinetic energy with complete precision in an infinitely short span of time. The uncertainty in our knowledge of energy, ΔE, and the duration taken to measure the energy, Δt, are related by the expression*

$$\Delta E \Delta t \sim h$$

The best accuracy we can ever hope to attain is that where the product of the energy and time uncertainties are equal to h. The more accurately we determine the energy of a photon, electron, or particle of any kind, the more uncertain we will be of the time it had that energy.

The uncertainty principle is relevant only to quantum phenomena. The inaccuracies in measuring the position and momentum of a baseball due to the interactions of observation, for example, are completely negligible. But the inaccuracies in measuring the position and momentum of an electron are far from negligible. The inaccuracies in measuring the energy and time of fall for the ram of a piledriver are negligible, but the inaccuracies in measuring simultaneously the energy release and time of de-excitation of an atom are not negligible. This is because the uncertainties in the measurements of these subatomic quantities are comparable to the magnitudes of the quantities themselves.

The Electron—Particle, Wave, or Particle-Wave?

The fundamental uncertainty of measurements impedes our efforts to pin down the exact nature of the electron. Whether it is a particle, a wave, or a particle-wave is not certain. As a particle we have a rough estimate of its "diameter," and we know its mass and charge with great accuracy. As a wave we can easily calculate its wavelength as a function of momentum from the interference fringes it produces.

We can produce an interference pattern with a beam of electrons by directing it through two narrow closely spaced slits. . . . But strangely enough, electrons directed toward the slits one at a time still produce an interference

*We can see that this follows from the uncertainty of momentum and position. Recall that momentum = force × time, and energy = force × distance. Then

$$\begin{aligned} h &= \Delta \text{ momentum } \Delta \text{ distance} \\ &= \Delta \text{ (force} \times \text{time) } \Delta \text{ distance} \\ &= \Delta \text{ (force} \times \text{distance) } \Delta \text{ time} \\ &= \Delta \text{ energy } \Delta \text{ time} \end{aligned}$$

pattern. This suggests that a single electron passes through two holes at the same time! On the basis of this evidence, it seems that the electron must be a wave, unless we allow that a particle can split in two just before it reaches the slits.

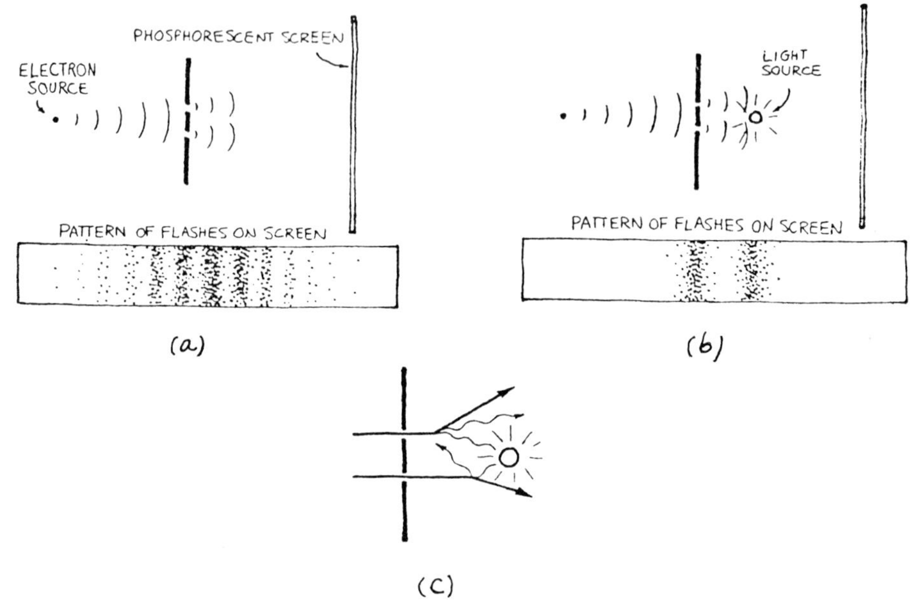

Interference experiment with electrons. (a) Electrons directed toward thin slit openings, one at a time, produce a wave interference pattern on the screen behind. (b) When any attempt is made to detect which hole(s) the electron passes through (with a light source, for example), a pattern characteristic of particles is produced on the screen. (c) The act of observing the behavior of an electron with a photon disturbs the behavior in question.

To further examine this idea, a lamp can be placed behind the slits, sending light past both openings. It is known that if a photon meets an electron it is scattered and can be observed as a tiny flash of light. Therefore, if light flashes at each slit simultaneously, it indicates that one electron has in fact passed through both holes simultaneously, thereby confirming the wave nature of the electron. When this experiment is performed, it is found that the electrons come through one slit or the other, but not both at the same time. But it is found that the pattern produced on the screen is the pattern that would be produced by *particles*. How can this be? The answer lies in the fact that we have disturbed the system we were attempting to measure. The act of observing the electron altered its momentum, thereby interrupting its synchronization with the "wave" from the other slit. The result of this experiment appears to be, if you don't "look at" the electrons they behave like waves, but if you do they behave like particles.

Many schemes designed to permit a particle to exhibit its full wave and particle characteristics at the same time have been tried. For more than forty years all such attempts have failed. The realm of the quantum is one of indeterminacy.

Determinism

The uncertainty principle has stirred much interest in a certain philosophical dilemma. With the laws of classical mechanics and a knowledge of the positions and velocities of the planets, future motions of the solar system can be accurately predicted. Before the development of quantum mechanics, scientists generally agreed that Newton's laws had a much larger application—they believed that if the exact positions and momenta of all the *particles* in the universe were known, one could evoke Newton's laws and, *in principle,* calculate the past and future course of all events. *In practice*, of course, it was realized that such calculations would not be possible because of the enormous number of particles involved—but this didn't undermine the basic idea.

Using the same broad application of Newton's laws, many philosophically inclined scientists concluded that all human actions are predetermined. Human beings, after all, are made up of tiny particles like everything else. If it was in principle possible to predict the future course of these particles, it was equally possible, in principle, to predict the course of a person composed of them. This implied that a person could not alter the future by changing his mind. Or rather, if he did change his mind, that change, along with its consequences, would be part of an inevitable predetermined chain. Our ability to choose rightly or wrongly, to decide on our actions, to control our destinies, seemed to be an illusion—more wishful thinking. There seemed to be an inescapable division between science and free will.

The advent of quantum mechanics caused many proponents of free will to feel that science itself had come to their aid. The basic indeterminacy of quantum mechanics makes the prediction of subatomic events impossible. We know of no way, for example, to determine which electron will absorb a photon in the photoelectric effect. All we can do is calculate the probability that the photon will be absorbed by a given electron. Neither can we say exactly where a given electron will strike the screen when it produces an electron wave-interference pattern. We can only state the probability of its striking a given area. With a radioactive substance, we cannot state when an individual atom will decay, but only the probability of its decaying within a certain time. In short, here is a realm into which the relentless Newtonian compulsion cannot reach.

This indeterminacy, however, does not give us proof of free will. Large aggregates of particles lend themselves to extremely accurate prediction. We can, for example, predict with certainty the electric current produced by a beam of light on a photosensitive material, or the relative intensities of a wave-interference pattern, or the rate of the decay of a radioactive substance. In most circumstances the macroscopic domain is quite adequately described by classical physics. Human behavior involves hordes of particles, not just one or two, and since statistical averages of these hordes can be determined with great accuracy, the indeterminacy at the quantum level can be discounted.

The dilemma about free will remains unresolved.

When the uncertainty principle is discussed by itself outside the area of quantum mechanics, it is often assigned a profoundness and depth that are unjustified. Some people cite the uncertainty principle as invalidating causality and determinism, and supporting free will. Others interpret the uncertainty principle as nature's shield of its innermost secrets. Critics of science use the uncertainty principle as evidence that science itself is uncertain. Determinism, nature's secrets, and the uncertainties of science have very little to do with Heisenberg's uncertainty principle. The profoundness of the uncertainty principle has to do with the probabilistic nature of the atomic world—mainly that there is no sharp boundary between particles and waves. The world of the atom is a fuzzy place.

Werner Heisenberg was a friend and contemporary of Niels Bohr, the Dane who revolutionized atomic theory. Heisenberg often visited Bohr at his country house, where they discussed, among other topics, the uncertainty principle and its applications to many scientific phenomena. Here Heisenberg recounts an evening spent with Niels Bohr.

Heisenberg and Bohr: A Conversation

Werner Heisenberg

In the evenings, we would often sit around the open fire, taking great pains to keep it alight. When the doors of the living room were closed, the chimney would smoke a good deal, and we had to leave at least one door open. The result was a mighty draft, and Niels, who loved paradoxical formulations, claimed that the fireplace had been especially put in as a cooling device. But hot or cold, the area round the fireplace was highly popular, and whenever other physicists came to visit us from Copenhagen, keen conversations on the problems that interested all of us would take place before it. I remember one evening particularly well. I think that Kramers and Oskar Klein were among those present, and, as so often, the discussion revolved around Einstein's refusal to accept the statistical character of the new quantum mechanics.

"Isn't it odd," Oskar Klein said, "that Einstein should have such great difficulties in accepting the role of chance in atomic physics? He knows more about statistical thermodynamics than most other physicists, and has himself produced a convincing statistical derivation of Planck's radiation law. Yet he still rejects quantum mechanics, simply because chance plays a fundamental part in it."

"It is precisely this fundamental aspect that upsets him," I tried to point out. "In a pot of water, we cannot possibly hope to tell how each water molecule moves. Hence no one should be surprised that physicists are applying statistics to it, making use of probability, much as life insurance companies must make actuarial computations of their clients' life expectancies. But we used to assume that, in principle at least, it was possible to describe the motion of every molecule in accordance with the laws of

From pp. 103–107 in *Physics and Beyond* by Werner Heisenberg, World Perspectives Series, Volume Forty-two, edited by Ruth Nanda Anshen.

Newtonian mechanics. In other words, nature was thought to have at any given moment an objective state from which one could deduce its state during the next moment. But this is no longer so in quantum mechanics. Here we cannot make observations without disturbing the phenomena—the quantum effects we introduce with our observation automatically introduce a degree of uncertainty into the phenomenon to be observed. This Einstein refuses to accept, although he knows the facts perfectly well. He thinks that our interpretation cannot possibly be complete, and hopes that the discovery of fresh data will help to close what he thinks are open gaps in our knowledge. But that is an idle hope."

"I don't entirely agree with you," Niels said. "There is, of course, a basic difference between classical thermodynamics and quantum mechanics, but you have exaggerated its importance. In any case, I find all such observations as 'observation introduces uncertainty into the phenomenon' inaccurate and misleading. Nature has taught us that the word 'phenomenon' cannot be applied to atomic processes unless we also specify what experimental arrangement or what observational instruments are involved. If a particular experimental setup has been defined and a particular observation follows, then we can admittedly speak of a phenomenon, but not of its disturbance by observation. And though the results of different observations can no longer be correlated as directly as they could in classical physics, this does not mean that the phenomena have been disturbed by observation; it simply means that we cannot objectify the observational results in the manner of classical physics or everyday experience. Different observational situations—by that I mean the over-all experimental setup, the readings, etc.—are often complementary, i.e., they are mutually exclusive, cannot be obtained simultaneously, and their results cannot be correlated without further ado. Hence I cannot see any very fundamental difference between quantum mechanics and thermodynamics. An observational situation involving a temperature reading also has an exclusive relationship with one in which the coordinates and velocities of all the participating particles can be determined. After all, the very concept of temperature may be defined as the degree of uncertainty about the behavior of fractions of the system characteristic of what we call 'canonical distribution.' Or, to put it less academically: if a system consisting of many particles exchanges heat steadily with the environment or with other macrosystems, then the energy of each particle will fluctuate continuously and so will that of the entire system. However, the mean values obtained from a large number of particles over long periods of time correspond very precisely to the mean values of this normal or canonical distribution. All this you can read in Gibbs. And temperature, after all, can only be defined by energy exchanges. It follows that a precise determination of temperature is incompatible with a precise determination of the positions and velocities of the molecules."

"But doesn't that mean," I asked, "that temperature is not an objective property? We have always thought that the statement 'The tea in this pot has a temperature of 70 degrees' refers to an objective fact; that everyone who measures the temperature in that particular teapot will get a reading of 70 degrees, regardless of how he performs the measurement. On the other hand, if temperature merely defines the degree of one's knowledge or ignorance of the molecular motions in the tea, it follows that different observers may obtain different temperature readings even if the real state of the system is identical; after all, different observers can have different levels of knowledge."

"No, you are quite wrong," Niels broke in. "The very word 'temperature' refers to an observational situation involving energy exchanges, say, between the tea and the thermometer, and this quite irrespective of the other properties of the thermometer. A thermometer is only a real thermometer if the molecular motions in the system to be measured, in our case the tea, and in the thermometer itself reflect the canonical distribution with the required degree of accuracy. If that is the case, all thermometers will give the same readings, and to that extent temperature is an objective quality. It all goes to show once again how problematical the concepts 'objective' and 'subjective,' which we normally use so glibly, really are."

Kramers seemed unhappy about this definition. "From the way you speak of processes inside the teapot," he said to Niels, "one might think that you were asserting some sort of uncertainty relation between the temperature and the energy inside. But surely you can't be thinking of applying such ideas in classical physics?"

"To a certain extent, I can," Niels replied. "Let us look at the properties of an individual hydrogen atom inside the teapot. Its temperature, if we can talk about it at all, is surely as high as that of the rest of the tea, in our case 70 degrees, because it exchanges heat with all the other tea molecules. Its energy, however, must fluctuate and this precisely because it exchanges heat; hence we can only define a probability curve for its energy. If, conversely, we had measured the energy of the hydrogen atom and not the temperature of the tea, then we could not deduce the latter unequivocally from the former; once again we could only draw a probability curve—for the temperature. The relative breadth of this curve, in other words the lack of certainty as to the precise temperature or energy values, is relatively large in so small an object as a hydrogen atom, and hence significant. In a much larger object, for instance a small quantity of tea within the pot, it becomes considerably smaller and can therefore be neglected."

"In the old thermodynamics as we teach it to our students," Kramers objected, "energy and temperature are attributed to an object simultaneously. Not a word is said about indeterminacy or uncertainty. How can you possibly reconcile that with your views?"

"The old thermodynamics," Niels replied, "is to statistical thermodynamics what classical mechanics is to quantum mechanics. With large objects we do not commit significant errors if we attribute certain values to their temperature and energy simultaneously, just as we can attribute simultaneous values to their position and velocity. With very small objects things are quite different. In thermodynamics, we used to say that the latter are endowed with energy but not with temperature. But this strikes me as a mistaken idea, if only because we don't know where to draw the line between small and large objects."

All of us now realized why Niels laid far less stress on the fundamental difference between the statistical laws of thermodynamics and those of quantum mechanics than Einstein. Niels felt that complementarity was a central feature of all attempts to describe nature, a feature inherent in, though insufficiently brought out by, statistical thermodynamics, especially in the form Gibbs had given it; Einstein, on the other hand, remained steeped in the conceptual world of Newtonian mechanics or of Maxwellian field theory and completely failed to notice these complementary features.

Out of the uncertainties of the years preceding World War II came an exodus of scientists from Germany and elsewhere in Europe, where the quest for knowledge had become adulterated with national socialism and death. Among the emigrés was Leo Szilard, an atomic physicist who eventually turned to geology. He spent the last months of his life at the Salk Institute in La Jolla, California, which he helped to establish. As his great friend and colleague Jacob Bronowski explains, Szilard was so convinced that the great power locked in the nucleus of the atom could be released by man in a chain reaction that he filed for a patent on the idea. Eventually Szilard found himself involved in drafting the historic letter—signed by Albert Einstein—which induced President Franklin D. Roosevelt to order the development of the atomic bomb. Following are Szilard's patent request and a section of *The Voice of the Dolphin* in which Szilard fantasizes "My Trial as a War Criminal."

Patent for a Chain Reaction
Leo Szilard

Application Date: June 28, 1934. No. 19157/34.

Application Date: July 4, 1934. No. 19721/34.

I, Leo Szilard, a citizen of Germany and subject of Hungary, c/o Claremont Haynes & Co. of Vernon House, Bloomsbury Square: London, W.C.1, do hereby declare the nature of this invention to be as follows:—

This invention has for its object the production of radio active bodies the storage of energy through the production of such bodies and the liberation of nuclear energy for power production and other purposes through nuclear transmutation.

In accordance with the present invention nuclear transmutation leading to the liberation of neutrons and of energy may be brought about by maintaining a chain reaction in which particles which carry no positive charge and the mass of which is approximately equal to the proton mass or a multiple thereof form the links of the chain.

I shall call such particles in this specification "efficient particles."

A way of bringing about efficiently transmutation processes is to build up transmutation areas choosing the composition and the bulk of the material so as to make chain reactions efficient and possible, the links of the chain being "efficient particles."

One example is the following. The chain transmutation contains an element C_2 and this element is so chosen that an efficient particle x when reacting with C may produce an efficient particle y, and the efficient particle y when reacting with C may produce either an efficient particle x or another efficient particle which in its turn is directly or indirectly when reacting with C capable of producing x. The bulk of the transmutation area, on the other hand, must be such that the linear dimensions of the area should sufficiently exceed the mean free path between two successive transmutations within the chain. For long chains composed of, say, 100 links the linear dimensions must be about ten times the mean free path.

I shall call a chain reaction in which two efficient particles of different mass number alternate a "doublet chain." An example for a doublet chain which is a neutron chain would be the following reaction, which might be set up in a mixture of a "neutron reducer element" (like lithium (6) or boron (10) or preferably some heavy "reducer" element), and a "neutron converter element" which yields $n(2)$ when bombarded by $n(1)$. An example for such a chain in which carbon acts as a reducer and beryllium acts as converter would be the following:

$$
\begin{aligned}
\mathrm{C}(12) + n(2) &= \mathrm{C}(13) + n(1) \\
\mathrm{Be}(9) + n(1) &= \text{“Be(8)”} + n(2)
\end{aligned}
$$

("Be(8)" need not mean an existing element, it may break up spontaneously).

One can very much increase the efficiency of the hitherto mentioned neutron chain reactions by having a "neutron multiplicator" O mixed with the elements which take part in the chain reaction. A neutron multiplicator is an element which either splits up $n(2)$ into $n(1) + n(1)$ or an element which yields additional neutrons for instance $n(1)$ when bombarded by $n(1)$. A multiplicator need not be a meta-stable element. Beryllium may be a suitable multiplicator

$$\mathrm{Be}(9) + n(1) = \text{“Be(8)”} + n(1) + n(1)$$

An efficient particle disappears (and a chain is therefore interrupted if this happens in a chain reaction) if a neutron reacts with a nucleus in such a way that the neutron disappears and a positive particle for instance a proton or an alpha particle is emitted. I can suppress the production of a positive particle when bombarding the element by neutrons by choosing the element and the neutron energy so that the positive particle, the creation of which has a potential possibility, should not have sufficient energy at its disposal to penetrate in the inverse process the nucleus of that element. In order to avoid such an occurrence in my chain reactions I shall use as reducers, converters and multiplicators the heaviest elements which are otherwise satisfactory.

My Trial as a War Criminal

Leo Szilard

I was just about to lock the door of my hotel room and go to bed when there was a knock on the door and there stood a Russian officer and a young Russian civilian. I had expected something of this sort ever since the President signed the terms of unconditional surrender and the Russians landed a token occupation force in New York. The officer handed me something that looked like a warrant and said that I was under arrest as a war criminal on the basis of my activities during the Second World War in connection with the atomic bomb. There was a car waiting outside and they told me that they were going to take me to the Brookhaven National Laboratory on Long Island. Apparently, they were rounding up all the scientists who had ever worked in the field of atomic energy.

Once we were in the car the young man introduced himself and told me that he was a physicist as well as a member of the Moscow Chapter of the Communist Party. I had never heard his name before and I was never able to remember it thereafter. He was obviously very eager to talk. He told me that he and the other Russian scientists were all exceedingly sorry that the strain of the virus which had been used had killed such a disproportionately large number of children. It was very fortunate, he said, that the first attack was limited to New Jersey and that the early cessation of hostilities made attacks of larger scope unnecessary. According to plan—so he said—stocks of this virus were merely held in reserve for an emergency. Another virus differing by five further mutational steps had been in the stage of pilot plant production, and it was this improved virus which was meant to be used in case of war. It would not affect children at all and would kill predominantly men between twenty and forty. Owing to the premature outbreak of the war, however, the Russian government found itself forced to use the stocks which it had on hand.

He said that all the scientists arrested would be given a chance to go to Russia, in which case they need not stand trial as war criminals; but that if I should elect to stand for trial he personally hoped that I would be exonerated and that afterward I would be willing to collaborate with the Russians here in the United States.

He said that the Russians were very anxious to get the support of people other than the American Communists for a stable political regime in the United States which would collaborate with them. Since they now had the support of the Communists anyway, he explained, they would rather bestow their favors on those whose co-operation was not yet assured. "We shall, of course, lean on the Communists for the next few months," he said, "but, in the long run, dissatisfied elements who are used to conspiracy would not be relied on by us. It is difficult to work with fellows who have no sense of humor," he added as an afterthought.

He told me that no scientist would be forced to go to Russia and that no one who was innocent need go there for fear of having to stand trial as a war criminal, because, he said, Russia would do everything in her power to make the trials fair and impartial. "The outcome of a bona fide trial," he added somewhat illogically, "is, of course, always something of a tossup."

He told me that he expected that Russia would, within a fortnight, change her position on the question of world government; that she would come out in favor of it, in principle, and that she would press for immediate strengthening of the United Nations. The tribunal which was being assembled to try war criminals would not be Russian-dominated, he said, but would, rather, be composed of representatives of all nations which were not at war with Russia.

I was surprised to hear him say that he expected Great Britain to delegate the Lord Chief Justice to sit on the tribunal, and, frankly, I did not believe him then, though of course this was technically not impossible, since the coalition Cabinet had declared Britain's neutrality twenty-four hours before the outbreak of the war. His prediction was confirmed, however, the following morning when the newspapers reported the speech of the British Prime

Minister, who had said that Great Britain, having participated in the Nuremberg trials, could not now refuse her participation without being guilty of displaying a double standard of morality. The information which I received from this young man proved to be most valuable to me, because it gave me time to make up my mind as to what line I would want to follow.

As far as going to Russia was concerned, my mind was made up. After having been raised in Hungary, I had lived in Germany and in England before I settled in the United States, and that is as much migration as is good for any man. Moreover, when you are above fifty you are no longer as quick at learning languages. How many years would it take me to get a sufficient command of Russian to be able to turn a phrase and to be slightly malicious without being outright offensive? No, I did not want to go to Russia.

Even less did I want to be put in the position of having favors bestowed upon me by the Russians or of having to refuse point-blank some position of importance which they might wish to offer to me. I did not want to incur the favor of the Russians, but I did not want to antagonize them, either. After devoting some thought to this dilemma, I decided that the best way for me to keep out of trouble was to stick to the truth and thereby to arouse the suspicion of the Russians.

I did not have to wait long for an opportunity to implement this plan of action. The next morning at Brookhaven I was interrogated by a Russian official. In the beginning his attitude was rather benevolent. Almost the first question he asked me was why I had not worked in the field of atomic energy prior to the Third World War. When I truthfully said that I had five good and valid reasons and named them one by one, he took them down in shorthand, but the longer I talked the more incredulous he looked. It was obvious that he felt himself unable to believe what I was saying to him. Realizing that my method worked, I answered all his questions as truthfully as I possibly could and then signed the transcript at the end of the interview.

I was called back for further interrogation in the afternoon; this time it was an older Russian scientist, who was known to me by name, but whom I had not previously met. He told me that he had asked to see me because he had read the transcript which I had signed in the morning. He said that the Russian scientists had followed with great interest the articles I had written before the war, and he quoted to me passages from articles entitled "Calling for a Crusade" and "Letter to Stalin" which I had published in the *Bulletin of the Atomic Scientists* in 1947. This pleased me very much. He went on to say, however, that these articles showed an almost incredible degree of naïveté and were models of un-Marxian writing. He acknowledged that they were free from any anti-Russian bias and told me that the Russian scientists had formed the opinion that I had not been working in the field of atomic energy before the Third World War because I had not wanted to make bombs that would be dropped on Russia. He said that he regretted that I had not given this as my reason, that he wanted to give me an opportunity to revise the answers which I had given, and that he was prepared to tear up my signed statement then and there, though by doing so he would be sticking his neck out, since he would be acting against regulations.

I thanked him for his kindness and told him that I had merely told the truth, which, unfortunately, it was not within my power to change; and there then ensued a most interesting and protracted conversation about the intrinsic value of truth. Since what he told me was told in confidence and might get him into trouble if revealed, I do not feel free to record it here.

The war crimes trials opened about one month later at Lake Success, and I was—apparently as a special favor—among the first to be tried. I was charged by the prosecutor, a Russian, first of all with having tried to induce the United States government to take up the development of atomic energy in a meeting held on October 21, 1939, i.e., at a time when the war in Europe was still an imperialist war, since Germany had not attacked Russia until 1941.

I was also charged with having contributed to the war crime of dropping an atomic bomb on Hiroshima. I thought at first that I had a good and valid defense against this latter charge, since I had warned against the military use of the bomb in the war with Japan in a memorandum which I had presented to Mr. Byrnes at Spartanburg, South Carolina, six weeks before the first bomb had been tested in New Mexico.

But unfortunately this memorandum, which Mr. Byrnes had put into a pocket of his trousers when I left him, could not be located by counsel for the defense either in the files of the State Department or in the possession of any of the Spartanburg cleaners who might have kept it as a souvenir. Mr. Byrnes was himself under indictment and was not called as a witness. Excerpts from the memorandum which were published in the fall of 1947 in the *Bulletin of the Atomic Scientists* were stricken from the record on the ground that the parts of this memorandum which were omitted from the publication for reasons of secrecy might have contained the opposite of what the published part of the document appeared to indicate.

Under these circumstances I had to fall back for my defense on a petition which I had circulated in the Uranium Project at the University of Chicago immediately after the testing of the bomb in New Mexico and which asked the President to withhold his approval of a military use of the bomb against the cities of Japan. The prosecutor moved, however, that this document be stricken from the record on the ground that it was not transmitted by me to the President directly, but was, rather, handed by me to the head of the project, who forwarded it through the Manhattan District of the War Department, headed by General Groves. The prosecutor said that I, Szilard, should have known better than to agree to such a method of transmittal.

Having rested my defense, I was now free on bail. Since I was not permitted to leave Lake Success, I was spending my time there listening to the trials of the statesmen and scientists. In spite of the seriousness of my own situation, I found it difficult sometimes to refrain from joining in the laughter which frequently interrupted the proceedings.

As a prelude to the Nuremberg trials, war crimes had been defined with the collaboration of the United States, represented by Justice Jackson of the United States Supreme Court. The "violation of the customs of war" had been defined as a war crime at that time. "Planning a war in violation of international agreements" had also been defined as a crime.

The first statesman to be tried on charges arising from the bombing of Hiroshima was Mr. Stimson, and he was tried on his own admission contained in an article which he had published in 1947 in *Harper's*. The prosecution pointed out that the "defense" put forward by Mr. Stimson in that article was untenable. Mr. Stimson's point was that, had the bomb not been used, millions would have perished in an invasion of Japan. The prosecutor, a Dutchman, quoted from a memorandum prepared after the surrender of Japan by the United States Strategic Bombing Survey which showed that the United States could have won the war against Japan without invasion, just by sitting tight, since Japan was essentially defeated before the bomb was dropped on Hiroshima. He further quoted passages from the book *Secret Mission*, by Ellis M. Zacharias, published in 1946, which showed that Japan's desperate position must have been known to Mr. Stimson, since it was fully disclosed in the reports prepared by the United States naval intelligence.

Counsel for the Defense, however, submitted a deposition obtained from the British Secretary of War in order to prove that secretaries of war never based decisions on reports prepared by naval intelligence. "Mr. Stimson," so counsel for the defense said, "should not be reproached for acting as all secretaries of war in all English-speaking countries have acted at all times."

The presiding judge, in summing up, disregarded the arguments presented by both the prosecution and the defense and took the line that prior to the Third World War it was not *customary* to drop atomic bombs on towns and cities, and that such a "violation of the customs of war" was a war crime which could not be justified on the ground that the government which committed it hoped that by doing so it would bring the war to a speedier conclusion.

It was expected that Mr. Stimson would be found guilty on his own admission, but that he would be reprieved primarily because of his article published in *Foreign Affairs* in 1947 in which he commented on the foreign policy of the Truman Administration. It was generally considered that in 1947 his was a voice of reason and moderation in the midst of general confusion.

Mr. Truman was charged with the "crime" of actually ordering the bombing of Hiroshima. At first, counsel for the defense took the line that at the time when the definition of war crimes was made public at Nuremberg, Mr. Truman was at sea—in the literal sense of the term. He was on board a battleship on his way back from Potsdam and did not have opportunity adequately to study the text of the Nuremberg Declaration prior to the bombing of Hiroshima. This plea was rejected by the court on the ground that those who were sentenced to death and executed at Nuremberg could—if they were alive—use much the same type of argument in their defense.

Subsequently, counsel for the defense took the line that Mr. Truman was not guilty because he had not acted on his own but had merely followed advice given to him and, so to speak, had been merely following orders. In proof of this the defense read into the record a magazine article published by Garbatov in Russia in 1947 which asserted that Mr. Truman had always been taking orders from one boss or another. This article had drawn a protest from the American ambassador at the time of its publication.

Having had little luck with any of his "lines," counsel for the defense raised the question why the use of an atomic bomb should be considered a "violation of the customs of war" any more than the use of a virus that killed children. But the presiding judge ordered his remark stricken from the record, saying that this was the trial of Harry S. Truman and not of Somebody Else, and that since Mr. Truman was not accused of having ordered the use of a virus in warfare, nothing relating to any virus could possibly be relevant to his defense.

It was generally expected that Mr. Truman would be found guilty, but it was rumored that there were powerful Russian influences at work to have him reprieved. There were all sorts of guesses as to what the reasons of the Russians may have been, and some thought that they favored Mr. Truman on account of his supposed Wall Street connections, since the Russians were known to nurture a secret admiration for Wall Street. I, myself, believe that the reason of the Russians may have been political and rather difficult to guess in detail without knowing on which of their misconceptions it was based.

The next to be tried was Mr. Byrnes, who was not only accused of being responsible for the decision of using the atomic bomb against Japan, but, above all, was accused of having advocated a war against Russia "in violation of international agreements" in his book *Speaking Frankly*, which appeared in 1947. The British prosecutor quoted from page 203 of the first edition:

... I do not believe the Red Army would try to hold permanently all of Eastern Germany. However, if I misjudge them, and they do go to the point of holding Eastern Germany and vetoing a Security Council Directive to withdraw occupation forces, we must be prepared to assume the obligation that then clearly will be ours. If our action is to be effective, we must be

clear in our minds and must make it clear to all that we are willing to adopt these measures of last resort if, for the peace of the world, we are forced to do so.

On this passage Mr. Byrnes was most severely cross-examined by the prosecutor. He was asked whether he was aware of the fact that the United States ratified the Charter of the United Nations at the time when Mr. Byrnes himself was Secretary of State. He was asked whether he was aware of the fact that by doing so the United States undertook the solemn obligation of refraining from war and that, under Article 51 of the Charter, the United States merely retained the right of waging war in case of an armed attack. He was asked whether the mere refusal of Russia to leave the territories which she had occupied after the Second World War could be construed as an armed attack. He was asked whether he could suggest any way of interpreting what he had been saying on page 203 of his book other than as advocating that the United States ought to violate her solemn obligation under the Charter and wage an illegal war against Russia in case Russia should refuse to settle on the terms set by the United States government.

Counsel for the defense replied that he wished to elucidate the meaning of the passage "measures of last resort" quoted by the prosecutor from Mr. Byrnes's book. At a press conference following shortly the publication of his book, Mr. Byrnes himself had explained this passage—so counsel for the defense said. " 'There is no suggestion as to whether such collective action should be persuasion, economic, or military action,' " counsel quoted. "Clearly," counsel said, raising his voice a little, "if Mr. Byrnes had had *military action* in mind, he would have spoken of 'measures of *very* last resort' and not merely of 'measures of last resort.' British statesmen," he said, looking sharply at the prosecutor, "may indulge in understatements, but that is no reason for accusing my client of one."

The prosecutor replied that Mr. Byrnes had condemned himself by the very words quoted by the defense, for by virtue of those words Mr. Byrnes had admitted that the term "measures of last resort" meant either persuasion or military action. "I am not conversant with American law," he said, "but surely in England a man who publicly proclaims that he is going to get hold of something that is in the possession of his neighbor either by persuasion or by pulling a gun on him is persuaded to go to jail."

At this point, counsel for the defense submitted evidence to show that, two weeks before the outbreak of the Third World War, Mr. Byrnes had sent a memorandum to the President of the United States warning against any aggressive act on the part of the United States armed forces that would result in war. The prosecutor's motion that this memorandum be ruled out as evidence was upheld by the presiding judge on the ground that if inconsistency were admissible as a defense at the trial of a statesman, then no statesman could ever be convicted as a war criminal and the statesmen would enjoy an immunity not shared by the other defendants.

All of us who attended his trial were unanimous in our praise of Mr. Byrnes for the patience and firmness he displayed. Of course, if sentence had been passed and executed, he would have lost his life; but as is generally known, no sentence was ever passed on Mr. Byrnes or any of the rest of us. The first Russian appeal for help reached the United States Public Health Service one week after Mr. Byrnes rested his defense.

Just what happened will never be known with certainty. This much is clear, that the vast quantities of vaccine which the Russians held in readiness to safeguard their own population against the virus were absolutely without any effect. In the laboratory tests such vaccine had proved to be 100 per cent effective; something must have gone wrong in the change-over from pilot plant operation to mass production, and *someone* must have forgotten to check the product for its effectiveness. Since the engineer in charge of the production plant at Omsk perished in the disorders which broke out after over half of the children of the town had died, and since all records of the production plants were destroyed in the fire, we shall never know just what had gone wrong.

The terms of the postwar settlement which had been reached within two weeks of the Omsk riots were in every respect very favorable to the United States and also put an end to all war crime trials. Naturally, all of us who had been on trial for our lives were greatly relieved.

"I am now convinced that theoretical physics is actual philosophy," said Max Born some years after Heisenberg proposed the Uncertainty Principle. It was his acknowledgment that man's vision of the world around him is limited by his own perceptions—and the limitations of his tools. Thus, as Heisenberg suggested, man's comprehension of the electron and its wave- and particlelike properties opens up a gap of uncertainty that the intellect cannot fully bridge. A look at the dual nature of light is provided in the following selections by James Perlman and Charlotte Ward.

Particles of Light

James S. Perlman

At the end of the 19th century, light seemed entrenched as electromagnetic radiation emitted in continuous waves by vibrating electric charges in luminous bodies. Newton's color spectrum was part of a far more extensive electromagnetic spectrum. But sooner or later, situations arise that the best of conceptual models cannot explain. Just when the electromagnetic wave theory seemed most firm, a *photoelectric effect* was observed that contradicted it. Light falling on a metal plate ejected electrons (negative charges) from the plate . . . to positively charge the plate. How is it possible for light waves to have impact on electrons? Water waves have impact on objects, but water is a material thing. Perhaps light is not "pure" energy after all!

The difficulty really began, however, when the intensity of the light on the photoelectric plate was increased. Instead of electrons leaving the plate with more speed, there were just more electrons leaving with the original speed. If light consists of waves, more intense light should mean higher waves

Excerpted from *The Atom and the Universe* by James S. Perlman, Wadsworth Publishing Company, Inc.

with nastier "wallops" on plate electrons. The electrons should move more quickly. If light consists of bulletlike particles, more intense light would mean more "bullets" at the same speed. A given electron hit by a light particle would not be affected by the additional "bullets." More "bullets" would just mean that more electrons would be hit and emitted. From the results, light seemed more like bullets. Maybe light is not a wave phenomenon after all!

In any case, Albert Einstein (1879–1955) assumed in 1905 that light is a stream of energy particles that he called *photons*. Einstein's photons are not minute bits of matter like Newton's corpuscles; they are particles of *energy* that act like matter. The energies of Einstein's photons are directly proportional to the color frequencies given in the electromagnetic spectrum. Violet light has a higher frequency than red light. Therefore, photons of violet light are larger than red light photons. The larger photons of violet striking metal should cause electrons to move faster than the smaller red photons do. Experiments showed this to be so. In his photon idea, Einstein was applying to light an energy packet, or *quantum*, idea that Max Planck (1858–1947) had successfully applied to heat phenomena. Einstein assumed that the energy E of each particle of energy was represented by:

$$E = hf,$$

where h is a now-famous universal constant of nature called Planck's constant (with a value determined at 6.62×10^{-34}), and where f is the frequency of the particular radiation in the electromagnetic spectrum.

In the transfer of a photon's energy to an electron in the photoelectric effect, Einstein emphasized that

$$hf = \tfrac{1}{2}mv^2 + w,$$

where hf equals the energy of the light photon, $\tfrac{1}{2}mv^2$ is the kinetic energy of the expelled electron, and w is the energy necessary to remove the electron from the metal. This famous photoelectric equation earned for Einstein a Nobel prize.

Today photoelectric cells (popularly called "electric eyes") open supermarket doors for customers, but the photon theory that explains the operation cannot explain diffraction phenomena. Bulletlike photons, ν, passing through a single pinhole, should create a spot on a screen directly in line with the hole. But alternate light and dark rings are also found. And photons through two pinholes should show two spots on a screen. Instead, a series of alternate light and dark stripes are found around a single spot midway between the two pinholes. The photon theory is also unable to explain how light bends around obstacles. But the opposing wave theory cannot explain the photoelectric effect. The nature of light thus seems relative to the experimental setup. Light is wavelike in a pinhole situation and particlelike in a photoelectric cell. But how can light be both a continuous wave and a discontinuous particle? Perhaps waves and particles are aspects of a bigger picture. Different situations may display light in its different aspects, just as at different angles "the neck of the peacock shows different colors."

Physical Optics
Charlotte R. Ward

The real question about the nature of light, whether it is a wave or a beam of particles, was apparently settled once and for all by an experiment done by Thomas Young about 1800. A single beam of light passing through a slit from a light source encounters a barrier that is opaque except for two narrow slits very close together, neither of which is aligned with the single slit. A pattern of alternating light and dark lines appears on the screen. The brightest line is the center one, which is aligned with neither slit, but is equidistant between them.

It is well nigh impossible to see how a stream of particles could produce this pattern of light and dark lines, called a *diffraction pattern*, but with waves, this is exactly the behavior expected. Two well-established principles of wave motion, one going back to Newton's contemporary, Huygens, serve to account thoroughly for diffraction phenomena.

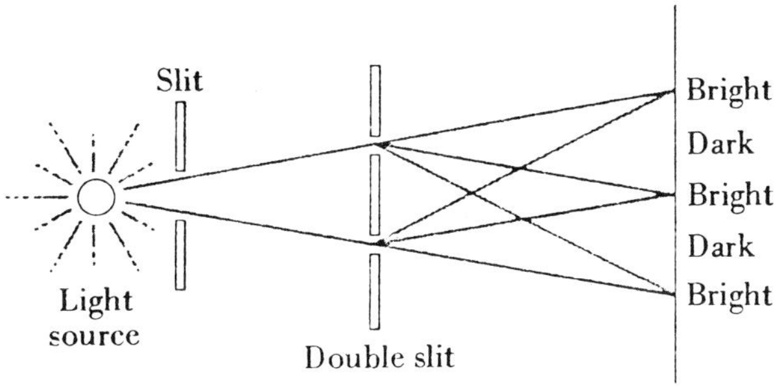

Young's double slit experiment.

Huygens' Principle: Every point on a wave front acts as a new source of waves.
Superposition Principle: When two waves travel together in a medium, the resultant wave is their algebraic sum.

The behavior of the ripples formed by dropping a pebble in water illustrates these principles clearly.

If we assume that light consists of waves, we can explain Young's experiment in this way: Wave fronts from the first slit spread out according to Huygens' principle. Eventually they fill the second set of slits, producing point sources there, from which waves spread out and reach the screen. If we use light of a single color (single wavelength) then waves that travel identical paths, or paths differing by a whole number of wavelengths, will arrive at the screen in phase, and when superimposed will reinforce each other to produce a bright spot. In between, waves will be out of phase when superimposed and will cancel each other. Of course, all degrees of phase mismatching are possible, so the pattern will shade from bright to dark over and over again.

Because each wavelength of light will produce its own diffraction pattern, white light can be dispersed by passing it through slits as well as through a prism. Actually, a diffraction pattern can be produced with a single slit, because the waves from different parts of the slit can interfere with each other, as anyone can see by looking at a light source through the narrow crack between two fingers. To disperse light for studying its component waves, however, the most effective slit system is the *diffraction grating*: a piece of glass ruled with a diamond point to contain as many as 13,000 or 14,000 lines (scratches) per inch. The clear glass areas between the scratches act as slits. The closer the spacing of the slits the greater the angle of diffraction, and so the greater the separation of the various wavelengths.

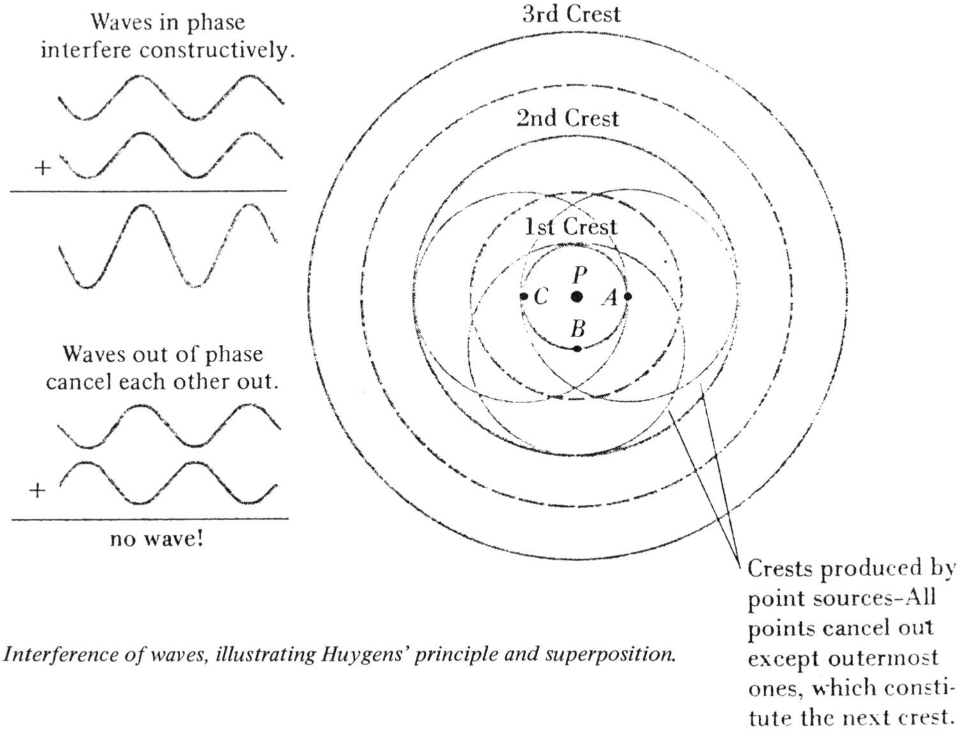

Interference of waves, illustrating Huygens' principle and superposition.

A final word on the uncertainty of science and its implications for the future is presented here in this reading from *Science Is Not Enough* by Vannevar Bush.

Science Is Not Enough

Vannevar Bush

Science proves nothing absolutely. On the most vital questions, it does not even produce evidence.

But is all the labor of science vain to the thinker, the seeker after a sure harbor, amid the mystery, evil, cruelty, majesty, that surrounds us? By no means. Science here does two things. It renders us humble. And it paints a universe in which the mysteries become highlighted, in which constraints on imagination and speculation have been removed, and which becomes ever more awe-inspiring as we gaze.

The first men who pondered did so on a small earth, which did not extend far beyond the horizon, for which the stars were mere lamps in the skies. Now, we are no longer at the center; there is no center. We look at congeries of stars by light that left them before the earth had cooled. Among the myriads of stars we postulate myriads of planets with conditions as favorable to life as is our earth. We puzzle as to whether the universe is bounded or extends forever; whether, indeed, it may be only one universe among many. We speculate as to whether our universe began in a vast explosion, whether it pulsates between utter compression and wide diffusion, whether it is self-renewing and thus goes on unchanged forever. And we are humble.

But science teaches more than this. It continually reminds us that we are still ignorant and there is much to learn. Time and space are interconnected in strange ways; there is no absolute simultaneity. Within the atom occur phenomena concerning which visualization is futile, to which common sense, the guidance from our everyday experience, has no application, which yield to studies by equations that have no meaning except that they work. Mass and energy transform one into another. Gravitation, the solid rock on which Newton built, may be merely a property of the geometry of the cosmos. Life, as its details unfold before us, becomes ever more intricate, emphasizing more and more our wonder that its marvelous functioning could have been produced by chance and time. The human mind, merely in its chemical and physical aspects, takes on new inspiring attributes.

And what is the conclusion? He who follows science blindly, and who follows it alone, comes to a barrier beyond which he cannot see. He who would tell us with the authority of scholarship a complete story of why we exist, of our mission here, has a duty to speak convincingly in a world where men increasingly think for themselves. Exhortation needs to be revised, not to weaken its power, but to increase it, for men who are no longer in the third century. As this occurs, and on the essential and central core of faith, science will of necessity be silent.

But its silence will be the silence of humility, not the silence of disdain. A belief may be larger than a fact. A faith that is overdefined is the very faith most likely to prove inadequate to the great moments of life. The late Mr. Justice Holmes said, ". . . the faith is true and adorable which leads a soldier to throw away his life in obedience to a blindly accepted duty, in a cause he little understands, in a plan of campaign of which he has no notion, under tactics of which he does not see the use." Young men, who will formulate the deep thought of the next generation, should lean on science, for it can teach much and it can inspire. But they should not lean where it does not apply.

12

Generation upon Generation

Make thee another self, for love of me,
That beauty still may live in thine or thee.

William Shakespeare

Beginning with Gregor Mendel, farm boy, unsuccessful teacher, and moderately successful monk, "the ascent of man quietly set off in a new direction," says Dr. Bronowski. Three years before Darwin published *The Origin of the Species*, in 1856, Mendel was already investigating heredity, using common garden peas as a model. The work of this gentle innovator laid the foundations of modern genetics. And yet his work was virtually ignored until the turn of the century. An account of his research is given in this selection from *The Coil of Life* by Ruth Moore.

The Discovery of Units of Heredity

Ruth Moore

Gregor Johann Mendel was born on July 22, 1822, into a long-settled peasant family living in the tiny village of Heinzendorf in what was then Austrian Silesia and is now Czechoslovakia. As a boy he aided his father in the grafting of trees and in cultivating the garden that was a part of "peasant holding no. fifty-eight." He received additional training in horticulture and beekeeping at the village school. Johann did so well in this work and in his academic studies that his teachers urged that he be sent on to the high school at Troppau. Mendel never lost his love of growing things as he struggled to make his way at Troppau and later at the Philosophical Institute at Olmütz. Gradually this early interest turned into a wider interest in science, although little science was included in the institute's course. . . .

Mendel was so excellent a teacher that the head of the school confidently urged him to try for a teacher's certificate. Mendel wrote the necessary essay and presented himself in Vienna for the oral examinations in the natural history and physics that he wanted to teach. He failed to pass the examination, partly, it seemed, because he lacked certain formal knowledge of his subjects. In part his failure may have been due to the conviction of the friendly chief examiner that the unusual young teacher should be given an opportunity for "higher scientific training" at the University of Vienna.

The abbot of the monastery agreed, and Mendel went to the university for four terms. At the end, he returned to teach in the Brünn "Modern School."

Soon after entering the monastery Mendel began trying to develop new colors in flowers. He also took up the breeding of gray and white mice. Although the latter experiments were dropped after a while, probably at a hint of ecclesiastical disapproval, they were a certain indication that the work with flowers was not a fancy but the first approach of a serious scientist to a fundamental problem.

Mendel's own account of how he happened to begin his studies is a simple one. As he tried to produce new colors in his flowers, he acquired considerable experience in artificial fertilization. His interest was aroused by the surprising and unaccountable results that he sometimes obtained. Whenever he crossed certain species, the same hybrid forms cropped up with striking regularity. But when he crossed one of his hybrids with another, some very different characters sometimes appeared among their progeny.

These phenomena sent Mendel to the literature. He searched all the books available in the monastery and elsewhere in Brünn. A number of scientists were working with the problem of hybrids, but no one had discovered any law governing their formation—if, indeed, anyone could conceive of a law capable of explaining the infinitude of forms in the multitude of plants. Inheritance in all its profusion seemed beyond any nailing down.

Mendel nevertheless had seen certain forms appearing with regularity. Others too had noted this phenomenon. They had not, as far as Mendel could find, counted the forms and classified them. Nor had anyone arranged hybrids according to their generations or worked out their statistical relationships. Mendel, in his remote Austrian town, did not realize that even his concept of counting and figuring the mathematical relationships of hybrid plants was original and entirely his own.

To study heredity properly, Mendel saw, a large number of generations would have to be bred. He also recognized that many plants would have to be included in each generation. A few might produce a misleading picture. "It indeed required some courage to undertake such far-reaching labors," Mendel wrote later in the introduction to one of his monographs. "It appears, however, to be the only way in which we can finally reach the solution of a problem which is of great importance in the evolution of organic forms."

The year was 1854. Many people were asking how the great number of species could have arisen, and yet Charles Darwin would not publish *The Origin of Species* for another five years. It seems unlikely that the Austrian monk was planning a direct study of the great problem of evolution. His interest was concentrated on heredity and how traits could be handed on from parent to offspring. Nevertheless, as his words show, Mendel was well aware of the evolutionary significance of the work he was about to launch.

Mendel planned the big experiment with care and foresight. His first requirement was a plant with varied characters, each of which bred true to form. If a plant produced green peas, he had to be sure that it would go on producing green peas if crossed with another green-pea plant. Mendel also had to have a plant that could be protected from foreign pollen during its flowering period. If even one bee carried in some pollen other than that he would dust on, the whole experiment would be upset. Equally important, the characters had to be easily observable and countable. It was a huge task Mendel was laying out for himself. . . .

. . . From the thirty-four varieties [of seeds] Mendel selected twenty-two, and from these, seeds with seven sharply contrasting pairs of characters. As he wanted to make no fine-line decisions about whether or not a part of a plant had changed, he selected the following clear-cut differences with which to work:

1. The form of the ripe seeds—round or wrinkled;
2. The color of the peas (the seed)—yellow or intense green;
3. The color of the transparent seed cover or skin—white or grayish;
4. The form of the ripe pods—inflated or constructed between the peas;

5. The color of the unripe pods—green or yellow;
6. The position of the flowers—distributed along the stem or bunched at the top: in other words, axial or terminal;
7. The length of the stem—tall (six or seven feet) or dwarf (three-fourths to one and a half feet).

Each kind of seed went into its own jar—to await the spring of 1856.

Mendel was at last ready to launch his experiment proper. In one section of his long narrow garden he planted round seeds, and in the next section, wrinkled ones. In other sections went the yellow and the green, and, in their own places, all of the others. Keeping the characters that he wanted to compare next to one another would simplify the work that lay ahead. . . .

As soon as the pollen had ripened in the . . . round peas, Mendel collected a bit on a fine camel's-hair brush and, removing the bags on the wrinkled peas, dusted it on their stigmas. The little bags then were tied on again to keep away the bees and other pollen-carrying insects. To make certain that his experiments were not affected by which plant served as the seed parent, Mendel also reversed his fertilization procedure. Some of the pollen from the wrinkled peas was deposited on the prepared stigmas of the round. . . .

His first indication of what was happening would come when the pods formed in the "unripe pod color" section of his garden. As the pods appeared and filled out and grew longer, Mendel saw with elation that all of them were green. Whether they grew in the "yellow" half of the plot and came from parents that had produced yellow pods or in the "green" half and sprang from parents with green pods, all of them hung green on the vines. Yellow and green parents alike had produced green offspring. Mendel searched his plants. He could not find a single yellow among the unripe pods. . . .

Confirmation of this striking result came as the summer drew to a close and the pods dried on the vines. . . .

During the winter, as he studied his jars of peas and the way all had completely taken on the character of one parent, it was evident that one characteristic in each pair had been entirely "dominant" over the other. Mendel decided to designate the hereditary trait that prevailed as a "dominant." Later he named the other factor, the one lost to sight, a "recessive."

Only the first step had been completed in the big experiment. The next, the crossing of hybrid with hybrid, had to await another spring. As the sun grew warm and the spring rains fell, Mendel planted his hybrid seeds. Each carefully marked group went into its own plot, but this time his procedure would be different. He would not operate on the buds. He would permit the peas to fertilize themselves in the natural way, and thus would obtain a cross of two identical hybrids.

The plantation was again filled with pea vines, and Mendel waited as patiently as he could for the pods and peas to form and dry.

As the young pods grew, Mendel saw a few yellows appearing among the greens. He counted both colors and entered the figures in his notebooks. When the pods dried in the "pod form" plot, some were crinkled around the peas they contained; they were, in the terms he was using, constricted. The others were inflated. Again Mendel counted both types, and again he reserved judgment. . . .

Mendel continued his plantings through six or seven generations in all cases. As he did so, he learned that the 3-to-1 ratio applied to the appearance of the peas only. When Mendel planted the hybrid parents, they produced both round and wrinkled seeds. One fourth were pure rounds and, as long as they were planted, would yield only round peas. Two fourths were round in appearance, but were actually hybrids and in the next generation would produce round and wrinkled peas in a steady 3-to-1 ratio. The wrinkled peas from the union were again pure wrinkled and never would yield anything but their own kind. The wrinkled peas from the first hybrids yielded only wrinkled peas and continued to produce only wrinkled as long as he planted them and permitted them to fertilize themselves. They were pure recessives also. . . .

"It can be seen," said Mendel, "how rash it may be to draw from the external resemblances conclusions as to their internal nature."

Appearance meant little or nothing. It was not surprising that human beings had eternally been bewildered when they tried to understand heredity. The surface differences of offspring were legion in themselves, and yet below the surface lay still other differences and possibilities for variation.

With the clarity and simplicity of genius, Mendel labeled the "dominant" in the union with a capital *A*, and the recessive with a small *a*. A constant dominant thus would be formed by the coming together of two *A*'s, and it would be described as *AA*; a hybrid by either *Aa* or *aA*; and a recessive by *aa*.

With this understanding, it was possible for him to chart this coming together and separation and the clear-cut results that it produced. Mendel drew, and the world has since followed, his chart of heredity. . . .

Charles Darwin also had experimented with peas and had noticed that the hybrid divided 3 to 1, but Darwin was no mathematician and did not pursue this revealing indication of order in heredity. Mendel, on the other hand, was an excellent mathematician. The new experiment confirmed what he had already glimpsed in his first experiments. He was simply obtaining every combination that could be formed by the separate factors present. If *A* and *a* were combined, only one combination could be formed: *Aa*. If *Aa* and *Aa* came together, four combinations could be made: *AA*, *Aa*, *aA*, *aa*. And this was exactly what had happened. . . .

Mendel had no way to look into the egg and pollen cells to search for the hereditary factors whose existence he inferred from his experiments. His results, however, were explainable in no other way. Just as Mendel had grasped the secrets of the combinations of these factors, so did he reason out the biological basis that had to underlie them. He formulated three laws of inheritance:

1. All living things are a complex of a large number of independent heritable units.
2. When each parent contributes the same kind of factor and the two come

together in the offspring, a constant character is produced. But if one parent contributes one kind of factor, say *A*, and the other another, say *a*, a hybrid results. When the hybrid forms reproductive cells, the two differentiating elements "liberate themselves" again and thus are free to form new combinations in the next union.

3. The factors are unaffected by their long association in the individual. They emerge from the union as distinct as when they entered it.

In a letter to a fellow scientist Mendel enlarged upon this radical idea: "The course of development consists simply in this: that in each successive generation the two primal characters issue distinct and unadulterated out of the hybridized pair, there being nothing whatever to show that either of them had inherited or taken over anything from the others."

This was Mendel's ultimately famous law of the purity of the gametes, or reproductive cells. Mendel regarded it only as a hypothesis, and felt that he must subject it to additional tests. . . .

At last the time had come to report on his eight years of unremitting work. During the fall and early winter of 1864, Mendel checked and rechecked his findings. In his fine copperplate script he wrote the paper that eventually would explain to the world the all-important phenomenon of heredity.

The Brünn Society for the Study of Natural Science was holding its February 1865 meeting at the Modern School, the school in which Mendel taught. The night was cold and snowy, but most of the forty members came to hear Mendel report on the work he had so long been carrying on at the monastery. Curiosity about it was keen.

As Mendel read, this curiosity gave way to incomprehension. The several botanists in the society were as much confused by the report on invariable hereditary ratios in peas as were the other members—a chemist, an astronomer, a geologist, and an authority on cryptograms. Mendel spoke for the hour allotted to him, and then announced that at the next meeting he would explain why the peculiar and regular segregation of characteristics occurred.

The next month, as Mendel presented his algebraic equations, attention wavered. The combination of mathematics and botany which Mendel was expounding was unheard of. And the idea that lay behind it, that heredity was a giant shuffling and reshuffling of separate and invisible hereditary factors, stood in such diametrical contrast to all that had been taught that it probably could not be grasped. No one rose to ask a question. The minutes recorded no discussion.

A number of the members spoke to Mendel afterward about his experimental work. "I encountered various views," Mendel told his friends. "No one undertook a repetition of the experiment."

The editor of the *Proceedings* of the society extended the usual invitation to Mendel to prepare his paper for publication in the society's journal. "I agreed to do so," said Mendel, "after I had once more looked through my notes relating to the various years of the experiment without being able to discover any sort of mistake."

Mendel's monograph—*Versuche uber Pflanzenhybriden*—appeared in 1866. According to the society's custom, copies of the *Proceedings* in which it was included were sent to Vienna, Berlin, Rome, St. Petersburg, and Uppsala. But the brief paper that could have altered all ideas of heredity, and at a time when Darwin was still at work on the role of heredity in evolution, attracted no attention. It sat all but unread on library shelves.

What Mendel did was to show that genetic information was passed on from parent to offspring. Nearly one hundred years later, Francis Crick and James Watson showed *what* information was passed on and that the information was contained in a double-helical structure made up of nucleic acids. The first reading by geneticist George Beadle interprets for us "the language of the genes." The reading that follows it is Watson and Crick's low-key sparse announcement of their discovery as it appeared in *Nature* on April 25, 1953.

The Language of the Genes

G. W. Beadle

Let me begin by reminding you that each of us begins development as a small almost microscopic single cell, the fertilized egg. It is about 1/300 of an inch in diameter—about one-tenth the diameter of the head of a pin. A human egg does not look very different from the egg of a rabbit, a guinea pig, or a chimpanzee. In fact it is not too unlike the egg cell from which spinach plant develops. Yet each of these cells obviously "knows" what its potential future is, for gross mistakes do not occur. The egg of a rabbit never gives rise to a guinea pig. Clearly, the egg cell must contain information that says what it is going to do. A human egg carries specifications for growing into a person—not just any person but a very specific and truly unique one.

It is these specifications about which I wish to speak. . . .

In confining my attention largely to the nucleus, I likewise do not wish to belittle the role of environmental factors. They, too, are absolutely necessary. For a human egg cell to develop into a person, for example, a highly specific physical and chemical environment must be maintained, especially in the early stages of foetal development. Raw material of the proper kind must be available in the right amounts and at the right times. Before and after birth the nervous system is being indelibly impressed with information. Especially after birth the information that is fed into the nervous system in massive amounts plays a large and important role in determining what we are. It includes a large input of cultural inheritance.

The Nucleus and Genes

Getting back to the nucleus, there is an impressive body of evidence showing that in it is contained a very large part of the primary biological infor-

Reproduced by permission of the author and the British Association for the Advancement of Science from *The Advancement of Science,* Vol. 17, 1961, pp. 511–521.

mation that directs development and function. Mendel demonstrated in his garden peas that this information is particulate. Unfortunately, he was far ahead of his time and was quite unable to convince such persons as the German botanist Nägeli that what he found should be taken seriously. Had Darwin known of his work, the history of science might have been profoundly changed.

Mendel postulated the units of heredity that we now call genes. He did not know they were in the nucleus of the cell or that they were carried in chromosomes. But he did know that they determine whether pea seeds will be round or wrinkled, whether green or yellow, whether the flowers will be purple or white, whether the pea plant will be tall or short, and so on. He had a remarkable understanding about how genes are transmitted from one generation to the next. Although modified in important ways, his laws of inheritance still hold as first approximations to the ways of nature in communicating specifications from parent to offspring.

We now know that the most basic features of the process of transmission of biological directions for development are common to viruses, bacteria, algae, protozoa, higher plants, multicellular animals and man. . . .

When sperms or eggs are produced by individuals of two types, each carries one or the other of the two messages, either the maternal or the paternal one. Thus from any pair of parents it is easy to predict statistically the types of offspring, just as Mendel did with his peas.

How many such units or genes are contained in the nucleus of a human egg cell? We do not know with any degree of certainty. But we estimate that it is probably more than 10,000 kinds and less than 1,000,000, each present in duplicate except for sex-linked genes which are for the most part present singly in the male. What do these genes specify? Many characteristics: eye color, hair color, hair form, skin color, and hundreds of other characters by which the species may differ.

You may remark, as many biologists did in the early part of the century just after Mendel's work was rediscovered and while the young science of genetics was struggling for recognition, that these are all trivial characters. So they are, as are most of the characteristics responsible for the uniqueness of normal individuals. But we now have abundant evidence that most genes carry information of vital importance. Thus two genes in man carry the instructions for making the hemoglobin molecules that play such an essential role in carrying oxygen and carbon dioxide. Hemoglobin is *not* trivial.

Soon after the rediscovery of Mendel's classical paper, it was suspected that genes were carried in chromosomes. The clinching proof came in 1915 from the laboratory of the late Thomas Hunt Morgan. His student, Calvin Bridges, showed that errors in transmission of genes are exactly paralleled by errors in chromosome behavior.

The Structure of DNA as Genetic Material

Chromosomes were known to be made up largely of protein and nucleic acid, the latter of the deoxyribose type. Deoxyribonucleic acid has now become widely known as DNA. . . .

What is the structure of DNA that enables it to carry hereditary information? A tremendous step was taken toward answering this question in 1953 by a young American biologist, James D. Watson, and the English chemist, Francis H.C. Crick, working together at Cambridge University. Making use of the information then available about DNA–nucleotide composition of DNA of various sources, X-ray diffraction observations of M.H.F. Wilkins and associates of King's College, London, general knowledge of the structural arrangements of atoms in nucleotide components, etc.–they succeeded in constructing a molecular model of DNA that seemed to satisfy essentially all requirements. Now, some seven years later, it is generally agreed that the Watson-Crick structure is essentially correct for the native DNAs of a number of organisms. It is now clear that their achievement is outstanding in modern-day biology. This is so because their model goes so far in suggesting plausible answers to the questions posed at the beginning of this lecture.

According to this model, DNA consists of a pair of antiparallel polynucleotide chains wound helically around a common axis and cross-linked through specific hydrogen bonding between purine and pyrimidine bases.

Letting A, T, C, and G represent the four nucleotides characterized by adenine, thymine, cytosine and guanine, a four-unit segment of DNA can be represented in two dimensions as . . . [shown]. Paired dots represent hydrogen bonds. The two chains of the DNA molecule run in opposite directions. This is determined by the orientations of nucleotides in the two chains.

Information in DNA

Genetic information must somehow be determined by sequence of nucleotides. Since only A:T and C:G nucleotide pairs are possible in the normal structure, the two chains obviously carry complementary information. As we shall see this is of special significance in hypotheses of DNA replication.

One can think of information being carried either as a sequence of nucleotide pairs in the double molecule or as nucleotide sequences in the two single chain components. . . .

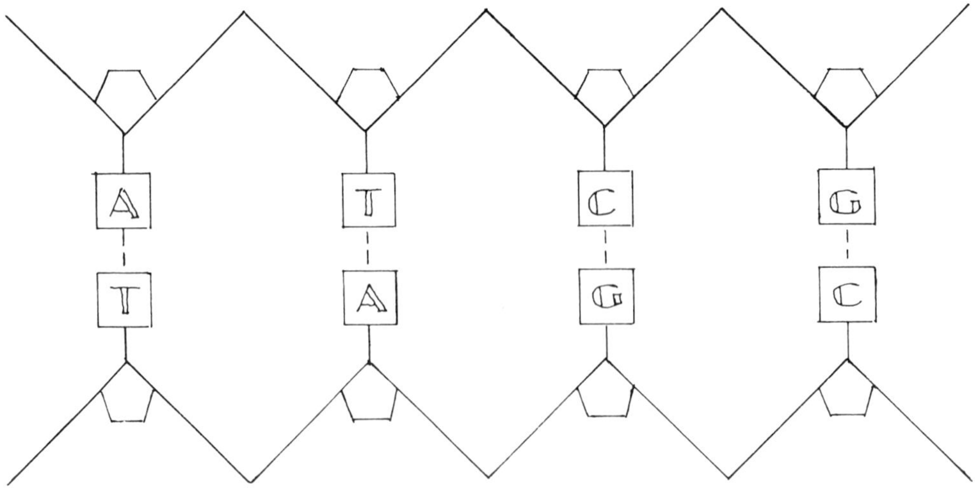

What about the physical size of DNA molecules? The diameter of the double helix is 20 Angstrom units. The base pairs are spaced at 3.4 A intervals. Thus a continuous linear double helix of the 5,000,000,000 base pairs of a single human egg would be somewhat more than 5 ft. in length—only slightly less than the height of an average person. This is a convenient way to remember the amount of DNA in a single human cell. If the strands were packed side by side in a monolayer on the head of a pin 1 mm. in diameter, this amount of DNA would cover less than 1/200 of the surface.

Replication of DNA

The Watson–Crick structure immediately suggested how DNA molecules might reproduce by separation of the double structure into single chains followed by synthesis of new partners against each of the old chains. Since the two originals are complementary, each carries the information necessary to direct the synthesis of its partner. It is presumed that single chains serve as templates against which free nucleotides are properly ordered by specific hydrogen pairing. Just how the problems of separation of paired chains helically coiled around a common axis is solved is not known. The forces required to "untwist" the double helix would not be great if it were to rotate around its axis in the manner of an automobile speedometer cable.

Use of DNA Information

How is information in the form of DNA made use of in the development of an organism like man? This is clearly a question of the most fundamental importance to biology. At the same time it is an enormously difficult one and we are a long way from knowing the complete answer.

All living systems contain DNA (or in some viruses a related form of nucleic acid, ribonucleic acid, called RNA) and protein. In cellular forms the proteins are of many kinds. We know that many genes, perhaps all, are somehow concerned with synthesis of specific proteins. Many of these serve as organic catalysts—enzymes—or as components of these catalysts. Enzymes accelerate vital reactions that would otherwise proceed at rates too low to sustain life. In cellular organisms there are thousands of kinds of enzymes, each owing its specificity to a particular protein. We can therefore narrow the problem of gene action, at least in some cases, to that of protein synthesis.

An hypothesis at present widely used as a working basis visualizes the process of protein synthesis in the following way: for each protein potentially capable of being formed, there is in the nucleus a specific segment of DNA that carries the information by which the twenty kinds of amino acid sub-units in that protein are properly ordered during its synthesis. Take human hemoglobin as an example. Each molecule of this vital oxygen-carrying red protein consists of four protein chains and an equal number of heme groups each containing an atom of iron. The protein chains are of two kinds, called alpha and beta chains. There are two of each, the members of a pair being identified and each made up of about 150 amino-acid units arranged in a precisely determined sequence. For each of the two kinds of protein chains, there is presumed—with some evidence—to be a gene consisting of a segment of DNA. This segment is, by the definition I shall use, a gene. It may be something of the order of 1000 nucleotide pairs in length.

How is the information in the gene for the alpha hemoglobin chain made use of? The hypothesis assumes the following sequence of events: from the gene in the nucleus, information is transferred to RNA, presumably by the DNA somehow acting as a template in the ordering of the four kinds of ribose nucleotides in RNA. This informational RNA then moves from the nucleus to the cytoplasm of the young red blood cell—before it loses its nucleus. There it is incorporated into microsomes, which are submicroscopic bodies made of structural protein and structural RNA. Once in microsomes, informational RNA molecules serve as templates against which amino-acids are arranged in proper order to make alpha hemoglobin protein chains.

Prior to this the amino-acids are activated and attached to small carrier segments of RNA, each specific for its own amino-acid. Thus for each of the twenty amino-acids there is a corresponding carrier RNA segment. Carrier RNA molecules are somehow coded to specific sites on the informational RNA in the microsome. In this way each amino-acid is carried to its proper position on the template. They are then joined through peptide linkages to form alpha chains and are released from the template and microsome. Whether association of alpha and beta chains with their hemes, and with each other, occurs in the microsome or outside is not known.

Not all proteins for which genetic directions are available in the nucleus are synthesized in any one cell. There are ingenious control mechanisms in operation which determine whether the information in a given gene will be used and, if so, when and for how long. In some cases it is clear that the presence or absence of substrate determines whether a given enzyme will be synthesized. In this way it is assumed that enzymes are not made in quantity when there are no substrates on which they can work. Through studies on such control mechanisms we are beginning to understand how it is that cells with identical genetic information in their nuclei may do quite different things, depending on their environmental context or on their previous history. . . .

Mutation

In terms of DNA structure gene mutation is believed to be the result of alterations in nucleotide sequence. Many mutations are thought to be the result of errors in DNA replication. In their original paper on DNA structure, Watson and Crick pointed out that if at the precise moment of partner selection a purine or pyrimidine were to exist in an improbable tautomeric form, it might form hydrogen bonds with a "wrong" nucleotide. At the next round of replication such a "wrong" nucleotide would direct that its own complementary nucleotide be inserted in the paired chain. Thus one of the two daughter DNA molecules would differ from its sister in one nucleotide pair. Its informational content would be modified accordingly.

Experimentally mutations can be increased in frequency in a number of ways. High-energy radiations—ultraviolet, X-rays, etc.—are effective in doing this. Transmutation of artificially incorporated radioactive isotopes such as phosphorus-32 is known to produce mutations. Since radioactive isotopes occur naturally in low frequencies, they are no doubt responsible for a small fraction of spontaneously produced mutations.

Many chemical substances are known to be mutagenic. Perhaps the one best known from the standpoint of mechanism of action is nitrous acid. This specifically oxidizes amino groups. In this way it may chemically change a natural DNA pyrimidine or purine in such a way that its hydrogen bonding specificity is modified. . . .

The Coding Problem

How are sequences of nucleotides in DNA related to amino-acid sequences in proteins? DNA and protein can be thought of as four- and twenty-symbol systems. If they are both equivalent to simple linear codes, it is clear that it requires at least three nucleotides to specify an amino-acid, for the maximum number of two-letter words is only sixteen. Several possible coding systems have been investigated but so far the problem has not been solved. Perhaps it will not be until nucleotide sequences in a gene can be compared with amino-acid sequences in the protein specified by that gene. As you know, complete amino-acid sequences have been worked out for only a very few proteins and unfortunately these are from organisms that are far from ideal for genetic study. There are now investigations under way in many laboratories designed to correlate genetic fine structure with protein fine structure.

Evolution

Presumably DNA replication bears no immediate relation to its informational content. Biologically useless DNA is replicated just as faithfully as that playing a vital role. Presumably, too, the mutation process is a random affair. If so, meaningful DNA sequences are much more likely to be made less useful through random mutation than they are to be converted into sequences more useful to the organism of which they are a part. In this sense they may be likened to random typographical errors in a useful message. Direct experiment verifies this expectation; most mutations are unfavorable in the context in which they occur.

If mutations are mostly unfavorable but are as faithfully reproduced as are their normal counterparts, why do they not accumulate with successive rounds of replication just as a typist would accumulate errors in the Lord's Prayer if she were to type copies in succession, each from the previous typing, in a purely mechanical way without proofreading or correcting errors? The answer is that biologically unfavorable mutations specify organisms with reduced reproductive fitness. The errors in their DNA are not corrected but the lines of descent carrying such errors are statistically bred out of existence over a shorter or longer number of generations depending on the degree of reduction in reproductive fitness. We call this natural selection.

Positive evolutionary progress depends on rare favorable mutations that increase reproductive fitness. They gradually replace the ancestral types from which they arose.

Modern concepts of organic evolution hold that all living systems have evolved gradually, mutation by mutation, from pre-existing organisms. Sometimes, as in parasites, it is advantageous to become simpler and to count increasingly on the host for raw materials, a proper environment and protection against enemies. But from the beginning, the overall trend in many lines such as our own has evidently been toward greater complexity.

Speculation as to how organic evolution began in the first place leads to the conclusion that there is no clear qualitative break in the sequence of events that spans the advent of life on earth. This is only another way of saying that we cannot clearly distinguish between living and non-living systems. No matter what system one contemplates, it is possible to imagine a closely related one, only slightly simpler, that could have given rise to the former by a single mutation-like step. . . .

As every organic chemist knows, organic molecules are formed through the interaction of inorganic molecules—again when the conditions are right. And as every inorganic chemist knows elements interact to produce inorganic molecules. Nuclear physicists tell us that elements themselves evolve from simpler elements as a result of processes that are both natural and inevitable, given the appropriate circumstances.

Thus it is clear that the sequence: hydrogen-helium-beryllium-8-carbon-oxygen-other elements-water-other inorganic molecules-simple organic molecules-more complex organic molecules like nucleotides, amino-acids, and small proteins-nucleic acids capable of replication-nucleic acids protected by protein coats-virus-like systems with protein coats serving catalytic functions-multigenic but subcellular organisms-simple cellular systems like bacteria-autonomous cellular forms like algae-protozoa-multicellular plants and animals and, in our line of descent, man himself—is a natural one that could have arisen by steps no one of which need have been larger than the individual mutational steps we know in today's living systems.

In the beginning there was a universe of hydrogen. How and when it was created—or whether it is and always has been in a steady state of continuous creation—science knows very little.[1] In whatever way the universe began, there must have been built into it from the very beginning the potentiality of essentially unlimited orderly evolution.

[1] See "Starfolk" by Carl Sagan in Chapter 10 for a discussion of one theory of the origin of the planets.—Ed.

The Double Helix

J.D. Watson and F.H.C. Crick

We wish to suggest a structure for the salt of deoxyribose nucleic acid (D.N.A.). This structure has novel features which are of considerable biological interest.

A structure for nucleic acid has already been proposed by Pauling and Corey. They kindly made their manuscript available to use in advance of publication. Their model consists of three intertwined chains, with the phosphates near the fibre axis, and the bases on the outside. In our opinion, this structure is unsatisfactory for two reasons: (1) We believe that the material which gives the X-ray diagrams is the salt, not the free acid. Without the acidic hydrogen atoms it is not clear what forces would hold the structure together, especially as the negatively charged phosphates near the axis will repel each other. (2) Some of the van der Waals distances appear to be too small.

Another three-chain structure has also been suggested by Fraser. In his model the phosphates are on the outside and the bases on the inside, linked together by hydrogen bonds. This structure as described is rather ill-defined, and for this reason we shall not comment on it.

We wish to put forward a radically different structure for the salt of deoxyribose nucleic acid. This structure has two helical chains each coiled round the same axis. . . . We have made the usual chemical assumptions, namely, that each chain consists of phosphate diester groups joining β-D-deoxyribofuranose residues with 3′,5′ linkages. The two chains (but not their bases) are related by a dyad perpendicular to the fibre axis. Both chains follow right-handed helices, but owing to the dyad the sequences of the atoms in the two chains run in opposite directions. Each chain loosely resembles Furberg's model No. 1; that is, the bases are on the inside of the helix and the phosphates on the outside. The configuration of the sugar and the atoms near it is close to Furberg's 'standard configuration', the sugar being roughly perpendicular to the attached base. There is a residue on each chain every 3.4 A. in the *z*-direction. We have assumed an angle of 36° between adjacent residues in the same chain, so that the structure repeats after 10 residues on each chain, that is, after 34 A. The distance of a phosphorus atom from the fibre axis is 10 A. As the phosphates are on the outside, cations have easy access to them.

The structure is an open one, and its water content is rather high. At lower water contents we would expect the bases to tilt so that the structure could become more compact.

The novel feature of the structure is the manner in which the two chains are held together by the purine and pyrimidine bases. The planes of the bases are perpendicular to the fibre axis. They are joined together in pairs, a single base from one chain being hydrogen-bonded to a single base from the other chain, so that the two lie side by side with identical *z*-co-ordinates. One of the pair must be a purine and the other a pyrimidine for bonding to occur. The hydrogen bonds are made as follows: purine position 1 to

Used with permission from James Watson and Francis Crick, "The Double Helix," *Nature*, Vol. 171, April 25, 1953, pp. 737–738.

pyrimidine position 1; purine position 6 to pyrimidine position 6.

If it is assumed that the bases only occur in the structure in the most plausible tautomeric forms (that is, with the keto rather than the enol configurations) it is found that only specific pairs of bases can bond together. These pairs are: adenine (purine) with thymine (pyrimidine), and guanine (purine) with cytosine (pyrimidine).

In other words, if an adenine forms one member of a pair, on either chain, then on these assumptions the other member must be thymine; similarly for guanine and cytosine. The sequence of bases on a single chain does not appear to be restricted in any way. However, if only specific pairs of bases can be formed, it follows that if the sequence of bases on one chain is given, then the sequence on the other chain is automatically determined.

It has been found experimentally that the ratio of the amounts of adenine to thymine, and the ratio of guanine to cytosine, are always very close to unity for deoxyribose nucleic acid.

It is probably impossible to build this structure with a ribose sugar in place of the deoxyribose, as the extra oxygen atom would make too close a van der Waals contact.

The previously published X-ray data on deoxyribose nucleic acid are insufficient for a rigorous test of our structure. So far as we can tell, it is roughly compatible with the experimental data, but it must be regarded as unproved until it has been checked against more exact results. Some of these are given in the following communications. We were not aware of the details of the results presented there when we devised our structure, which rests mainly though not entirely on published experimental data and stereochemical arguments.

It has not escaped our notice that the specific pairing we have postulated immediately suggests a possible copying mechanism for the genetic material.

The discovery of the double helix opened the study of heredity to chemists. Now, like any other more or less complex chemical compound, the mechanism that passes on genetic information from generation to generation can be studied in a test tube. If that is so, it should also be possible to manipulate the genetic machinery, to insert chemical information that would correct genetic defect or even to make carbon copies of an individual without the need of the normal reproductive processes. That possibility was foreseen in a somewhat different context by the good Dr. Bokanovsky in Aldous Huxley's prophetic *Brave New World.* Dr. Bokanovsky perfected a method for mixing eggs and sperm in a test tube so that they would eventually grow into scores of identical adults. In this section, the Director of Hatcheries and Conditioning describes the Bokanovsky process.

Brave New World—an Excerpt

Aldous Huxley

Still leaning against the incubators he gave them, while the pencils scurried illegibly across the pages, a brief description of the modern fertilizing process; spoke first, of course, of its surgical introduction—"the operation undergone voluntarily for the good of Society, not to mention the fact that it carries a bonus amounting to six months' salary"; continued with some amount of the technique for preserving the excised ovary alive and actively developing; passed on to a consideration of optimum temperature, salinity, viscosity; referred to the liquor in which the detached and ripened eggs were kept; and, leading his charges to the work tables, actually showed them how this liquor was drawn off from the test-tubes; how it was let out drop by drop onto the specially warmed slides of the microscopes; how the eggs which it contained were inspected for abnormalities, counted and transferred to a porous receptacle; how (and he now took them to watch the operation) this receptacle was immersed in a warm bouillon containing free-swimming spermatozoa—at a minimum concentration of one hundred thousand per cubic centimetre, he insisted; and how, after ten minutes, the container was lifted out of the liquor and its contents re-examined; how, if any of the eggs remained unfertilized, it was again immersed, and, if necessary, yet again; how the fertilized ova went back to the incubators; where the Alphas and Betas remained until definitely bottled; while the Gammas, Deltas and Epsilons were brought out again, after only thirty-six hours, to undergo Bokanovsky's Process.

"Bokanovsky's Process," repeated the Director, and the students underlined the words in their little notebooks.

One egg, one embryo, one adult—normality. But a bokanovskified egg will bud, will proliferate, will divide. From eight to ninety-six buds, and every bud will grow into a perfectly formed embryo, and every embryo into a full-sized adult. Making ninety-six human beings grow where only one grew before. Progress.

"Essentially," the D.H.C. concluded, "bokanovskification consists of a series of arrests of development. We check the normal growth and, paradoxically enough, the egg responds by budding."

Responds by budding. The pencils were busy.

He pointed. On a very slowly moving band a rack-full of test-tubes was entering a large metal box, another, rack-full was emerging. Machinery faintly purred. It took eight minutes for the tubes to go through, he told them. Eight minutes of hard X-rays being about as much as an egg can stand. A few died; of the rest, the least susceptible divided into two; most put out four buds; some eight; all were returned to the incubators, where the buds began to develop; then, after two days, were suddenly chilled, chilled and checked. Two, four, eight, the buds in their turn budded; and having budded were dosed almost to death with alcohol; consequently burgeoned again and having budded—bud out of bud out of bud—were thereafter—further arrest being generally fatal—left to develop in peace. By which time the original egg was in a fair way to becoming anything from eight to ninety-six embryos—a prodigious improvement, you will agree, on nature. Identical twins—but not in piddling twos and threes as in the old viviparous days, when an egg would sometimes accidentally divide; actually by dozens, by scores at a time.

"Scores," the Director repeated and flung out his arms, as though he were distributing largesse. "Scores."

But one of the students was fool enough to ask where the advantage lay.

"My good boy!" The Director wheeled sharply round on him. "Can't you see? Can't you *see?*" He raised a hand; his expression was solemn. "Bokanovsky's Process is one of the major instruments of social stability!"

Major instruments of social stability.

Standard men and women; in uniform batches. The whole of a small factory staffed with the products of a single bokanovskified egg.

"Ninety-six identical twins working ninety-six identical machines!" The voice was almost tremulous with enthusiasm. "You really know where you are. For the first time in history." He quoted the planetary motto. "Community, Identity, Stability." Grand words. "If we could bokanovskify indefinitely the whole problem would be solved."

The more up-to-date version of Dr. Bokanovsky's technique is the process of cell cloning, that is, the process of growing a colony or clone of identical animals from cells of a

single parent. It has already been done in the axolotl and there is anticipation among some scientists that it might also be done with the cells of human beings. The prospect of cloning humans, or "tissue-culture reproduction," raises a number of questions, theological and otherwise. Albert Rosenfeld wants to know, for instance, whether cultured organisms could be said to have separate souls:

If a new being were grown from the cell of a man still living, a possibility to be elaborated upon later, would it too have a soul? If so, it cannot be the man's own soul because his still presumably remains in residence. Where did the new soul come from? If he has none, is he human? Can he be saved? And suppose one hundred people are grown from the still living cells of a dead man. Do they all have souls? Did the dead man whose tissue provided the cells have one hundred souls to spare? Where were they meanwhile? Perhaps in the DNA of the cell nucleus? If so, do the millions of cells still kept alive in the culture also have souls?

It has been demonstrated that, even in hamburger meat, there are still live cells which, transferred to a congenial medium, will begin multiplying. If a new cow were grown from such a cell, would it mean that the old cow, though butchered and ground up, had never really been killed at all? Applied to people, this means that a corpse mangled beyond recognition in some horrible accident would still yield up intact cells from which a new being might be grown. Or that a frozen cadaver which turned out to be not entirely revivable as a whole human being might still contain cells that would thrive in culture. Dr. Elof Carlson, a U.C.L.A. zoologist, has suggested that even the mummified cells of people long dead might be viable for tissue-culture growth. If any one of these cells has the capacity to become a human being, is there a soul resident in each one?

There are more serious moral and ethical questions, too, questions that touch on many areas of the rapidly growing fields of biomedical research. Along with cloning, we face the implications of organ transplantation, legalized abortion, even chemical and surgical manipulation of the mind. In an address delivered a few years ago Nobel-prize-winning biologist Salvador Luria gave his views on the ethical constraints of biomedical research.

From Al Rosenfeld, *The Second Genesis*. Used by permission of the author and the Ann Elmo Agency, 52 Vanderbilt Avenue, New York, N.Y. 10017.

As for cloning of human beings, Luria believes that it would be working against man's uniqueness as well as against natural evolution.

Ethical Aspects of the New Perspectives in Biomedical Research

Salvador E. Luria

Ethical problems arise, in the biomedical field as in all other areas, when decisions are influenced by conflicting sets of values. Such conflicts are often unrecognized or at least not explicitly formulated, especially when one set of values is so prevalent in society that it is taken as an absolute while conflicting values are considered as "deviant". This is often observed in sociopolitical affairs: for example, some of the values of liberal democracy, such as free speech and due process, proclaimed originally by a revolutionary bourgeoisie and widely assumed since then to be "intrinsically good", are sometimes experienced as artificial restraints by radical groups bent on changing again the structure of society and, more frequently, are violated by the instruments of a now dominant class.

Another example, this one closer to medicine, is the conflict between the demand for optimal care for individual patients, including costly major surgery, and the demand for adequate care on a broad social front—a conflict that may persist as long as medical services continue to be available in scarcity rather than in plenty. Both examples illustrate the simple truth that, in political as well as in medical affairs, values change as the situation changes. Conflicts arise, at first implicit and unrecognized, then seen as oppositions between "valid" and "deviant" values, finally as the expression of new human needs or aspirations or opportunities. Nowhere is this more evident than in the influence of modern technology on ethical values.

Can this view of ethics, that is, the awareness of the existence of conflicting sets of values, provide guidance for the practical resolution of specific conflicts, especially those that may face the biologist in his experimentation, or the pharmaceutical chemist in drug testing, or the physician in his daily practice? Are there any grounds for choice between guidelines, except one's personal convictions or prejudices or interests?

One thought I would like, as a biologist, to offer is that ethical values are not only the product of social and cultural history, but exert in turn a profound effect on the rate and direction of change in human affairs. Man is a unique species, whose current evolution, cultural as well as biological, is in good part directed by its own activities, which alter the natural, social, and intellectual environment and thereby change the human species itself. But this quasi self-domestication is, of course, experienced at the conscious level. Man's reactions to the situations created by his own activities are embodied in sets of values, which in turn exert regulatory functions. These functions are sometimes homeostatic (negative feedback), tending to prevent or delay the exploration of new options but also to avoid premature commitment to dubious courses of action. At other times, the regulatory functions of ethical principles are evolutionary (positive feedback), tending to accelerate the rate of change by removing cultural constraints that impede new explorations.

Reprinted with permission of the author and publisher from S.E. Luria, "Ethical Aspects of the New Perspectives in Biomedical Research," *The Challenge of Life,* Birkhauser Verlag, 1972.

My argument is that not only does ethics evolve but it has an *adaptive* evolutionary history, in the sense that the "validity" of its principles is tested by their consequences as regulatory mechanisms on the course of human events in a changing world. Ethical rules are part and parcel of human culture at a given stage, tested and validated by their effects on that culture. This argument rejects, of course, any metaphysical justification of ethical principles; it also rejects the relativistic or anarchistic views of ethics, by which any set of values is equally tenable and suitable as a guideline. (The existentialist view, of unlimited individual freedom coupled with individual responsibility, falls outside this discussion because it is not concerned with guidelines for collective choice in social situations.). . .

Still different is the situation with regard to what has been called genetic intervention—the modification of the genetic endowment of individuals through the introduction of foreign cells (chimerism) or of genetic materials such as chromosomes or genes. The techniques to carry out such operations are not yet at hand and, according to some, may not be forthcoming. Even more distant appears to be the possibility of genetic intervention on the germ line, whether on the intact human being or on eggs fertilized *in vitro* and then implanted into the womb of a mother or foster-mother. Yet, despite the lack of immediate prospects, discussion of promises and dangers has been heated, if inconclusive. The conflicting values are not easily discerned. On the one hand, the extent of prospective medical benefits is difficult to evaluate; on the other hand, the fears are based mainly on a distrust of powerful technologies becoming available in a society whose power structure and institutional mechanisms make a misuse of such technologies only too likely—as was the case for nuclear fission or fusion. In fact, the possible creation of genetic weapons, especially through virus research, is one of the fears most often expressed. The misgivings reflect the fact that our society cannot trust its own mechanisms to provide adequate guidelines for handling the new power that may be offered to it. Here we encounter another ethical question, that of whether not only the development of some technologies, but even the pursuit of certain lines of scientific research should be discouraged or stopped, either voluntarily or by legislation. Again no solid guidelines are available for choosing between the "traditional" (but only three centuries old!) ethics of freedom of research and what may prove to be a new ethical demand for restraint till the social implications of certain kinds of research become clear. Note how in this case it could be the prevalent values rather than the "deviant" ones that play the role of troublemakers.

Even more heated, because of the less remote feasibility of the practice, has been the discussion concerning the "cloning" of human beings by transplant of nuclei into enucleated human eggs, followed by implantation of the eggs into the womb. The case against cloning and the objection to experimentation on cloning of humans—an objection in which I personally concur—have often been presented in a confusing way because the arguments have been based on ethical misunderstanding. Theologians and medical scientists alike have upheld against cloning *the right of the individual to be unique*, that is, to be the product of the genetic lottery of meiosis and fertilization, as if this were an intrinsic biological and metaphysical right. But this argument, even when presented with a garnish of psychological jargon, reflects only a special set of values, that of the individualistic, puritan ethics of capitalist democracy. I suspect that anthropologists could cite a number of societies in which identity with other individuals, if possible, would be a highly prized value. Would the champions of uniqueness deny the members of those societies the right to be cloned?

A sounder argument is that cloning would reduce the genetic variability of human populations. This, however, would hardly be a significant effect unless the practice were to become widespread. The real ground on which many people, including myself, would like to prevent cloning of humans (and even regulate research in this and other similarly delicate areas of reproductive biology) is a concern for the social and biological consequences of untested interference with the process of human reproduction. This judgment is not based on any metaphysical belief in the sacredness of certain genetic rights, but on the ethical principle that steps that may affect masses of human beings ought not to be taken until one has some idea of what they entail. Like all principles, this is no doubt a debatable principle, but one defensible in terms of the concept of adaptive ethics. It simply means that when it comes to tampering with the genetic endowment of mankind (and also of individual men) a great deal of conservatism is wise and justified.

In relation to the series of problems discussed above—abortion and birth control, genetic engineering, and cloning—we must entertain, however, the possibility that future generations, bred under new conditions and new stresses, may evaluate the choices differently and develop and prize alternative values. Thus population pressures may tend to diminish the respect for individual life and may even make palatable certain forms of coercive restraints on procreation which are now abhorrent to most of us (incidentally, this seems to me a very good reason for trying to prevent the development of such pressures by voluntary limitation of the birth rate). On the other hand, any coercively enforced limitation of population size may increase the appeal and danger of a "selective" ethics, encouraging the preferential breeding of certain human types or social groups. In opposition to this we have—and, I believe, should uphold and defend—the value of individual freedom tightly coupled with individual reproductive responsibility.

To counterbalance the suspicion that there are evil scientists off in the shadows formulating malevolent plans for the genetic destiny of man comes a suggestion from Sir Peter Medawar, another Nobel-prize-winning biologist. He proposes that although it is generally believed that nature always knows best, she has been guilty of some very serious genetic mistakes.

Nature Does Not Always Know Best

Peter B. Medawar

It is a profound truth—realized in the nineteenth century by only a handful of astute biologists and by philosophers hardly at all (indeed, most of those who held any views on the matter held a contrary opinion)—a profound truth that nature does *not* know best; that genetical evolution, if we choose to look at it liverishly instead of with fatuous good humour, is a story of waste, makeshift, compromise, and blunder.

I could give a dozen illustrations of this judgment, but shall content myself with one . . . the immunological defences of the body, the reactions that are set in train by the invasion of the tissues by foreign substances. Reactions of this kind are more than important: they are essential. We can be sure of this because some unfortunate children almost completely lack the biochemical aptitude for making antibodies, the defensive substances upon which so much of resistance to infectious disease depends. Until a few years ago these children died, because only antibiotics like penicillin can keep them alive; for that reason, and because the chemical methods of identifying it have only recently been discovered, the disease I am referring to was only recognized in 1952. The existence of this disease confirms us in our belief that the immunological defences are vitally important; but this does not mean that they are wonders of adaptation, as they are so often supposed to be. Our immunological defences are also an important source of injury, even of mortal injury.

For example: vertebrate animals evolved into the possession of immunological defences long before the coming of mammals. Mammals are viviparous: the young are nourished for some time within the body of the mother: and this (in some ways) admirable device raised for the first time in evolution the possibility that a mother might react immunologically upon her unborn children—might treat them as foreign bodies or as foreign grafts. The haemolytic disease that occurs in about one newborn child in 150 is an error of judgment of just this kind: it is, in effect, an immunological repudiation by the mother of her unborn child. Thus the existence of immunological reactions has not been fully reconciled with viviparity; and this is a blunder—the kind of blunder which, in human affairs, calls forth a question in the House, or even a strongly worded letter to *The Times*.

But this is only a fraction of the tale of woe. Anaphylactic shock, allergy, and hypersensitivity are all aberrations or miscarriages of the immunological process. Some infectious diseases are dangerous to us not because the body fails to defend itself against them but—paradoxically—because it does defend itself: in a sense, the remedy *is* the disease. And within the past few years a new class of diseases has been identified, diseases which have it in common that the body can sometimes react upon its own constituents as if they were foreign to itself. Some diseases of the thyroid gland and some inflammatory diseases of nervous tissue belong to this category; rheumatoid arthritis, lupus erythematosus, and scleroderma may conceivably do so too. I say nothing about the accidents that used to occur in blood transfusions, immunological accidents; nor about the barriers, immunological barriers, that prevented our grafting skin from one person to another, useful though it would so often be; for transfusion and grafting are artificial processes, and, as I said in an earlier lecture, natural evolution cannot be reproached for failing to foresee what human beings might get up to. All I am concerned to show is that natural devices and dispositions are highly fallible. The immunological defences are dedicated to the proposition that anything foreign must be harmful; and this formula is ground out in a totally undiscriminating fashion with results that are sometimes irritating, sometimes harmful, and sometimes mortally harmful. It is far better to have immunological defences than not to have them; but this does not mean that we are to marvel at them as evidences of a high and wise design.

We can, then, improve upon nature; but the possibility of our doing so depends, very obviously, upon our continuing to explore into nature and to enlarge our knowledge and understanding of what is going on. If I were to argue the scientists' case, the case that exploration is a wise and sensible thing to do, I should try to convince you of it by particular reasoning and particular examples, each one of which could be discussed and weighed up; some, perhaps, to be found faulty. I should not say: Man is driven onwards by an exploratory instinct, and can only fulfil himself and his destiny by the ceaseless quest for Truth. As a matter of fact, animals do have what might be loosely called an inquisitiveness, an exploratory instinct; but even if it were highly developed and extremely powerful, it would still not be binding upon us. We should not be *driven* to explore.

Contrariwise, if someone were to plead the virtues of an intellectually pastoral existence, not merely quiet but acquiescent, and with no more than a pensive regret for not understanding what could have been understood; then I believe I could listen to his arguments and, if they were good ones, might even be convinced. But if he were to say that this course of action or inaction was the life that was authorized by Nature; that this was the life Nature provided for and intended us to lead; then I should tell him that he had no proper conception of Nature. People who brandish naturalistic principles at us are usually up to mischief. Think only of what we have suffered from a belief in the existence and overriding authority of a fighting instinct; from the doctrines of racial superiority and the metaphysics of blood and soil; from the belief that warfare between men or classes of men or nations represents a fulfilment of historical laws. These are all excuses of one kind or another, and pretty thin excuses. The inference we can draw from an analytical study of the differences between ourselves and other animals is surely this: that the bells which toll for mankind are—most of them, anyway—like the bells on Alpine cattle; they are attached to our own necks, and it must be *our* fault if they do not make a cheerful and harmonious sound.

"Sex produces diversity and diversity is the propeller of evolution." If that is so, as Dr. Bronowski strongly suggests, the propeller began whirling about three billion years ago, and the vehicle that carried it was probably a single-celled microorganism. This possibility is discussed more fully in the following selection from *Geology Today*.

Sexual Reproduction in Bacteria

from *Geology Today* (CRM Books, publisher)

In 1954, Stanley Tyler of the University of Wisconsin and Barghoorn reported the discovery of a large community of fossil microorganisms, that had been preserved in chert of the Gunflint Iron Formation of Ontario, Canada.

Geologic evidence indicated that the stromatolitic chert was somewhere between 1.6 and 2.0 billion years in age. This discovery of microscopic organisms preserved in sedimentary rocks that had been deposited at a point in geologic time approaching half the age of the earth stimulated much of the recent search for Precambrian fossils that had previously been considered either nonexistent or undetectable. The discovery also motivated many of the later fossil searches already mentioned and pointed to an excellent source of additional traces of very ancient life: fine-grained cherts rich in organic compounds. This rock type, which is composed of microcrystalline quartz, may bear organisms embedded in silica along with fossilized cellular residue.

Tyler and Barghoorn, in 1965, summarized results from their line of research and grouped the twelve species of microscopic fossil plants then known into four categories. The Gunflint assemblage is predominantly—perhaps entirely—procaryotic, that is, made up of cells lacking a true nucleus. It offers no evidence, as yet, that the nucleated cell had evolved by 1.6 to 2.0 billion years ago.

The *eucaryotic*, or nucleated, cell type differs from the procaryotic form that preceded it in several aspects. . . . The eucaryotes being fundamentally aerobic, that is, depending on free oxygen for existence, could not have evolved until after oxygen became fairly plentiful in the environment. It is an interesting fact that the modern procaryotes, bacteria and blue-green algae, are still more resistant to ultraviolet irradiation than the eucaryotes, an advantage preserved from when ozone was not abundant to serve as an ultraviolet shield.

The appearance of the eucaryotic cell was a milestone in biological evolution: each part within the integrated cell became contained separately. Intracellular differentiation substantially increased metabolic efficiency and resulted in a new and highly effective means of distributing genetic material.

In 1965, Barghoorn and Schopf reported identifying new microbiota in cherts in Bitter Springs Formation of central Australia, dated at some 900 million years old. Detailed studies of this amazingly rich deposit have identified fifty new species of Precambrian microorganisms. The Bitter Springs microbiota is the most diverse and best preserved early assemblage of microfossils known. Like the Gunflint assemblage, it is composed chiefly of blue-green algae, many of them strikingly similar in morphology to living species.

Like the Gunflint biota, the Bitter Springs assemblage is dominated by procaryotes. Unlike the Gunflint samples, it also includes eucaryotes, in some of which remnants of nuclei have been preserved. The state of preservation of cells in the Australian cherts is remarkable—careful examination has revealed that some cells were apparently "frozen" while dividing. . . . Although the appearance of eucaryotes cannot yet be established precisely, there is evidence suggesting that they may have originated between 1.8 and 1.4 billion years ago.

Without these basic, biological innovations at the cellular level it is doubtful that life forms on this planet would ever have evolved beyond microscopic size or that the crowning event in early evolution—the appearance of sexual reproductivity—could ever have occurred.

The major significance of sexual reproduction is that it makes possible a fantastic number of genetic combinations and recombinations in the population. For example, the occurrence of ten mutations in a population of *asexual* organisms could result in no more than eleven different types of individuals—the original type and the ten mutants. In contrast, ten mutations in a *sexual* population could combine to produce 59,049 genetically distinct types of the same species. Although not every potential combination would occur in an actual population, the message is clear that sexual reproduction, by vastly increasing the overall variability among the members of a species, proportionately increases the number of potentially *advantageous* variations in the population. In this way, sexual reproduction greatly increases the speed and efficiency of the evolutionary process.

The next important question is when did sexuality arise? The substantial diversity and high rates at which fossil eucaryotes evolved indicates that they reproduced sexually. A large number of evolutionary changes—thus, considerable time—must have been required for the transition from sexual unicells to multicellular eucaryotes. The diverse and rapidly evolving invertebrate fauna and algal flora of the early Phanerozoic, some 600 million years ago, implies that sexual reproduction had become well established during the Precambrian. In contrast, sexuality could not have originated prior to the appearance of asexual eucaryotes—around 1.8 to 1.4 billion years ago. The Bitter Springs microflora, about 900 million years old, is a likely place to begin searching for evidence of primitive sexual cells.

It is difficult to detect sexuality in microfossils, however. Some of the higher algae, fungi, and other advanced eucaryotic microorganisms have distinctive reproductive structures that could be preserved, but the Bitter Springs eucaryotes apparently are too primitive to have such features. Researchers have little knowledge of what a truly primitive sexual system was like or how to detect evidence of sexuality in microfossils. But these mysteries are not unsolvable. All known sexual reproduction involves a specific process, *meiosis*, that in most plant groups results in a characteristic tetrahedral configuration of sexual cells. The fossils to seek, therefore, would be those exhibiting this characteristic.

After the proper type of fossil to hunt and the appropriate rocks in which to look for it had been determined, tetrahedral spores were discovered in the Bitter Springs cherts in 1971. These cells are the earliest geologic evidence of life cycles apparently involving sex.

This discovery indicated that eucaryotic sexuality, perhaps the most consequential of all evolutionary innovations, was probably established as early as 900 million years ago. The decidedly primitive nature of these eucaryotes suggests that sexual reproductivity may still have been a novel evolutionary experiment, not then widespread in the biota. Nevertheless, the advent of sexual reproduction provided a fresh potential that soon resulted in diverse new stocks of algae, fungi, and single-celled animals. Before the Precambrian ended multicellular life was well established on earth. It was a long period in which microscopic organisms were constantly modifying their biochemical functioning, their modes of reproduction, and their environment. Finally, the atmosphere attained a critical level of oxygen content that created many new niches for organisms—in the open sea and on the land. In these new habitats, through mutation, the forces of evolution and natural selection enabled an enormous diversity of organic function and form to generate. Invertebrates, amphibians, reptiles, mammals, and finally man all evolved in only the last 600 to 800 million years—a relatively short span of geologic history.

"Ours is a cultural species, and I believe that our unique attention to sexual choice has helped to mould it," says Dr. Bronowski, echoing his earlier suggestion that sex is the propeller of evolution. This concept was discussed by Socrates and Glaucon in *Plato's Republic* (Book 5). Their topic was animal husbandry, the selective breeding of animals.

"And do you breed from them all indifferently or do you take care to breed from the best only?"

"From the best."

"And do you take the oldest or the youngest, or only those of ripe age?"

"I choose only those of ripe age."

"And if care was not taken in breeding, your dogs and birds would greatly deteriorate?"

"Certainly."

"And the same of horses and animals in general?"

"Undoubtedly."

"Good heavens! My dear friend," I [Socrates] said, "what consummate skill will our rulers need if the same principle holds of the human species."

"Certainly, the same principle holds. . . ."

Neither was the importance of sexual selection lost on Charles Darwin:

This [sexual selection] depends on the advantage which certain individuals have over others of the same sex and species solely in respect of reproduction. When . . . the two sexes differ in structure in relation to different habits of life, they have no doubt been modified through natural selection, and by inheritance limited to one and the same sex. So again the primary sexual organs, and those for nourishing or protecting the young, come under the same influence; for those individuals which generated or nourished their offspring best would leave, coeteris paribus, *the greatest number to inherit their superiority, which those which generated or nourished their offspring badly would leave but few to inherit their weaker powers.* (The Descent of Man)

Darwin, like Dr. Bronowski, saw evidences of the same selection process among male and female human beings: the genetic and biological basis for a good match. He concludes:

Man scans with scrupulous care the character and pedigree of his horses, cattle, and dogs before he matches them; but when he comes to his own marriage he rarely, or never, takes any such care. He is impelled by nearly the same motives as the lower animals, when they are left to their own free choice, though he is in so far superior to them that he highly values mental charms and virtues. On the other hand he is strongly attracted by mere wealth or rank. Yet he might by selection do something not only for the bodily constitution and frame of his offspring, but for their intellectual and moral qualities. Both sexes ought to refrain from marriage if they are in any marked degree inferior in body or mind; but such hopes are Utopian and will never be even partially realised until the laws of inheritance are thoroughly known. Everyone does good service, who aids towards this end. When the principles of breeding and inheritance are better understood, we shall not hear ignorant members of our legislature rejecting with scorn a plan for ascertaining whether or not consanguineous marriages are injurious to man.

The advancement of the welfare of mankind is a most intricate problem: all ought to refrain from marriage who cannot avoid abject poverty for their children; for poverty is not only a great evil, but tends to its own increase by leading to recklessness in marriage. On the other hand, as Mr. Galton has remarked, if the prudent avoid marriage, whilst the reckless marry, the inferior members tend to supplant the better members of society. Man, like every other animal, has no doubt advanced to his present high condition through a struggle for existence consequent on his rapid multiplication; and if he is to advance still higher, it is to be feared that he must remain subject to a severe struggle. Otherwise he would sink into indolence, and the more gifted men would not be more successful in the battle of life than the less gifted. Hence our natural rate of increase, though leading to many and obvious evils, must not be greatly diminished by any means. There should be open competition for all men; and the most able should not be prevented by laws or customs from succeeding best and rearing the largest number of offspring. Important as the struggle for existence has been and even still is, yet as far as the highest part of man's nature is concerned there are other agencies more important. For the moral qualities are advanced, either directly or indirectly, much more through the effects of habit, the reasoning powers, instruction, religion, etc., than through natural selection; though to this latter agency may be safely attributed the social instincts, which afforded the basis for the development of the moral sense. (The Descent of Man)

13

The Long Childhood

We are the music-makers,
And we are the dreamers of dreams,
Wandering by the lone sea-breakers,
And sitting by desolate streams;
World-losers and world-forsakers,
On whom the pale moon gleams;
Yet we are the movers and shakers
Of the world for ever, it seems.

Arthur O'Shaugnessy

In the ascent of man, justice, forethought, and creativity have become "part of the biological equipment of man," Dr. Bronowski points out. The remarkable ability of the brain to channel man's behavior in so many different directions sets him apart from all of his evolutionary ancestors, perhaps from everything the other side of *Australopithecus*. Man has become a "social solitary" who charts his own course, free from many of the chemical, physical, and biological constraints that have kept other animals where they are, having essentially the same evolutionary future as their evolutionary past. The human brain has become a selective advantage, as Harry J. Jerison observes in the following reading.

Language and Brain Deficiency

H.J. Jerison

The evolutionary viewpoint leads us to the position that the capacity for language evolved with the evolution of the human brain. There are, of course, difficulties in defining language precisely, and one may demonstrate symbolic, languagelike behavior in various nonhuman species. It is, nevertheless, obvious that the use of language is an especially and perhaps, peculiarly, human adaptation and that it is only with difficulty that analogies or possible identities with language may be demonstrated in the behavior of other species.

An issue has occasionally been raised about when in human evolution and the evolution of brain size it became possible for hominids to use language. This is an unanswerable question. For one thing, language in living men is learned, and the evolutionary heritage refers only to the capacity for learning language. There is no reason to correlate that capacity with a particular amount of hominid brain, except in instances of extreme pathology. If it is analogous to other behavioral capacities in other species, language should be identified with a particular type of brain. Thus, a microcephalic living human would not be expected to be completely incapable of the use of language (although in some cases such individuals may not have demonstrated the capability) any more than a decerebrate bird would be expected to be incapable of flight.

A useful way to think of language in evolutionary terms is as a species-specific adaptation, such as the flight mechanism of the hummingbird or stalking behavior in cats. We know that mammals may suffer brain damage in the laboratory or in nature, yet we would scarcely expect them to revert to reptilian or piscine adaptive mechanisms. In the same sense, we should not be surprised when human beings with pathologically small brains retain the capacity to learn speech. The use of language by deficient humans may be unusual, but it must be gauged relative to the normal human use of language, not to the studies on "language" in other species in which ingenious experimental manipulations are used to force us to sharpen our definitions of the nature of language.

Were it not that so spectacular a human capacity is involved, we could omit this discussion. Yet language is what seems to make us most obviously different from other animals, and many of us would go to some length to protect our identities and emphasize the difference. This is a valid aim for the evolutionist because language as we know it is certainly a characteristic that could be used to distinguish man from other species in the taxonomic as well as theological sense. We should conclude that deficiencies in human development will not produce nonhuman individuals, merely deficient human beings.

Plasticity and the Capacity for Culture

Plasticity, or modifiability, is one of the fundamental capacities of nervous systems. In the construction of a complex "real" world, in which each sense modality contributes to the cognitive image, it would be difficult to imagine a prewired central nervous system that was prepared, ready-made, for all the capacities. Among the consequences of the elaboration of functionally modifiable nervous systems, one must include the capacity for acculturation, the development of adult–child relations that lead to social and emotional dependence, the use of artifacts, communication by languages, and so forth. Each of these (or other categories, such as play, duration of infancy, love) can be used as a point of departure for speculations about the sources of hominization. But in every instance, the question of the evolution of the brain has to be tied to the work of the brain.

The simplest intuitive description of the brain's work (for me) is that it creates a "real" world. Within that real world all the events of a lifetime take place. (The realities differ, of course, depending on the nervous systems that do the work of building a real world and on the experiences—the information— to which the nervous systems are exposed during the lifetimes of the individuals.) The perception of others in social roles can often be handled at the level of organization of "fixed action patterns," as discussed in Lorenz's famous analysis of the companion in the bird's world. The perception of the self, on the other hand, may be a peculiarly human development of the capacity for creating "objects" in a "real" world. The reality of the self as an object (a person) is one of the most compelling of intuitions. Our capacity for imagery and imagination, though still poorly understood, may be related to our models of our "selves" and is clearly a kind of information processing that goes beyond simple model building based on the accommodation to stimulation through a variety of sense modalities.

Individuals capable of constructing elaborate multisensory "real" worlds might construct a reality that seems more fundamental than the immediate information from the senses. The capacity for imagery, in which one manipulates a possible real world in one's imagination, must early have led the hominids, by the time these capacities were well developed, to reach an appreciation of a past prior to one's lifetime and a sense of a future after one's death.

These have become singularly human hallmarks, understandable as cognitive structures with temporal as well as spatial extent and with some symbolic superstructures such as self-awareness or consciousness added to them. We can carry these speculations no further if we limit ourselves to the organic evolution of the brain. But it is well for us to recognize that when it became selectively advantageous to construct a reality beyond any immediate sense modality, the path to hominization could be followed. The capacity for abstraction is inherent in the process of construction of such a reality as a synthesis of information from many sources. The development of language in the process was, for this perspective at least, a reasonable extension of a conscious perceptual world, given the existence of visual and nonvisual imagery and auditory and vocal capacities. We model "reality" linguistically just as we model it visually or tactilely. Languages may be media for communication because different human brains construct essentially the same "reality." And we can share that reality because it is already shared in the linguistic structures we have in common and in the linguistic factors that are part of our fundamental image of the real world. Speech and reading, in this sense, provide a genuine shared consciousness for members of the human species, given a common cultural background.

In all the animal world, no animal other than man can claim the gift of speech. The simple reason behind that awesome fact is that the machinery of the brain of man is wired and engineered in such a way that the processes of the brain can be converted into words and understood by other human beings. Wilfred Penfield from the Montreal Neurological Institute literally pinpointed the speech area in the left-hand side of the brain. Penfield's findings are discussed here in a selection from Dean E. Wooldridge's *The Machinery of the Brain.*

The Speech Center

Dean E. Wooldridge

This important work started thirty years ago, when Penfield and his associates embarked on the treatment of focal epilepsy by radical surgical excision of abnormal areas of brain. In those days, based upon the earlier discoveries of Broca and others, it was the general belief that the portion of the brain controlling speech was located along the side of the dominant hemisphere—the left side of the brain for right-handed persons, the right side for left-handed persons. Because of the extensive involvement of speech mechanisms with higher intellectual activities, it was the practice at that time to refuse radical operation upon the dominant hemisphere unless the tumor or other brain damage was located far forward in the frontal lobe or far to the rear in the occipital lobe.

But, by accepting such a large portion of the brain as "forbidden territory," the surgeons denied the benefits of their techniques to many patients with damage or tumors that produced chronic epileptic symptoms and, not infrequently, by spreading led to progressive degeneration and death. In the meantime, increasing numbers of patients were observed with focal epilepsy resulting from lesions and small tumors that happened to fall within the general area that earlier work had indicated was vital to the control of speech, and many of these patients did not have speech difficulties. This led Penfield and his associates to suspect that not all the territory that had been proscribed by earlier research workers should really be considered to be forbidden, and that a more precise delineation of the speech areas might lead to the possibility of extending the region within which operations could be performed for the removal of defective tissue and the relief of epileptic symptoms.

Mapping the Speech Areas

The tool chosen by Penfield for the more accurate mapping of the cortical speech areas was the modern technique of electric stimulation. Penfield's reasoning was about as follows: Insertion of an electrode and injection of a train of current pulses into a portion of the cortex that is responsible either for the receipt of sensory information or for the transmission of motor commands typically produces behavioral symptoms that permit the normal function of that portion of cortex to be identified. If the spot chosen for electric stimulation is in the visual cortex, the injected current in effect preempts a portion of the subject's field of view and fills it with flashes of light. Stimulation of the auditory cortex similarly blocks the normal receiving channels from the ears and produces an uncoordinated noise or clashing sensation in the brain; stimulation of a portion of the postcentral sensory cortex may produce a sensation of numbness in the right thumb, which at the same time is precluded from performing its normal function of touch discrimination. Penfield reasoned that something similar should happen if he were to stimulate the part of the cortex involved in the control of speech. Because of the complexity of the speech processes, he did not expect to be able to cause the subject to produce specific words or phrases, but he thought that the other, negative result commonly produced by stimulation might reasonably be expected to occur—the incapacitation of the particular aspect of the speech process controlled by the stimulated portion of cortex. By performing the brain operation under local anesthesia, with the patient fully conscious and cooperating, he hoped that the specific areas of the cortex where electric stimulation would interfere with speech could be accurately delineated.

In such a description as this one, in which the development of a new technique for obtaining improved understanding of brain mechanisms is the subject of immediate interest, there is danger of conveying an improper impression of the approach of such workers as Penfield and his associates to their patients. These men are physicians, not experimenters. Their objective in each operation is to apply their skill and knowledge in such a way as to maximize the probability of improving the condition of the one patient they are working on. When they insert probes in his exposed cortex, apply electric stimulation and record behavioral effects, it is always done for the purpose of localizing his specific abnormality, so that subsequent surgery may be most effective. For this reason, little can be done with one patient that has not been previously successfully checked out on others. Nevertheless, individual patients are different enough, both in inherent make-up and with respect to their lesions, that fresh application of a well-proved electrical exploration technique frequently leads to information about a previously undetermined aspect of brain function. When the surgeon is also a scientist, and when throughout the years he performs operations to relieve the suffering of many hundreds of patients, much new information can be secured. It was in this way that the Montreal Neurological Institute group slowly and cautiously developed what turned out to be a powerful new technique. For Penfield's method worked. Specific regions of the cortex were indeed found where application of the stimulus while the patient was talking would produce prompt and dramatic disturbance in his speech. The accurate mapping of the regions of the cortex within which electric stimulation produces such disturbances, and determination of the nature of the disturbances themselves, have provided important clues as to how the brain controls the processes of speech.

Nature of Electrically Induced Speech Disturbances

Penfield and his associates knew from the beginning that there was one kind of speech disturbance that they would certainly be able to produce and that would be of little interest to them. This was interference by incapacitation of the part of the brain that controls the muscles used in speech. Such a result, it was known, could easily be obtained by stimulation through an electrode inserted in the part of the precentral motor strip employed in the control of the muscles of lips, tongue, and throat. Such stimulation would indeed interfere with, or entirely prevent, talking by the patient. But it was purely a muscular effect; it would frequently be accompanied by uncontrolled twitching of the muscles and, upon termination of the stimulation, the patient would confirm that the only interference with speaking that he had been aware of was the inability to control his muscles. This kind of interference was observed only as a result of stimulation in the well-established cortical motor regions. It was also only in such regions that occasionally stimulation produced positive effects. With the electrode in certain positions the injection of current would cause the subject to "vocalize," that is, to emit long-sustained vowel sounds. With such understandable exceptions, the effect of electrode stimulation was always negative; it interfered with the processes of speech, just as Penfield's preliminary working hypothesis had contemplated.

But electric interference with the thought processes underlying speech was Penfield's main interest. The inability of a patient to speak properly, even though he possesses adequate muscular control, is called *aphasia* by the workers in the field. And electrically induced aphasia was vividly demonstrated in many of Penfield's tests. On being asked to name the object in a picture while the speech area of his cortex was being stimulated, one patient said: "Oh, I know what it is. That is what you put in your shoes." After withdrawal of the electrode, he said, "Foot." A little later he was unable to name the picture of a tree, although he knew what it was, naming it properly as soon as the stimulus was turned off. Another patient under electric stimulation could not think of the word for a comb, although when asked its use he said, "I comb my hair." When asked again to name it, he couldn't until the electrode was removed.

From descriptions by the patients of their sensations when these tests are under way, it is clear that what is involved is interference with only a part of the speech mechanism. The patient recognizes the object he is trying to name, can speak effectively in most respects, but for some reason no longer has access to the common words that ordinarily would come to him easily. He frequently tries to think of synonyms when he finds he cannot remember the word itself. On one occasion, the patient struggled unsuccessfully to answer a question during the electric discharge; afterward he said, "I couldn't get that word 'butterfly' and then I tried to get the word 'moth'." The test is evidently a puzzling and frustrating experience for the subject.

The blocking of access by the patient to the names of objects is typical of the effects produced by electric stimulation in suitable cortical zones, but it is by no means the only type of aphasia that can be induced. Hesitation, slurring, and distortion frequently occur. These effects can be produced by stimulation in the motor cortex, but they can also result from interference with the ideational speech processes as a result of stimulation in other areas of the brain. Sometimes repetition of words and syllables is caused by the electric stimulus. Confusion of numbers is frequently observed when the current is turned on while the patient is counting. In such circumstances the patient may jump from "six" to "twenty" and then back to "nine." Interestingly, the confusion is apparently of a limited nature, for nonnumerical words are not used; the patient displays the proper psychological "set" but is simply unable to give the correct numbers. In another kind of misnaming the patient may use words closely related in sound such as "camel" for "comb"; or he may use synonyms such as "cutters" for "scissors," "hay" for "bed," and "moth" for "butterfly." Sometimes an entirely unrelated word is used, such as "rink" for "scissors," or "comb" for "hammer."

While all of these results were obtained by the artificial injection of electric current into the exposed cortex of the patient, the behavioral effects were quite similar to those produced in the same and other patients when the stimulating current was injected into the speech cortex as a result of spontaneous epileptic attacks. Lesions in the cortical speech areas also were found to produce similar speech defects, although there was a tendency for a wider variety of symptoms to appear from natural causes than from artificial stimulation. As we shall see shortly, one of the cortical areas important for speech extends back from the temporal lobe almost to the associative area of the visual cortex in the occipital lobe. Lesions occurring in this region, close to the visual cortex, were sometimes found to produce much more severe disturbance of writing than of speech. Other cases were observed in which the patient could express himself adequately but could not understand what was said to him. Interesting though these special cases may be, they appear to be relatively rare. Until better understanding exists of the basic nature of aphasia, it is not likely that attempts to interpret these special situations will be fruitful. In most cases of aphasia, whether produced naturally or artificially, the phenomena are more general; they suggest a breakdown of the mechanism that we ordinarily use in going from a concept to a word. . . .

Penfield's results . . . show that all the speech areas ordinarily are situated in the left hemisphere. He did not confirm the earlier belief that the dominant hemisphere in terms of left- or right-handedness, controls speech. Some cases of speech control by the right hemisphere were observed, but these were rare and always appeared to be the consequence of special circumstancces (to be discussed below). In the normal brain, the left hemisphere was found to control speech.

The existence of so nearly a standardized assignment of cortical tissue for the control of speech processes is rather remarkable. Speech is clearly an acquired function, with no hereditary overtones. The fact that the process of learning to speak, which must be a new and special experience for each individual, should automatically result in the involvement of a certain standardized portion of the cortex is probably replete with implications as to the detailed construction and operation of the brain. It is to be hoped that one day our knowledge will permit us to form the proper deductions from this significant evidence of speech localization.

Throughout *The Ascent of Man,* Dr. Bronowski stressed that man's unique faculties, especially the brain, have enabled him to create an aggregate of human values that impinge on every phase of his culture. At the same time, as man learns more precisely how the brain functions, the possibilities for altering or controlling its operation point to the need for a new set of values. This need is discussed in considerable detail by Roger W. Sperry.

Mind, Brain, and Humanistic Values

Roger W. Sperry

Any mixing of values and science is a red flag in some quarters. Value judgments, we are told, lie outside the realm of science. Values are for popes and prophets, for philosophers and perhaps scout leaders and civic planners, but not for science or scientists. As a student of brain and behavior, I have never quite been able to accept this. It seems like saying that value judgments lie outside the realm of knowledge and understanding, or that the best method of applying the human brain to problems of understanding must be discarded when it comes to values. It is like saying that science is able to deal only with those phenomena that appeared prior to the emergence of higher brains, with their wants, needs, and other goal-directed properties and, of course, the corresponding value systems that these latter impose.

Values have natural and logical origins. They are interdependent and interrelated in logical, hierarchical systems. These value systems and their

From Roger Sperry, "Mind, Brain, and Humanistic Values," *Bulletin of the Atomic Scientists*, September 1966. Reprinted by permission of the Bulletin of the Atomic Scientists.

perturbations ought to be subject to study, analysis, and prediction—and perhaps even some experimentation on a model basis these days, with computer assistance.

Humanist Impacts of Behavioral Science

We turn to our main topic now to consider some of the major impacts on human values that stem from recent developments in the sciences that deal with mind and brain. At first glance the record achieved by the brain-behavior sciences during the past half century must appear, to the humanist, to read less like a list of contributions and advancements than like a list of moral offenses and major setbacks. The accusations that anti-science can raise in this area are not exactly trivial. For example, prior to science, man had reason to believe that he possessed a mind that was potent and replete with something called consciousness. Modern experimental objective psychology and the neurosciences in general would dispense not only with the conscious mind but also with most other spiritual components in human nature. Before science man used to think himself a free agent possessing free will. Science gives us, instead, causal determinism wherein our every act is seen to follow inevitably from preceding patterns of brain excitation. Where we used to see purpose and meaning in human behavior, science now shows us a complex biophysical machine composed entirely of material elements, all of which obey inexorably the universal laws of physics and chemistry.

Science, abetted by Freud with an assist from astro-physics, stands accused of depriving the thinking man of his Father in heaven, along with heaven itself. Man's inner nature and his heritage also seem to have fared poorly. Since Darwin, and again Freud, man now enters this life, not "trailing clouds of glory" and divinity but clouds of jungleism and bestiality with a predisposition to oedipal and other complexes. The veneer of civilization is seen to be superficial, and when it rubs thin or cracks the basic animal within quickly shows.

In the face of these and related onslaughts of science on the worth and meaning of human nature and existence, one can understand why humanist thinkers look for other roads to truth. For the scientist himself, the current dim picture puts a rather severe test to his credo that it is better to know and live by the truth, however ugly, than by false premises and illusory values.

I find that my own conceptual working model of the brain leads to inferences that are in direct disagreement with many of the foregoing; especially I must take issue with that whole general materialistic-reductionistic conception of human nature and mind that seems to emerge from the currently prevailing objective analytic approach in the brain-behavior sciences. When we are led to favor the implications of modern materialism in opposition to older, more idealistic values in these and related matters, I suspect that science may have sold society and itself a somewhat questionable bill of goods.

The Nature of Consciousness: The Central Issue

Most of the disagreements I refer to relate to a central point of controversy that comes out of the following question: Is it possible, in theory or in principle, to construct a complete, objective explanatory model of brain function without including consciousness and mental phenomena in the causal sequence?

If the prevailing view in neuroscience is correct—that consciousness and mental forces in general must be excluded from any objective model—then we write off all that inner subjective world from science and come out with materialism and all its implications. Conversely, if it should prove true, when the facts are in, that conscious mental forces do in fact govern and direct the nerve-impulse traffic and other biochemical and biophysical events in the brain and, hence, do have to be included as important features in the objective chain of control, we then come out on the side of mentalism, and with quite a different and more idealistic set of values.

Some 99.9 per cent of those of us who work in brain research have held firmly during the past fifty years to the conviction that conscious mental forces have no place in any explanatory model or theory of brain function. The inner sensations, feelings, percepts, concepts, mental images, and the like cannot be weighed or measured, photographed, spectrographed, or chromatographed, or otherwise recorded or dealt with objectively by any known scientific methodology. The dictates of the scientific method, requiring experimental demonstration and proof, demand that these introspective, private, inaccessible, will-o'-the-wisp unknowns must be excluded from any scientific explanation. Furthermore, the neuroscientist of today feels he has a pretty fair idea anyway about the kinds of things that excite and fire the cellular elements of the brain: Membrane changes, ion flow, chemical transmitters, pre-synaptic and post-synaptic potentials, and the like may be on his list of acceptable causal influences, but not consciousness.

Science can see the brain as a complex electrochemical communications network full of nerve excitations, all governed by respectable scientific laws of biophysics, biochemistry, and physiology; but few investigators, and none that I know, have been ready to tolerate an interjection into this causal machinery of any mental or conscious forces. This then in brief is the general stance of modern science out of which has come today's prevailing objective, mechanistic, materialistic, behavioristic, reductionistic, fatalistic view of the nature of mind and psyche. This kind of thinking is not confined to our laboratories and classrooms, of course. It leaks and spreads, and though never officially imposed on the societies of our Western world, we nevertheless see, on all sides, the pervasive influence of creeping materialism.

An Alternative Mentalist Position

I am going to align myself in a counterstand, along with that approximately 0.1 per cent mentalist minority, in support of a hypothetical brain model in which consciousness and mental forces generally are given their due representation as important features in the chain of control. These appear as active operational forces and dynamic properties that interact with and upon the physiological machinery. Any model or description, that leaves out conscious forces, according to this view, is bound to be pretty sadly incomplete and unsatisfactory. The conscious mind in this scheme, far from being put aside and dispensed with as an "inconsequential by-product," "epiphenomenon," or "inner aspect," as is the customary treatment these days, gets located, instead, front and center, directly in the midst of the causal interplay of cerebral mechanisms.

Mental forces in this particular scheme are put in the driver's seat, as it were. They give the orders and they push and haul around the physiology and physicochemical processes as much as or more than the latter control them. This is a scheme that puts mind back in its old post, over matter, in a sense—not under, outside, or beside it. It's a scheme that idealizes ideas and ideals over physicochemical interactions, nerve impulse traffic—or DNA. It's a brain model in which conscious, mental, psychic forces are recognized to be the crowning achievement of some five hundred million years or more of evolution.

Let us now examine more closely this seemingly ridiculous notion, this "water-on-the-brain" contention that ideas and other mental entities push around, control, and direct the biophysical and biochemical events in the nervous system. The basic reasoning is simple: First, we contend that conscious or mental phenomena are dynamic, emergent, pattern (or configurational) properties of the living brain in action—a point accepted by many, including some of the more tough-minded brain researchers. Second, the argument goes a critical step further, and insists that these emergent pattern properties in the brain have causal control potency—just as they do elsewhere in the universe. And there we have the answer to the age-old enigma of consciousness.

To put it very simply, it becomes a question largely of who pushes whom around in the population of causal forces that occupy the cranium. There exists within the human cranium a whole world of diverse causal forces; what is more, there are forces within forces within forces, as in no other cubic half-foot of universe that we know. At the lowermost levels in this system are those local aggregates of subnuclear particles confined within the neutrons and protons of their respective atomic nuclei. These individuals, of course, don't have very much to say about what goes on in the affairs of the brain. Like the atomic nucleus and its associated electrons, the subnuclear and other atomic elements are "moleculebound" for the most part, and get hauled and pushed around by the larger spatial and configurational forces of the whole molecule.

Similarly the molecular elements in the brain are themselves pretty well bound up, moved, and ordered about by the enveloping properties of the cells within which they are located. Along with their internal atomic and subnuclear parts, the brain molecules are obliged to submit to a course of activity in time and space that is determined very largely by the overall dynamic and spatial properties of the whole brain cell as an entity. Even the brain cells, however, with their long fibers and impulse conducting elements, do not have very much to say either about when or in what time pattern, for example, they are going to fire their messages. The firing orders come from a higher command.

The Mental Entities

The flow and the timing of impulse traffic through any cell, or nucleus of cells, in the brain is governed very largely by the overall encompassing properties of the whole cerebral circuit system, and also by the relationship of this system to other circuit systems. Even the circuit properties of the cerebral system as a whole, and the way in which these govern the flow pattern of impulse traffic throughout—that is, the circuit properties of the whole brain—may undergo radical and widespread changes with just the flick of a cerebral facilitatory "set." This set is a shifting pattern of central excitation that will open or prime one group of circuit pathways while at the same time closing, repressing, or inhibiting endless other circuit potentialities. Such changes of set are involved in a "shift of attention," "a turn of thought," "a change of feeling," or "a new insight," etc. In short, if one climbs upward through the chain of command within the brain, one finds at the very top those overall organizational forces and dynamic properties of the large patterns of cerebral excitation that constitute the mental or psychic phenomena.

Let us now illustrate one of these "power-packed" mental entities. For simplicity, consider an elemental subjective sensation—and for reasons that will become evident let us use the sensation of pain instead of philosophy's old favorite, the color red. More specifically, make it pain in the wrist and fingers of the left hand of an arm that was amputated above the elbow some months previously. Suffering caused by pain localized in a phantom limb is no easier to bear than if the limb were still there. It is easier, however, with this example, to infer where our conscious awareness must reside.

With regard to this conscious sensation of pain, the contention is that any groans it may evoke—and any other response measures the patient may take as a result of the pain sensation—are indeed caused, not by the biophysics, chemistry, or physiology of the cerebral nerve impulses as such, but by the pain quality, the pain property, per se. This brings us to the real crux of the argument. Nerve excitations are just as common to pleasure, of course, as to pain, or any other sensation. What is critical is the unique patterning of cerebral excitation that produces pain instead of something else. It is the overall functional property of this pain pattern that is critical in the causal sequence of brain affairs. This pattern has a dynamic entity, the qualitative effect of which must be conceived functionally and operationally, and in terms of its impact on a living, unanesthetized cerebral system. This overall pattern effect in brain dynamics is the pain quality of inner experience.

Above simple pain and other elemental sensations in brain dynamics, we find, of course, the more complex but equally potent forces of perception, emotion, reason, belief, insight, judgment, and cognition. In the onward flow of conscious brain states, one state calling up the next, these are the kinds of dynamic entities that call the plays. It is exactly these encompassing mental forces that direct and govern the inner flow patterns of impulse traffic, including their physiological, electro-chemical, atomic, subatomic, and subnuclear details. It is important to remember in this connection that all of the simpler, more primitive, elemental forces remain present and operative; none has been cancelled. These lower-level forces and properties, however, have been superseded in successive steps, encompassed or enveloped as it were, by those forces of increasingly complex organizational entities. For the transmission of nerve impulses, all of the usual electrical, chemical, and physiological laws apply, of course, at the level of cell, fiber, and synoptic junction. Proper function in the uppermost levels depends to a large extent upon normal operation at the subsidiary levels. It is a special characteristic of these larger functional patterns in the brain, however, that they have a coherence and organization that enables them to carry on orderly function in the presence of considerable disruptive damage in the lower-level components.

Ideas as Causal Forces

Near the apex of this compound command system in the brain we find ideas. In the brain model proposed here the causal potency of an idea or an ideal becomes just as real as that of a molecule, a cell, or a nerve impulse. Ideas cause ideas and help evolve new ideas. They interact with each other and with other mental forces in the same brain, in neighboring brains, and in distinct, foreign brains. And they also interact with real consequence upon the external surroundings to produce in toto an explosive advance in evolution on this globe far beyond anything known before, including the emergence of the living cell.

Problems of complexity and adequate technology aside, there would seem to be no great obstacle to the eventual objective, scientific treatment of mental phenomena. Statements in the literature discourage the hope that the mind is capable of explaining itself in terms of its own ideas. The argument is that no machine, living or otherwise, can logically embody within itself a complete description of itself. But underline that word "complete," and then consider the extent of the explanatory possibilities that still remain. Also underline "itself" and remember that this logic does not prevent a man's mind from acquiring a complete description of his neighbor's mind nor from passing on this description to other neighbors, excepting only the one being described.

Looking back at this point, one may see that the earlier dichotomy between mentalism and materialism is resolved in this interpretation. The former polar differences with respect to human values, when recast in the present scheme, become mainly errors of reductionism. For a theory of mind the new twist here, if any, lies in the attempt to make the emergent properties of inner experience conform to the brain code rather than isomorphs of the outside world—combined, of course, with the critical interjection of these mental qualities into the causal sequence.

The present scheme would put the conscious mind back into the brain of objective science and in a position of top command. It would eliminate the old dualistic confusions, dichotomies, and paradoxes, proposing instead a single unified system extending from subnuclear forces near the bottom up through ideas at the top. It would provide a long-sought unifying view on which to base our conception of human nature. Moreover it suggests a possible answer not only for the relation between mind and brain but also for that between the outside world and its inner cerebral representation. As for the older materialist-behaviorist movement, it may be said in retrospect that the denial or downgrading of conscious mental forces in objective experimental psychology during the past half-century has had value as a tactical expedient for a developing science and remains appropriate for much analytic research aimed at lower levels of brain function. It is hardly something, however, on which to build societal philosophy and cultural values.

Free Will

Another serious threat to cherished images of human nature is the scientific rejection of free will. Every advance in the science of behavior, whether it comes from implanted electrodes, psychomimetic drugs, the psychiatrist's couch, brain surgery, imprinting, or Skinner boxes, seems only to reinforce the old suspicion that free will is merely an illusion. The more we learn about the brain and behavior, the more deterministic, lawful, and causal it appears. Like most others in brain research, I assume that every apparently free mental choice must in fact have been causally predetermined in the preceding brain states and related events. This means that any decision any of us has ever made could not possibly have had any other outcome in the given situation. Attempts to restore free will to the human brain by recourse to various forms of indeterminacy—physical, logical, emergent, or others—have failed, so far as I can see, to do much more than perhaps introduce a bit of unpredictable caprice into our comportment that most of us would prefer to be without. Neither science nor philosophy seems able as yet to find in the brain any satisfying exceptions to the classical onward flow of causal determinism.

Before we become overly disturbed by all this, however, there are a few more points we should keep in mind. These add up to the conclusion that if we were given freedom of choice in this whole matter, we might well prefer not to have it; that is, we would probably prefer to leave determinism in control exactly as science postulates. It should be clear that the kind of determinism proposed is not that of the atomic, molecular, or cellular level, but rather the kind that prevails at the level of cerebral mentation, involving the interplay of ideas, reasoning processes, judgment, emotion, insight, and so forth.

The proposed brain model provides in large measure the mental forces and abilities to determine one's own actions. It provides a high degree of freedom from outside forces as well as mastery over the inner molecular and atomic forces. In other words it provides plenty of free will provided we think of free will as self-determination. A person does indeed determine with his own mind what he is going to do and often from among a large series of alternative possibilities.

This does not mean, however, that there are cerebral operations that occur without antecedent cause. Man is not free from the higher forces in his own decision-making machinery. In particular, our model does not free a person from the combined effects of his own thought, his own impulses, his own reasoning, feeling, beliefs, ideals, and hopes, nor does it free him from his inherited makeup or his lifetime memories. All these and more, including unconscious desires, exert their due causal influence upon any mental decision, and the combined resultant determines an inevitable but nevertheless self-determined, highly special, and highly personal outcome. Thus the question: Do we really want free will, in the indeterministic sense, if it means gaining freedom from our own minds?

There may be worse fates, perhaps, than causal determinism. Maybe after all it is better to be an integral part of the causal flow of cosmic forces than to be out of contact with these—free-floating, as it were, with behavioral possibilities that have no antecedent cause, and hence no reason nor any reliability relative to future plans, predictions, or promises. If one were assigned the task of trying to design and build the perfect free-will model, consider the possibility that the aim might be not so much to free the machinery from causal contact as the opposite, that is, to try to incorporate into the model the potential value of universal causal contact. In other words, contact with all related information in proper proportion—past, present, and future.

At any rate it is clear that the human brain has come a long way in evolution in exactly this direction, when you consider the amount and the kind of causal factors that this multidimensional, intracranial vortex draws into itself, scans, and brings to bear in turning out one of its preordained decisions; potentially included, through memory, are the events and wisdom of most of a human lifetime. Potentially included, also, with a visit to the library, is the accumulated knowledge of all recorded history. And we can add, thanks to reason and logic, much of the forecast and predictive value extractable from all these data as well as creative insights newly conceived. Maybe the total falls a bit short of universal causal contact; maybe it is not even up to the kind of thing evolution has going for it over on galaxy nine; and maybe, in spite of all, any decision that comes out is still predetermined. Nevertheless it certainly represents a very long jump in the direction of freedom from the primeval slime mold, the Pleistocene sand dollar, or even the latest model orangutan.

It will be evident that our current view does not deny the animalistic in human nature—any more than it denies the molecular or atomistic. It does deny, however, that the higher human properties in the mind and nature of man are the same as, or are reducible to, the components from which they are fashioned. On the debit side, there is little in our proposed model for consciousness to bolster one's hopes either for extrasensory perception or for postmortem perception. Similarly prepartum perception in the embryo would presumably be negligible until after the requisite cerebral machinery for conscious awareness begins to attain functional maturity in the later months of fetal life, and in subsequent postnatal development.

Plasticity of Human Nature and Inheritance of Behavior Traits

Finally, in connection with development, I should mention briefly certain other advances in the brain-behavior sciences that have resulted in important revisions in our general conception of human nature. These concern the extent to which behavior traits can be inherited and the extent to which human nature is plastic and subject to shaping by experience and environment.

The objective, materialist movement in psychology, established first in the Soviet Union partly under Pavlov, and pioneered in this country by Watson as behaviorism, has been identified almost as much with the promotion of the conditioned response as it has with the demotion of consciousness. Personality and behavior generally were attributed to a life-long chain of conditioned reflex associations. The whole idea of the genetic inheritance of behavior patterns was renounced until the term "instinct" became highly discredited in behavioral science. The embryonic growth of brain pathways was believed to be by nature nonselective and diffuse. Specific anatomical hook-ups in the brain were held to be unimportant anyway for orderly function, and subject to radical, wholesale disarrangements by surgery, injury, and regeneration without causing much functional disturbance. In the scientific thinking of those times the brain was endowed with an almost mysteriously omnipotent plasticity and readaptation capacity. In general, science seemed to be telling us as late as the early forties that the human brain and human nature in general are extreme in their malleability. It seemed at that time a scientifically sound conclusion that it would be possible, through an appropriate program of training and environmental conditioning, to shape human nature and hence society within wide limits into a desired mold.

Much of the basic scientific thinking and evidence behind these views has since suffered a series of severe upsets leading to current stands on these same matters that are almost diametrically opposed to the earlier doctrines. Instead of a loose, universal plasticity in brain hook-ups we see today a basic built-in wiring diagram that is characteristic of the species and functionally rather rigid. Instead of diffuse nonselective growth of nerve connections in brain development, we see now a very precise and highly ordered patterning of brain fiber pathways and connections—all strictly preregulated through cytochemical affinities under genetic control. Where there used to be an outright denunciation of the whole instinct concept, we now accept the idea that an entire evolutionary tree can be set up on the basis of inherited behavior patterns. The conditioned response, along with other forms of learning, continues to be recognized as a highly powerful modeling influence, and especially so in man, but only within limits that are much narrower than previously believed.

Within the specialized fields of scientific inquiry involved here the pendulum of opinion continues to swing, at this date, in the direction of inheritance. How far it will go can only be guessed. It is still too soon for the implications of these changes to have fully permeated even the neighboring scientific disciplines. In any case, to return to our central theme here: It would seem that the evidence available today demands that we renounce, along with other aspects of the behaviorist-materialist approach, the old "Pavlovian–Watsonian conditioned reflex theory of the psyche" with its radical environmentalism that used to tell us, literally, that 99 per cent of human nature and mind is a product of experience, environment, and training.

Sperry's call to reevaluate human values in brain research has not gone completely unheeded. Scientists are aware of the diabolical directions that such research can take, and they seem to be able to weigh the values involved. "The ascent of man," Dr. Bronowski proposes "is always teetering in the balance." Just how seriously brain researchers view the problem is evident in a selection from a symposium on physical manipulation of the brain held at the Hastings Center in 1973. Participants were José M.R. Delgado, Rudolph Ehrensing, Willard Gaylin, Perry London, and Ann Orlov.

Physical Manipulation of the Brain

DELGADO: There are two ways to control behavior: One is through the environment, the second (our subject) is through the brain. And in the brain we have mainly three types of controls: (1) mechanical, through surgery, (2) electrical, through brain stimulation, and (3) chemical, by placing drugs inside the brain (I'm not talking about drugs taken by mouth, only direct application of chemicals in the brain). Then, there is a fourth type of control, the effect of the environment on the electrical and chemical activity of the brain. Through the environment, we *are* really modifying the structure and substance of the brain electrically and chemically. . . .

Experimental research of the brain may change the classical question of philosophy, which was, What is man? This is perhaps the residue of the Aristotelian conception of the soul. The question rather than, What is man?, should be, What kind of man are we going to construct? Through sensory inputs, and through brain manipulation, we are really constructing and modifying the materials that form the neurological basis of man. We have the power and the knowledge to understand how "man" is constructed inside the working brain, and to manipulate these building blocks that construct man. This conception may be dangerous because it changes the conceptions of freedom, identity, and integrity. This is an important question to which we should address ourselves.

LONDON: On the one hand, you're telling us about the immediate potential of the technology of brain implantation and about the philosophical implication—the practical philosophical implication—that we can define man as a constructible entity. On the other hand, this technology by itself doesn't do anything; it only does things in interaction with the environment. In your own writing you make a strong point about the need to reconstruct man by essentially *educational, informational* means. There's a great distance between being able to control behavior via brain stimulation, even with the most elegant internal computers, and restructuring society, its educational systems and so forth, by educating children. Where do you see the technology moving in the next 5, 10, 35 years? How do you see this technology progressing in such a way that it orients us toward that longer-range goal, and what are the dangers that you see in connection with that?

Excerpted from "Physical Manipulation of the Brain," *Hastings Center Report,* Special Supplement, May 1973. Institute of Society, Ethics and the Life Sciences, Hastings-on-Hudson, New York.

DELGADO: The main implication of brain research for education is that the research will make available unique information about the neurological mechanisms. Present technology may give us a new understanding about how education modifies children's brains. It is not possible, however, to educate children with electrodes.

This new knowledge may change the philosophy of our educational system, which is based on the sanctity of the individual, and on the assumption that an individual exists at the moment of birth. This belief may be questioned. We should also try to clarify the biological bases of the potentials of the individual and what effect they have on democratic beliefs. We are *forming* the individual; we are *constructing his* brain.

GAYLIN: Dr. Delgado, you said in effect that for the first time man has the capacity to enter into or to control the mind. I am not convinced that *communication* is less "entering into the mind" than insertion of an electrode. I'm not sure that we are more into the mind when we're into the brain with a cannula or an electrode than we are with an emotion.

DELGADO: You're absolutely right. Sensory inputs have intracerebral correlates that were unknown in the past but may be investigated in the present. In the past, the brain was considered an unreachable entity, but today we can look at the working neurons. We need to understand man from his cultural outside and also from the inside of his central nervous system. We need a convergence between both approaches.

ORLOV: Do you really believe that we should try to restructure society and reeducate man in this fashion? Save a lot of money by putting electrodes, at $1,000 each, into the brains of children?

DELGADO: Let me repeat what I've said. We cannot educate children with electrodes. The answer is, No. However, the information obtained about learning, memory, emotionality, etc., may be used to restructure a different kind of education, and therefore to create a different kind of child. I'm not satisfied with the way in which our present civilization is behaving. I do not believe that we must be slaves of natural chance. Up to now, man has been structured in a natural way by nature. And the results are that we are full of flaws. We should improve the society in which we live.

EHRENSING: I'm asking about your (Delgado's) concept of freedom. First, what happens to the concept of informed consent in the context of asking what kind of man *we* want to construct? And second, do you see any validity in or a future for the ideal of freedom?

DELGADO: Freedom is only a relative concept. Are you free to speak Chinese, or Russian? . . . Not unless experiential inputs modify our brain and give us that freedom.

EHRENSING: I'm defining freedom very broadly as self-determination.

DELGADO: No, this is the wrong notion. I would like to redefine freedom as the use of our intelligence within frames of reference given by culture and by experience. We are a conglomerate of cultural values, with the symbols determined by culture and experience. Therefore, we are limited by the building blocks at our disposal; we do not invent these building blocks, they are given from the outside and they change the structure of our brain. Education provides a frame of reference for each child.

If indeed the special refinements in the convoluted twists and turns of man's brain have contributed so strongly to man's ascent, an obvious question is: Is he still ascending, or has his brain taken him as far as he can go? Julian Huxley gives his answer to this question in this selection from *Evolution: the Modern Synthesis.*

Is Man Still Ascending?

Julian Huxley

Conscious and conceptual thought is the latest step in life's progress. It is, in the perspective of evolution, a very recent one, having been taken perhaps only one or two and certainly less than ten million years ago. Although already it has been the cause of many and radical changes, its main efforts are indubitably still to come. What will they be? Prophetic phantasy is a dangerous pastime for a scientist, and I do not propose to indulge it here. But at least we can exclude certain possibilities. Man, we can be certain, is not within any near future destined to break up into separate radiating lines. For the first time in evolution, a new major step in biological progress will produce but a single species. The genetic variety achieved elsewhere by radiating divergence will with us depend primarily upon crossing and recombination.

We can also set limits to the extension of his range. For the planet which he inhabits is limited, and adventures to other planets or other stars are possibilities for the remote future only. . . .

More basic, however, though much slower in operation, are changes in the genetic constitution of the species, and it is evident that the main part of any large genetic change in the biologically near future must then be sought in the improvement of the fundamental basis of human dominance—the feeling, thinking brain, and the most important aspect of such advance will be increased intelligence, which, as A. Huxley has stressed, implies greater disinterestedness and fuller control of emotional impulse. . . .

Meanwhile there are many obvious ways in which the brain's level of performance could be genetically raised—in acuteness of perception, memory, synthetic grasp and intuition, analytic capacity, mental energy, creative power, balance, and judgment. If for all these attributes of mind the average of our population could be raised to the level now attained by the best endowed ten-thousandth or even thousandth, that alone would be of far-reaching evolutionary significance. Nor is there any reason to suppose that such quantitative increase could not be pushed beyond its present upper limits. . . .

The problem immediately poses itself whether man's muscular power and urges to hunting prowess may not often be a handicap to his new modes of control over his environment, and whether some of his inherited impulses and his simpler irrational satisfactions may not stand in the way of higher values and fuller enjoyment. The poet spoke of letting ape and tiger die. To this pair, the cynic later added the donkey, as more pervasive and in the long run more dangerous. The evolutionary biologist is tempted to ask whether the aim should not be to let the mammal die within us, so as the more effectually to permit the man to live. . . .

The ordinary man, or at least the ordinary poet, philosopher, and theologian, is always asking himself what is the purpose of human life, and is anxious to discover some extraneous purpose to which he and humanity may conform. Some find such a purpose exhibited directly in revealed religion; others think that they can uncover it from the facts of nature. One of the commonest methods of this form of natural religion is to point to evolution as manifesting such a purpose. The history of life, it is asserted, manifests guidance on the part of some external power; and the usual deduction is that we can safely trust that same power for further guidance in the future.

I believe this reasoning to be wholly false. The purpose manifested in evolution, whether in adaptation, specialization, or biological progress, is only an apparent purpose. It is as much a product of blind forces as is the falling of a stone to earth or the ebb and flow of the tides. It is we who have read purpose into evolution, as earlier men projected will and emotion into inorganic phenomena like storm or earthquake. If we wish to work towards a purpose for the future of man, we must formulate that purpose ourselves. Purposes in life are made, not found. . . .

Obviously the formulation of an agreed purpose for man as a whole will not be easy. There have been many attempts already. Today we are experiencing the struggle between two opposed ideas—that of the subordination of the individual to the community, and that of his intrinsic superiority. Another struggle still in progress is between the idea of a purpose directed to a future life in a supernatural world, and one directed to progress in this existing world. Until such major conflicts are resolved, humanity can have no single major purpose, and progress can be but fitful and slow. Before progress can begin to be rapid, man must cease being afraid of his uniqueness, and must not continue to put off the responsibilities that are really his on to the shoulders of mythical gods or metaphysical absolutes. . . .

Progress is a major fact of past evolution; but it is limited to a few selected stocks. It may continue in the future, but it is not inevitable; man, by now become the trustee of evolution, must work and plan if he is to achieve further progress for himself and so for life.

In his ascent, man has had a "long childhood" that has stretched from the time of the Taung baby to the modern era of the sophisticated adult, blending emotion, intellect, art, science, and culture. And as Julian Huxley wrote in the preceding selection, man, if he plays his cards right, has not yet reached the end of that cultural and biological ascent. In going through Dr. Bronowski's notebooks, shortly after his death, I found a fable that he had copied from T.H. White's *The Sword in the Stone*. It is the badger's version of how man was created, and its bright and optimistic conclusion mirrors Dr. Bronowski's conviction that "Every man, every civilisation has gone forward because of its engagement with what it has set itself to do. The personal commitment of a man to his skill, the intellectual commitment and the emotional commitment working together as one, has made the Ascent of Man."

Genesis According to the Badger

T. H. White

"It's not very good," explained the badger coyly. "It's just a rough draft, you know. I shall alter a lot before I send it in."

"I am sure it must be interesting," said the Wart.

"Oh, no, it isn't a bit interesting. It's just an odd thing I threw off in an odd half-hour, just to pass the odd time, you know. But still, this is how it begins."

"Hem!" said the badger. Then he put on an impossibly high falsetto voice and began to read as fast as possible.

"People often ask as an idle question whether the process of evolution began with the chicken or the egg. Was there an egg out of which the first chicken came, or did a chicken lay the first egg? I am in a position to state that the first thing created was the egg.

"When God had manufactured all the eggs out of which the fishes and the serpents and the birds and the mammals and even the duck-billed-platypus would eventually emerge, he called the embryos before him, and saw that they were good."

"Perhaps I ought to explain," added the badger, lowering his papers nervously and looking at the Wart over the top of them, "that *all embryos look very much the same.* They are what you are before you are born, and, whether you are going to be a tadpole or a peacock or a cameleopard or a man, when you are an embryo you just look like a peculiarly repulsive and helpless human being. I continue as follows:

"The embryos stood up in front of God, with their feeble hands clasped politely over their tummies and their heavy heads hanging down respectfully, and God addressed them.

"He said: 'Now, you embryos, here you are, all looking exactly the same, and We are going to give you the choice of what you are going to be. When you grow up you will get bigger anyway, but We are pleased to grant you another gift as well. You may alter any parts of yourselves into anything which you think would be useful to you in after life. For instance, at the moment you can't dig. Anybody who would like to turn his hands into a pair of spades or garden forks is allowed to do so. Or, to put it another way, at present you can only use your mouths for eating with. Anybody who would like to use his mouth as an offensive weapon, can change it by asking, and be a corkindrill or a sabre-toothed tiger. Now then, step up and choose your tools, but remember that what you choose you will grow into, and will have to stick to.'

"All the embryos thought the matter over politely, and then, one by one, they stepped up before the eternal throne. They were allowed two or three specialisations, so that some chose to use their arms as flying machines and their mouths as weapons, or crackers, or drillers, or spoons, while others selected to use their bodies as boats and their hands as oars. We badgers thought very hard and decided to ask three boons. We wanted to change our skins for shields, our mouths for weapons, and our arms for garden forks. These boons were granted to us. Everybody specialised in one way or another, and some of us in very queer ones. For instance, one of the lizards decided to swap his whole body for blotting-paper, and one of the toads who lived in the antipodes decided simply to be a water-bottle.

"The asking and granting took up two long days—they were the fifth and sixth, so far as I remember—and at the very end of the sixth day, just before it was time to knock off for Sunday, they had got through all the little embryos except one. This embryo was Man.

" 'Well, Our little man,' said God. 'You have waited till the last, and slept on your decision, and We are sure you have been thinking hard all the time. What can We do for you?"

" 'Please God,' said the embryo. 'I think that You made me in the shape which I now have for reasons best known to Yourselves, and that it would be rude to change. If I am to have my choice I will stay just as I am. I will not alter any of the parts which You gave to me, for other and doubtless inferior tools, and I will stay a defenceless embryo all my life, doing my best to make unto myself a few feeble implements out of the wood, iron and other materials which You have seen fit to put before me. If I want a boat I will endeavor to construct it out of trees, and if I want to fly I will put together a chariot to do it for me. Probably I have been very silly in refusing to take advantage of Your kind offer, but I have done my best to think it over carefully, and now hope that the feeble decision of this small innocent will find favor with Yourselves.'

" 'Well done,' exclaimed the Creator in delighted tones. 'Here, all you embryos, come here with your beaks and what-nots to look upon Our first Man. He is the only one who has guessed Our riddle, out of all of you, and We have great pleasure in conferring upon him the Order of Dominion over the Fowls of the Air, and the Beasts of the Earth, and the Fishes of the Sea. Now let the rest of you get along, and love and multiply, for it is time to knock off for the week-end. As for you, Man, you will be a naked tool all your life, though a user of tools: you will look like an embryo till they bury you, but all others will be embryos before your might; eternally undeveloped, you will always remain *potential* in Our image, able to see some of Our sorrows and to feel some of Our joys. We are partly sorry for you, Man, and partly happy, but always proud. Run along then, Man, and do your best. And listen, Man, before you go . . .'

" 'Well?' asked Adam, turning back from his dismissal.

" 'We were only going to say,' said God shyly, twisting Their hands together. 'Well, We were just going to say, God bless you.' "

Bibliography

Chapter One

H.G. Bandi and J. Maringer, *Art in the Ice Age*, New York, Praeger, 1953.

W.W. Bishop, "The Tempo of Human Evolution," *Nature*, Vol. 244, 1973, pp. 405-407.

J. Bronowski and W. Long, "Statistical Methods in Anthropology," *Nature*, Vol. 168, 1951, p. 794; and "The Australopithecine Milk Canines," *Nature*, Vol. 172, 1953, p. 251.

M.Y. Coppens, "Les restes d'Hominidés des formations pliopléistocenes de l'Omo en Ethiopie," *Comptes rendus des séances de l'Academie des Sciences*, November 30, 1970, December 21, 1970, and January 4, 1971, Paris.

G.B. Dalrymple and M.A. Lanphere, *Potassium-Argon Dating*, San Francisco, Freeman, 1969, pp. 43–48.

R.A. Dart, "Taungs and its Significance," *Natural History*, Vol. 26, No. 3, 1926, pp. 315-327.

T.G. Dobzhansky, *Genetics of the Evolutionary Process*, New York, Columbia University Press, 1970.

H. Faul, *Ages of Rocks, Planets, and Stars*, New York, McGraw-Hill, 1966.

R.L. Holloway, "The Casts of Fossil Hominid Brains," *Scientific American*, July 1974, pp. 106–115.

J. Kingdon, *East African Mammals: An Atlas of Evolution in Africa*, Vol. 1, New York, Academic Press, 1971, pp. 104–107.

G.B. Kolata, "!Kung Hunters-Gatherers: Feminism, Diet, and Birth Control," *Science*, Vol. 185, 1974, pp. 932–934.

R.E.F. Leakey, "Evidence for an Advanced Plio-Pleistocene Hominid from East Rudolf, Kenya," *Nature*, Vol. 242, 1973, pp. 447–450.

W.F. Libby, *Radiocarbon Dating*, Chicago, University of Chicago Press, 1955.

V J. Maglio, "Vertebrate Faunas and Chronology of Hominid-Bearing Sediments East of Lake Rudolf, Kenya," *Nature*, Vol. 239, 1972, pp. 379-385.

H.V. Merrick, "Archaeological Occurrences of Early Pleistocene Age from the Shungura Formation, Lower Omo Valley, Ethiopia," *Nature*, Vol. 242, 1973, pp. 572–575.

J. Napier, "The Evolution of the Hand," *Human Variation and Origins*, San Francisco, Freeman, 1967, pp. 155-161.

K.P. Oakley, "Tools Makyth Man," *Antiquity*, Vol. 31, 1957, pp. 199–209.

C.E. Parker, "Responsiveness, Manipulation, and Implementation Behavior in Chimpanzees, Gorillas and Orang-utans," *Proceedings of the Second International Congress on Primates* (1968), Vol. 1, Basel, Karger, 1969, pp. 160–166.

D. Pilbeam, *The Evolution of Man*, London, Thames and Hudson, 1970.

C.M. Turnbull, *The Mountain People*, New York, Simon and Schuster, 1972.

J.S. Weiner, *The Piltdown Forgery*, London, Oxford University Press, 1955.

Chapter Two

R. Bigelow, *The Dawn Warriors*, Boston, Little, Brown, 1969.

J. Bronowski, "Technology and Culture in Evolution," *Philosophy of the Social Sciences*, Vol 1, 1971, pp. 195–206; *The American Scholar*, Vol. 41, No. 2, 1972, pp. 197–211; and abridged in *Cambridge Review*, Vol. 91, No. 2196, 1970, pp. 140–144.

G. Clark, *The Stone Age Hunters,* New York, McGraw-Hill, 1967, pp. 67-90.

I. Cornwall, *Ice Ages, Their Nature and Effects*, London, John Baker, 1970.

R.S. Dietz and J.C. Holden, "The Breakup of Pangaea," *Scientific American*, October 1970, pp. 30–41.

R.S. Dietz and J.C. Holden, "Continental Drift and Plate Tectonics," *Journal of Geophysical Research*, Vol. 75, No. 26, 1970, pp. 4939–4955.

I.M. Drew, D. Perkins, and P. Daly, "Prehistoric Domestication of Animals: Effects on Bone Structure," *Science*, Vol. 171, 1971, pp. 280–282.

G.E. Evans, *The Farm and the Village*, London, Faber and Faber, 1969.

J. Garstang and J.B.E. Garstang, *The Story of Jericho*, London, Marshall, Morgan & Scott, 1948.

J.H. Hulse and D. Spurgeon, "Triticale (Hybrid of Wheat and Rye)," *Scientific American,* August 1974, pp. 72–91.

E. Isaac, *Geography of Domestication*, Englewood Cliffs, N.J., Prentice-Hall, 1970, pp. 57–64.

Joshua, Book of, *Holy Bible,* New Scofield Reference Edition, New York, Oxford University Press, 1967.

K. Kenyon and A.D. Tushingham, "Jericho Gives Up its Secrets," *National Geographic*, Vol. 104, No. 6, pp. 853–870.

B. Kurtén, *The Ice Age*, New York, Putnam, 1972.

D.P. McKenzie, D. Davies, and P. Molnar, "Plate Tectonics of the Red Sea and East Africa," *Nature*, Vol. 226, 1970, pp. 243–248.

E.B. McNeil, ed., *The Nature of Human Conflict*, Englewood Cliffs, N.J., Prentice-Hall, 1965.

P.C. Manglesdorf, *Corn, its Origin, Evolution and Improvement*, Cambridge, Mass., Belknap (Harvard University Press), 1974.

A.C. Moore, *The Grasses*, New York, Macmillan, 1960.

S. Piggott, "The Beginnings of Wheeled Transport," *Scientific American*, July 1968, pp. 82–89.

G. Schultz, *Glaciers and the Ice Age*, New York, Holt, Rinehart and Winston, 1963.

P.J. Ucko and G.W. Dimbleby, *The Domestication and Exploitation of Plants and Animals*, Chicago, Aldine, 1969.

A.M. Wall, Ralph Riley, and M.D. Gale, "The Position of a Locus on Chromosome 5B of *Triticum aestivum* Affecting Homoeologous Meiotic Pairing," *Genetical Research,* Vol. 18, 1971, pp. 329–339.

S.L. Washburn, ed., *Classification and Human Evolution*, Chicago, Aldine, 1963.

H.T. Waterbolk, "Food Production in Prehistoric Europe," *Science*, Vol. 162, 1968, pp. 1093–1102.

Sir Charles L. Wooley, *Excavations at Ur*, Ernest Benn, 1954.

Chapter Three

J.L. Bada, R.A. Schroeder, and G.F. Carter, "New Evidence for the Antiquity of Man in North America Deduced from Aspartic Acid Racemization," *Science*, Vol. 184, 1974, pp. 791–793.

E. Baller, "The Ancient Ones," *Westways*, December 1972, pp. 37–41.

H. Bingham, *Lost City of the Incas*, New York, Duell, Sloan and Pearce, 1948.

W.C. Boyd, "Genetics and the Human Race," *Science*, Vol. 140, 1963, pp. 1057–1064.

J. Bronowski, "The Enigma of Leonardo," *Realités*, No. 20, July 1952.

A.M. Bues, "Stochastic Tests of Selection in the ABO Blood Groups," *American Journal of Physical Anthropology*, **21** (1963), pp. 287–299.

B. Bunting, *Taos Adobes*, Fort Burgwin, N. Mex., Museum of New Mexico Press, 1964.

G.H.S. Bushnell, *The First Americans*, New York, McGraw-Hill, 1968.

G.H.S. Bushnell, *Peru*, London, Thames and Hudson, 1963.

F. Guerra, "Aztec Science and Technology," *History of Science*, Vol. 8, Cambridge, Crombie and Hoskin, Heffer, 1969.

D.M. Hopkins, *The Bering Land Bridge*, Palo Alto, Calif., Stanford University Press, 1967.

F. Johnson, "Radiocarbon Dating and Archaeology in North America," *Science*, Vol. 155, 1967, pp. 165–169.

D. Knoop and G.P. Jones, *The Medieval Mason*, New York, Barnes and Noble, 1967.

T. Kroeber, *Ishi in Two Worlds: A Biography of the Last Wild Indian in North America*, Berkeley, Calif., University of California Press, 1961.

W.S. Laughlin, "Eskimos and Aleuts: Their Origins and Evolution," *Science*, Vol. 142, 1963, pp. 633–645.

E. MacCurdy, *The Notebooks of Leonardo Da Vinci,* New York, George Braziller, 1954.

R.S. MacNeish, "Early Man in the Andes," *Scientific American*, April 1971, pp. 36–46.

R.S. MacNeish, R. Berger, and R. Protsch, "Megafauna and Man from Ayacucho, Highland, Peru," *Science*, Vol. 168, 1970.

P.S. Martin and H.E. Wright, *Pleistocene Extinctions: The Search for a Course*, New Haven, Conn., Yale University Press, 1967.

B.S. Meyers, *Art and Civilization*, New York, McGraw-Hill, 1967.

John Ruskin, *Seven Lamps of Architecture*, London, Smith, Elder, 1849.

B. Russell, *The Philosophy of Leibniz*, London, Allen and Unwin, 1964.

J.H. Seward and L.C. Xavan, *Native Peoples of South America*, New York, McGraw-Hill, 1959.

H.M. Wormington, *The Prehistoric Indians of the Southwest,* Denver, The Colorado Museum of Natural History, 1947.

Chapter Four

R. Boyle, *The Sceptical Chymist,* New York, E.P. Dutton, 1949.

D.R. Brothwell, "The Real History of Syphilis," *Science Journal,* September 1970, pp. 27–34.

F.P. Bundy, "Superhard Materials," *Scientific American,* August 1974, pp. 62–70.

K.S. Davis, *The Cautionary Scientists,* New York, Putnam, 1966.

E. Farber, ed., *Great Chemists,* New York, Interscience, 1961.

M.I. Finley, *Aspects of Antiquity: Discoveries and Controversies,* London, Chatto and Windus, 1968.

G.J. Gruman, "A History of Ideas about the Prolongation of Life," *Transactions of the American Philosophical Society,* Vol. 56, Pt. 9, 1966.

A.R. Hall, *The Scientific Revolution,* London, Longmans, 1954.

Antoine L. Lavoisier, "Mémoire sur l'affinité du principe oxygene avec les différentes substances auxquelles il est susceptible de s'unir," *Hist. Acad. Roy. Sci.,* 1785, pp. 530–540.

Antoine L. Lavoisier, *Traité Elementaire de Chimie,* Vols. 1 and 2, Paris, 1789.

R. Legame, trans., *The Autobiography of Benvenuto Cellini,* New York, Dodd Mead, 1961.

A.M. Lewis, *The Struggle Between Science and Superstition,* Chicago, Charles H. Kerr, 1916.

D.P. Mellor, *The Evolution of the Atomic Theory,* Amsterdam, Elsevier, 1971.

J.G. Parr, *Man, Metals and Modern Magic,* Cleveland, Ohio, and Ames, Iowa, American Society for Metals and the Iowa State College Press, 1958.

H. Partington, *A Short History of Chemistry,* Third Edition, New York, St. Martins Press, 1965.

C. Patterson, "Native Copper, Silver and Gold Accessible to Early Metallurgists," *American Antiquity,* 36 (1971):3, pp. 286–321.

C.S. Smith, *A History of Metallography,* Chicago, University of Chicago Press, 1960.

C.S. Smith, "Matter Versus Materials: A Historical View," *Science,* Vol. 162, 1968, pp. 637-644.

C.S. Smith, "Metallurgical Footnotes to the History of Art," *Proceedings of the American Philosophical Society,* Vol. 116, No. 2, 1972, pp. 104-107.

R.A. Smith, *Memoir of John Dalton,* London, H. Bailliere, 1856.

A. Thackray, *John Dalton: Critical Assessments of His Life and Science,* Cambridge, Mass., Harvard University Press, 1972.

S. Toulmin and J. Goodfield, *The Architecture of Matter,* New York, Harper & Row, 1962.

R. Vallery-Radot, *The Life of Pasteur,* New York, Dover, 1960.

M.E. Weeks and H.M. Leicester, *Discovery of the Elements,* Seventh Edition, Easton, Pa., Journal of Chemical Education, 1968.

Chapter Five

Alhazen (Abu-'Ali Al-Hasen ibn Al-Haytham), *Opticae Thesaurus Alhazeni: Arabis Libri Septem, nunc Primum Editi,* Basel, 1572.

J. Aubrey, "Thomas Hobbes," in *Brief Lives and Other Selected Writings,* Anthony Powell, ed., London, Cresset Press, 1949, pp. 237-262.

L. Carroll, *Euclid and His Modern Rivals,* New York, Dover, 1973.

E. Cassirer, *An Essay on Man,* New York, Doubleday Anchor Books, 1953.

A. Dürer, *Vier Bücher von menschlicher Proportion,* Nürenberg, Germany, J. Formschneyder, 1528.

C.C. Gillespie, *The Edge of Objectivity; an Essay in the History of Scientific Ideas,* Princeton, N.J., Princeton University Press, 1960.

S. Goudsmit, "Symmetry of Symbols," *Nature,* March 6, 1937.

F. Hoyle, *Astronomy,* London, Macdonald, 1962.

M. Jammer, *Concepts of Space: The History of Theories of Space in Physics,* Second Edition, Cambridge, Mass., Harvard University Press, 1969.

H. W. Janson, ed., *Key Monuments of the History of Art*, Englewood Cliffs, N.J., Prentice-Hall, 19[illegible].

A. Koyré, *Metaphysics and Measurement*, London, Chapman and Hall, 1968.

P. Le Corbeiller, "Crystals and the Future of Physics," *Scientific American*, Vol. 50, January 1953.

P. Le Corbeiller, "The Curvature of Space," *Scientific American*, Vol. 80, November 1954.

G. W. Leibniz, *The Early Mathematical Manuscripts of Leibniz*, James Mark Child, trans., Chicago, Open Court, 1920.

Edith A. Müller, *Gruppentheoretische und Strukturanalytische Untersuchunger der Maurischen Ornamente aus der Alhambra in Granada*, Rüschlikon, Buchdruckerei Baublatt AG, 1944.

J. R. Newman, *The World of Mathematics*, Vols. 1–4, New York, Simon and Schuster, 1956–1960.

E. Panofsky, *The Life and Art of Albrecht Dürer*, Princeton, N.J., Princeton University Press, 19[illegible]5.

G. Polya, "Über die Analogie der Kristallsymmetrie in der Ebene," *Zeitschrift fur Krystallographie*, Vol. 60, 1924, pp. 278–282.

A. N. Whitehead, and B. Russell, *Principia Mathematica*, Vols. 1–3, Cambridge, Cambridge University Press, 1910–1913.

Chapter Six

K. Åkerblom, *Astronomy and Navigation in Polynesia and Micronesia: A Survey*, [illegible] Publication 14. Stockholm, The Ethnographical Museum (Etno-[illegible]

[illegible]erse: The Astrarium of Giovanni de [illegible]l Society, New Series, Vol. 56, [illegible]

[illegible]onian Magazine, Vol. 4, No. 1, April 1973, pp. [illegible] *[illegible]ure of Scientific Discovery*, Owen Gingerich, ed., The National Academy of Sciences (in press).

H. Butterfield, *The Origins of Modern Science*, London, Bell, 1949.

J. Dobrzycki, *The Reception of Copernicus' Heliocentric Theory,* Boston, D. Reidel, 1972.

S. Drake, ed., *Discoveries and Opinions of Galileo,* New York, Doubleday, 1957.

P. Duhem, *Le système du monde: histoire des doctrines cosmologiques de Platon à Copernic,* Vols. 1-10, Paris, A. Hermann, 1913-1959.

Galileo Galilei, *Dialogue on the Great World Systems* (Salusbury Translation), revised by Giorgio de Santillana, Chicago, University of Chicago Press, 1957.

L. Geymonat, *Galileo Galilei: A Biography and Inquiry into His Philosophy of Science,* New York, McGraw-Hill, 1965.

H.C. King, *The History of the Telescope,* Cambridge, Mass., Sky, 1955.

A. Koyré, *La révolution astronomique,* Paris, A. Hermann, 1961.

D.J. de Solla Price, "Leonardo da Vinci and the Clock of Giovanni de Dondi," *Antiquarian Horology,* June 1958.

W. Shumaker, *The Occult Sciences in the Renaissance,* Berkeley, Calif., University of California Press, 1972.

J.E.S. Thomson, *A Commentary on the Dresden Codex: A Maya Hieroglyphic Book,* Memoirs of the American Philosophical Society, Vol. 93, Philadelphia, 1972.

R. Wittkower, *Architectural Principles in the Age of Humanism,* New York, Random House, 1965.

F.A. Yates, *Giordano Bruno and the Hermetic Tradition,* London, 1964.

Chapter Seven

G.N. Clark, *Science and Social Welfare in the Age of Newton,* Oxford, Clarendon Press, 1937.

I.B. Cohen, ed., *Introduction to Newton's Principia,* Cambridge, Mass., Harvard University Press, 1970.

Albert Einstein, "Electrodynamik bewegter Korper," *Annalen der Physik* (4), Vol. 17, 1905, p. 891.

Albert Einstein, "Über einen die Erzeugung und Verwandlung des Lichtes betreffenden heuristischen Gesichtspunkt," *Annalen der Physik* (4), Vol. 17, 1905, p. 132.

Albert Einstein, *The Meaning of Relativity,* Princeton, N.J., Princeton University Press, 1953.

G. Holton, *Thematic Origins of Scientific Thought,* Cambridge, Mass., Harvard University Press, 1973.

S.L. Jaki, "Goethe and the Physicists," *American Journal of Physics,* Vol. 37, No. 2, 1969, pp. 195-203.

M. Jammer, *Concepts of Space: The History of Theories of Space in Physics,* Second Edition, Cambridge, Mass., Harvard University Press, 1969.

A. Koyré, "The Significance of the Newtonian Synthesis," *Archives internationales d'histoire des sciences,* Vol. 29, 1950, pp. 291-311.

B. Kuznetsov, *Einstein and Dostoevsky,* London, Hutchinson Educational, 1972.

M. Leon-Portilla, *Time and Reality in the Thought of the Maya,* Boston, Beacon Press, 1968.

A.A. Michelson, and E.W. Morley, "On the Relative Motion of the Earth and the Luminiferous Ether," *American Journal of Science,* Series 3, Vol. 34, 1887, p. 333.

J. Needham, W. Ling, and D.J. de Solla Price, *Heavenly Clockwork,* Cambridge, Cambridge University Press, 1960.

Isaac Newton, "A Letter of Mr. Isaac Newton, Professor of the Mathematicks in the Univeristy of Cambridge; containing his New Theory about Light and Colours: sent by the Author to the Publisher from Cambridge, February 6, 1671/72 , *Philosophical Transactions,* No. 80, February 19, 1671/72, pp. 3075-3087.

Isaac Newton, *Opticks, or a Treatise of the Reflections, Refractions, Inflections and Colours of Light,* E.T. Whittaker, ed., reprinted from the Fourth Edition (1730), London, G.Bell and Sons, 1931.

N. Reingold, ed., *Science in Nineteenth Century America,* New York, Macmillan, 1966.

A.I. Sabra, *Theories of Light from Descartes to Newton,* London, Oldbourne, 1967.

D.T. Whiteside, ed., *The Mathematical Papers of Isaac Newton,* Vols. 1-5, Cambridge, Cambridge University Press, 1967-1972.

G. Wick, "The Clock Paradox Resolved," *New Scientist,* February 3, 1972, pp. 261-263.

Chapter Eight

R. Calder, *Science in Our Lives,* East Lansing, Mich., Michigan State College Press, 1954.

S.N.L. Carnot, *Réflexions sur la puissance notrice du feu et sur les machines propres à développer cette puissance,* Paris, Bachelier, 1824 and A. Blanchard, 1953.

I.B. Cohen, *Franklin and Newton: An Inquiry into Speculative Newtonian Experimental Science and Franklin's Work in Electricity as an Example Thereof,* Revised Edition, Cambridge, Mass., Harvard University Press, 1973.

D. Defoe, *A Tour Through England and Wales; Divided into Circuits or Journies,* George Douglas Howard, ed., London, J.M. Dent and Sons, 1928.

H.W. Dickinson, and R. Jenkins, *James Watt and the Steam Engine,* Oxford, Clarendon Press, 1927.

Benjamin Franklin, *Experiments and Observations on Electricity, Made at Philadelphia in America,* London, 1751-1754.

J.L. Hammond, and B. Hammond, *The Village Labourer 1760-1832,* London, Longmans, Green and Company, 1911.

J.P. Joule, "On the Calorific Effects of Magneto-Electricity, and on the Mechanical Value of Heat," London, 1843.

F.D. Klingender, *Art and the Industrial Revolution,* London, N. Carrington, 1947.

P. Mantoux, *The Industrial Revolution in the Eighteenth Century,* London, Jonathan Cape, 1928.

I.J. Parker, *Dissenting Academies in England,* New York, Octagon Books, 1969 (a reprint of the 1914 edition).

G. Rudé, *Paris and London in the Eighteenth Century,* London, Fontana, 1970.

S.P. Thompson, *Life of William Thomson, Baron Kelvin of Largs,* London, 1910.

A.S. Turberville, ed., *Johnson's England,* Vols. 1 and 2, London, Oxford University Press, 1933.

Chapter Nine

N. Barlow, ed., *Darwin and Henslow—The Growth of an Idea,* Berkeley, Calif., University of California Press, 1967.

H.W. Bates, "Contributions to an Insect Fauna of the Amazon Valley," *Transactions of the Linnaean Society,* Vol. 23, 1862, pp. 495-566.

H.W. Bates, *The Naturalist on the River Amazon,* London, John Murray, 1863.

M. Calvin, *Chemical Evolution,* Eugene, Ore., University of Oregon Press, 1961.

Charles Darwin, *The Descent of Man,* London, John Murray, 1871.

Charles Darwin, *On the Origin of Species by Means of Natural Selection,* New York, Heritage, 1963.

Charles Darwin, *The Voyage of the Beagle,* London, J.M. Dent and Sons, 1965.

E. Darwin, *Zoönomia, or the Laws of Organic Life,* London, J. Johnson, 1794-1796.

J.C. Kendrew, ed., "Evolution of the Genetic Apparatus," *Journal of Molecular Biology,* Vol. 38, 1968, pp. 381-393.

D.H. Kenyon, and G. Steinman, *Biochemical Predestination,* New York, McGraw-Hill, 1969.

C. Lyell, *Principles of Geology,* Vols. 1-3, London, John Murray, 1830-1833.

C. Lyell, *The Geological Evidences of the Antiquity of Man,* London, G.W. Childs, 1863.

A.I. Oparin, *The Chemical Origin of Life,* Springfield, Ill., Charles C. Thomas, 1964.

Louis Pasteur, "La Dissymetrie moleculaire," lecture delivered before the Société chimique de Paris, December 22, 1883, *Oeuvres,* Vol. 1, pp. 369-380.

C. Ponnamperuma, ed., *Exobiology,* North-Holland Research Monographs Frontiers of Biology, Vol. 23, Amsterdam, North-Holland, 1972.

C. Sagan, "On the Origin and Planetary Distribution of Life," *Radiation Research,* Vol. 15, 1961, pp. 174-191.

I.S. Shklovskii and C. Sagan, *Intelligent Life in the Universe,* San Francisco, Holden-Day, 1966.

R. Vallery-Radot, *The Life of Pasteur,* New York, Doubleday, Page, 1923.

A.R. Wallace, *Darwinism, Theory of Natural Selection,* New York, Macmillan, 1889.

A.R. Wallace, *My Life, A Record of Events and Opinions,* Vol. 1, New York, Dodd, Mead, 1906.

A.R. Wallace, and C. Darwin, "On the Tendency of Species to Form Varieties; and on the Perpetuation of Varieties and Species by Natural Means of Selection," *Journal of the Linnaean Society (Zoology),* Vol. 3, 1858, p. 35.

A. Williams-Ellis, *Darwin's Moon: a Biography of Alfred Russel Wallace,* London, Blackie, 1966.

L.G. Wilson, ed., *Sir Charles Lyell's Scientific Journals on the Species Question,* New Haven, Conn., Yale University Press, 1970.

Chapter Ten

H.A. Bethe, "Energy Production in Stars," *Physical Review,* Vol. 55, 1939, p. 434.

Niels Bohr, "On the Constitution of Atoms and Molecules," *Philosophical Magazine,* Vol. 26, No. 1, 1913, pp. 476 and 857.

Niels Bohr, "Light and Life," *Nature,* Vol. 131, 1933, p. 421.

Ludwig Boltzmann, "Der zweite Hauptsatz der mechanischen Warmetheorie" (1886), *Populäre Schriften,* 1925.

J. Chadwick, "The Existence of a Neutron," *Proceedings of the Royal Society,* Vol. A. 136, 1932, p. 692.

R. Clausius, "Über die bewegende Kraft der Wärme, und die Gesetze, welche sich daraus für die Wärmelehre selbst ableiten Passen," *Poggendorff's Annalen,* March and April 1850.

L. Fermi, *Atoms in the Family,* Chicago, University of Chicago Press, 1954.

G. Gamow, *Thirty Years that Shook Physics: The Story of Quantum Theory,* New York, Doubleday Anchor Books, 1966.

W.I. Homer, *Seurat and the Science of Painting,* Cambridge, Mass., M.I.T. Press, 1964.

B. Jaffe, *Moseley and the Numbering of the Elements,* New York, Doubleday, 1971.

E. Mach, *Die Mechanik in ihrer Entwickelung historis chkritisch dargestellt,* 1883.

Dmitri Mendeleev, "The Atomic Volumes of the Simple Bodies," (in Russian), *Transactions of the Second Congress of Russian Naturalists in Moscow,* April 23, 1869 and *Annalen der Chemie und Pharmacie,* Suplementebande 8, 1871, pp. 133 ff.

Dmitri Mendeleev, *Journal of the Chemical Society, Transactions,* Vol. 55, 1889, pp. 634–639.

Max Planck, "Über das Gesetz der Energieverteilung im Normalspectrum," *Verhandlungen der Deutschen Physikalischen Gesellschaft,* Vol. 2, 1900, pp. 202 and 237.

Max Planck, *Vorträge und Erinnerungen,* Stuttgart, 1949.

R. Rosenblum, *Cubism and Twentieth-Century Art,* Revised Edition, New York, Abrams, 1966.

J. Rowland, *Ernest Rutherford–Atom Pioneer,* New York, Philosophical Library, 1957.

S. Rozental, ed., *Niels Bohr, His Life and Works as Seen by His Friends and Colleagues,* New York, Wiley, 1964.

Lord Ernest Rutherford, *The Newer Alchemy,* Cambridge, Cambridge University Press, 1937.

Lord Ernest Rutherford, "The Scattering of α and β Particles by Matter and the Structure of the Atom," *Philosophical Magazine,* Vol. 21, No. 669, 1911.

D. Schroeer, *Physics and Its Fifth Dimension: Society,* Reading, Mass., Addison-Wesley, 1972.

M.H. Shamos, *Great Experiments in Physics,* New York, Holt, Rinehart and Winston, 1964.

Leo Szilard, "Reminiscences," *Perspectives in American History,* Gertrud Weiss Szilard and Kathleen R. Winsor, eds., Vol. 2, 1968.

G.P. Thomson, *J.J. Thomson, Discoverer of the Electron,* New York, Doubleday Anchor Books.

J.J. Thomson, "Cathode Rays," *Philosophical Magazine,* Vol. 44, 1897, p. 293.

J.J. Thomson, *Recollections and Reflections,* New York, Macmillan, 1937.

Chapter Eleven

Max Born, *My Life and My Views,* New York, Charles Scribner's Sons, 1968.

J. Bronowski, *Nature and Knowledge: The Philosophy of Contemporary Science,* Eugene, Ore., University of Oregon Books, 1969.

M. Brown, ed., *The Social Responsibility of the Scientist,* New York, Free Press, 1971.

A. Burroughs, *Art Criticism from a Laboratory* (use of x-rays), Boston, Little, Brown, 1938.

P. Dirac, "The Quantum Theory of the Electron," *Proceedings of the Royal Society,* Vol. A. 117, 1928, p. 610.

G.W. Dunnington, *Carl Friedrich Gauss–Titan of Science,* New York, Hafner, 1955.

Albert Einstein, and L. Infeld, *The Evolution of Physics,* New York, Simon and Schuster, 1938.

G.W.F. Hegel, *Dissertatio Philosphica de Orbitis Planetarum,* Jena, Typis Prageri *et soc.,* 1801.

Werner Heisenberg, *Natural Law and the Structure of Matter,* London, Rebel Press, 1970.

Werner Heisenberg, M. Born, E. Schrödinger, and P. Auger, *On Modern Physics,* New York, Clarkson N. Potter, 1961.

J.J. Lagowski, *The Structure of Atoms,* Boston, Houghton Mifflin, 1964.

Wilhelm K. Röntgen, *Eine neue Art von Strahlen,* Wurzberg, 1895-1896.

Erwin, Schrödinger, *Science and Humanism,* Cambridge, Cambridge University Press, 1951.

Leo Szilard, "Über die Entropieverminderung in einem thermodynamischen System bei Eingriffen intelligenter Wesen," *Zeitschrift für Physik,* Vol. 53, 1929, p. 840.

Chapter Twelve

G. Beadle, and M. Beadle, *The Language of Life,* New York, Doubleday, 1966.

B. Campbell, ed., *Sexual Selection and the Descent of Man,* Chicago, Aldine, 1972.

H.C. Cannon, *Lamarck and Modern Genetics,* Springfield, Ill., Charles C. Thomas, 1959.

Francis H.C. Crick, *Of Molecules and Men,* Seattle, Wash., University of Washington Press, 1966.

Francis H.C. Crick, "The Origin of the Genetic Code," *Journal of Molecular Biology,* Vol. 38, 1968, pp. 367-379.

R.A. Fisher, *The Genetical Theory of Natural Selection,* New York, Dover, 1958.

H. Iltis, *Life of Mendel,* New York, Hafner, 1966.

Jean Baptiste Lamarck, *Philosophie Zoologique,* Vols. 1 and 2, Paris, J.B. Baillière, 1809.

Gregor Mendel, "Versuche über Pflanzenhybriden," *Verhandlungen des Naturforschenden Vereins in Brunn,* 4, 1865.

Jacques Monod, *Chance and Necessity,* New York, Random House, 1971.

A.G. Motulsky, "Brave New World?" *Science,* Vol. 185, 1974, pp. 653-663.

R.C. Olby, *Origins of Mendelism,* New York, Schocken Books, 1966.

A.I. Oparin, *Genesis and Evolutionary Development of Life,* New York, Academic Press, 1968.

Linus Pauling, R.C. Corey, and B.R. Branson, "The Structure of Proteins: Two Hydrogen Bonded Helical Configurations of the Polypeptide Chain," *Proceedings of the National Academy of Sciences,* Vol. 37, 1951, p. 205.

J.W. Schopf, "Precambrian Micro-Organisms and Evolutionary Events Prior to the Origin of Vascular Plants," *Biological Review,* Vol. 45, 1970, pp. 315-352.

Chapter Thirteen

J. Bronowski, "The Creative Process," *Scientific American,* Vol. 199, No. 3.

J. Bronowski, "Language in a Biological Frame," *Current Trends in Linguistics, Vol. 12,* Thomas A. Sebeok, Ed., The Hague, Mouton, in press.

J. Bronowski, "A Twentieth Century Image of Man," *Leonardo,* Vol. 7, No. 2, pp. 117-121.

F.J.J. Buytendijk, "Considérations de psychologie comparée à propos d'expériences faites avec le singe cercopithecus," *Archives Neerlandaises de Physiologie de l'Homme et des Animaux,* Vol. 5, 1921, pp. 42-88.

John Cairns, Günther S. Stent, and James D. Watson, eds., *Phage and the Origins of Molecular Biology,* Cold Spring Harbor Laboratory of Quantitative Biology, Cold Spring Harbor, New York, 1966.

N. Chomsky, *Language and Mind,* New York, Harcourt Brace Jovanovich, 1968.

T.H. Clutton-Brock, "Primate Social Organization and Ecology," *Nature,* Vol. 250, 1974, pp. 539-542.

K. von Frisch, "Decoding the Language of the Bee," *Science,* Vol. 185, 1974, pp. 663-668.

F. Galton, *Inquiries into Human Faculty and Its Development,* New York, Macmillan, 1883.

C.F. Hockett, and R. Ascher, "The Human Revolution," *Current Anthropology,* Vol. 5, No. 3, 1964, pp. 135-167.

W.S. Hunter, "The Delayed Reaction in Animals and Children," *Behavior Monographs,* Vol. 2, 1913, pp. 1-86.

C.F. Jacobsen, "Functions of the Frontal Association Area in Primates," *Archives of Neurology and Psychiatry,* Vol. 33, 1935, pp. 558-569.

C.F. Jacobsen, "Studies of Cerebral Functions in Primates. I. The Function of the Frontal Areas in Monkeys," *Comparative Psychology Monographs,* Vol. 13, No. 63, 1936, pp. 3-60.

C.A. Lawson, *Brain Mechanisms and Human Learning,* Boston, Houghton Mifflin, 1967.

J. Levy, "Possible Basis for the Evolution of Lateral Specialization of the Human Brain," *Nature,* Vol. 224, 1969, pp. 614-615.

J. von Neumann, and O. Morgenstern, *Theory of Games and Economic Behavior,* Princeton, N.J., Princeton University Press, 1943.

W. Penfield, and L. Roberts, *Speech and Brain Mechanisms,* Princeton, N.J., Princeton University Press, 1959.

N. Tinbergen, *Social Behavior in Animals,* New York, Wiley, 1965.

N. Tinbergen, *The Study of Instinct,* Oxford, Clarendon Press, 1951.

N. Tinbergen and M. Cullen, "Ritualization Among Black-Headed Gulls," *Royal Society Conference on Ritualization of Behaviour in Animals and Man,* London, 1965.

J.M. Warren and K. Akert, eds., *The Frontal Granular Cortex and Behavior,* New York, McGraw-Hill, 1964.

R.H. Weir, *Language in the Crib,* The Hague, Mouton, 1962.